G. Grioli • C. Truesdell (Eds.)

Non-linear Continuum Theories

Lectures given at a Summer School of the
Centro Internazionale Matematico Estivo (C.I.M.E.),
held in Bressanone (Bolzano), Italy,
May 31-June 9, 1965

Springer

FONDAZIONE CIME ROBERTO CONTI

C.I.M.E. Foundation
c/o Dipartimento di Matematica "U. Dini"
Viale Morgagni n. 67/a
50134 Firenze
Italy
cime@math.unifi.it

ISBN 978-3-642-11032-0 e-ISBN: 978-3-642-11033-7
DOI:10.1007/978-3-642-11033-7
Springer Heidelberg Dordrecht London New York

Reprint of the 1st ed. C.I.M.E., Ed. Cremonese, Roma 1966
With kind permission of C.I.M.E.

Printed on acid-free paper

Springer.com

CENTRO INTERNAZIONALE MATEMATICO ESTIVO

(C.I.M.E.)

1° Ciclo - Bressanone - dal 31 maggio al 9 giugno 1965

"NON-LINEAR CONTINUUM THEORIES"

Coordinatori : Prof. C. Truesdell
Prof. G. Grioli

CENTRO INTERNAZIONALE MATEMATICO ESTIVO

(C.I.M.E.)

B. COLEMAN

M.E. GURTIN

THERMODYNAMICS AND WAVE PROPAGATION IN ELASTIC AND VISCOELASTIC MEDIA

Corso tenuto a Bressanone dal 31 maggio al 9 giugno 1965

THERMODYNAMICS AND WAVE PROPAGATION IN ELASTIC AND VI-SCOELASTIC MEDIA

by

Bernadrd D. Coleman

Mellon Institute, Pittsburgh

Morton E. Gurtin

Brown University Providence

In continuum physics the word wave is used with several distinct meanings. To some a wave is a sinusoidal disturbance, to some it is any member of a certain class of solutions to a hyperbolic equation, and to others it is a propagating singular surface. Here we follow (Christoffel, Hugoniot, Hadamard, and Duhem and use the word in the last sense; thus, we define a wave to be a surface , moving with respect to the material , across which some kinematical variable, such as the acceleration or the velocity, suffers a jump discontinuity. In the present age of sonic booms and nuclear explosions , even the layman is familiar with "shock waves" . To find, however, applications for a general theory of propagating singular surfaces, one does not have to turn to the latest accomplishments of physics ; it suffices to think carefully of the motion of an object struck with a hammer.

Here we review some aspects of the classical theory of wave propagation in elastic materials and discuss recent extensions of the classical theory to materials with memory.

Let $x_i(X_j, t)$ give the spatial position at time t of the material point which occupies the position X_j in the reference configuration.

---

*We use Cartesian tensor notation.

According to the Duhem-Hadamard classification scheme, a surface $\Sigma = \Sigma(t)$ is a wave of order N if the N' th-order derivatives of $x_i(X_j, t)$ exhibit jump discontinuities at Σ, but all lower derivatives are continuous across Σ. A shock is a wave of order 1; that is, $x_i(X_j, t)$ is continuous, but the velocity $\dot{x}_i = \frac{\partial}{\partial t} x_i(X_j, t)$ and the deformation gradient $F_{ij} = \frac{\partial}{\partial X_j} x_i(X_k, t)$ show jumps across Σ. At an <u>acceleration wave</u> second derivatives, such as the acceleration $\ddot{x}_i = \frac{\partial^2}{\partial t^2} x_i(X_j, t)$, are the first to suffer jumps; hence, an acceleration wave is a singular surface of order two. Our interest here is in waves with $N \geq 2$.

In the terminology of Truesdell and Toupin [1960, 2] and Truesdell [1961, 2], the <u>speed of propagation</u> V of a wave Σ is the rate of advance of Σ along its unit normal relative to the particles instantaneously situated on Σ. A convenient measure of the <u>amplitude</u> of wave of order N is a vector s_i defined by

$$(1) \qquad (-V)^N s_i = [\overset{(N)}{x}_i]$$

where $[\overset{(N)}{x}_i]$ is the jump in $\overset{(N)}{x}_i = \frac{\partial^N}{\partial t^N} x_i(X_k, t)$ across Σ. for an acceleration, $V^2 s_i = [\ddot{x}_i]$.

Let the <u>Piola - Kirchhoff stress tensor</u> S_{ij} be defined by the formula*.

---

*Whenever an index is repeated in a product of two terms, summation over that index is understood.

(2) $$\rho S_{ik} F_{jk} = T_{ij}$$

where T_{ij} is the familiar stress tensor of Cauchy and ρ is the present mass density. For an elastic material, when thermodynamic influences are ignored, S_{ij} is a function of the deformation gradient :

(3) $$S_{ij} = S_{ij}(F_{k\ell})$$

The classical Fresnel-Hadamard theorem asserts that for any material obeying (2) the amplitude s_i and speed of propagation V of an acceleration wave traveling in the direction n_k must obey the propagation condition

(4) $$Q_{ij}(n_k)s_j = V^2 s_i \;,$$

where the tensor $Q_{ij}(n_k)$, called the acoustic tensor, is given by[**]

(5) $$Q_{ij}(n_k) = F_{am} F_{b\ell}\, n_a n_b \frac{\partial}{\partial F_{jm}} S_{i\ell}(F_{pq}).$$

Ericksen [1953, 1], working in the theory of isotropic incompressible hyperelastic materials, and Truesdell [1961, 2], working in the theory of compressible elastic materials, have shown that all elastic waves of order $N > 2$ must also obey the propagation condition (4) with $Q_{ij}(n_k)$ given by (5).

Even in elasticity theory one should include thermodynamic incluences and allow S_{ij} to depend not only on $F_{k\ell}$ but also

---

** For the most general statement of the Fresnel-Hadamard Theorem, and for a detailed discussion of its consequences, see Trusdell [1961, 2]

on a thermodynamic variable such as the temperature θ or the specific entropy η. In the thermodynamic theory of acceleration waves, $\dot{x}_i$, F_{ij}, θ, and η are taken to be continuous across the wave:

(6) $$[\dot{x}_i] = [F_{jk}] = [\theta] = [\eta] = 0.$$

An acceleration wave is said to be homothermal if, in addition to to (6), one has

(7) $$[\dot{\theta}] = [\frac{\partial\theta}{\partial X_j}] = 0.$$

On the other hand, if, in place of (7),

(8) $$[\dot{\eta}] = [\frac{\partial\eta}{\partial X_j}] = 0,$$

then the wave is called homentropic. It is the content of the first theorem of Duhem, that homothermal (homentropic) acceleration waves in elastic materials obey (4), proveded that the derivative $\frac{\partial}{\partial F_{jm}}$ in (5) is taken at constant temperature (entropy). The second theorem of Duhem gives the physical circumstances in which acceleration waves are homothermal or homentropic: Every acceleration wave in an elastic material obeying Fourier's law of heat conduction with positive-definite thermal conductivity is homothermal: every acceleration wave in an elastic material which does not conduct heat is homentropic*.

---

* The two theorems of Duhem are explained and given modern proofs by Truesdell [1961, 2].

An elastic material is said to be <u>hyperelastic</u> if the stress-strain function of (3) is obtained by differentiating a stored energy function Φ; i.e., if

$$S_{ij}(F_{k\ell}) = \frac{\partial}{\partial F_{ij}} \Phi(F_{k\ell}). \tag{9}$$

It is a familiar assertion of classical thermodynamics that every elastic material is hyperelastic in the sense that

$$S_{ij}(F_{k\ell}, \theta) = \frac{\partial}{\partial F_{ij}} \psi(F_{k\ell}, \theta), \tag{10}$$

$$S_{ij}(F_{k\ell}, \eta) = \frac{\partial}{\partial F_{ij}} (F_{k\ell}, \eta), \tag{11}$$

where ψ is the specific Helmoltz free energy $\psi = \epsilon - \theta\eta$, and ϵ the specific internal energy.

A <u>theorem of Hadamard</u> asserts that for a hyperelastic material the acoustic tensor (5) must be symmetric :

$$Q_{ij}(n_k) = Q_{ji}(n_k). \tag{12}$$

It follows from this result, then first theorem of Duhem, and the classical equations (10) and (11), that the acoustic tensor of an elastic material must be symmetric in homothermal and homentropic waves.

The recent years have seen the development of general theories of non-linear materials with memory*. In these studies it is assumed that the present stress depends not only on the present value of the strain but also on the past history of the strain, and one attempts to solve problems without specializing the functional expressing this dependence. Let us define a function $F^{(t)}_{kl}$ over the half-closed interval $[0, \infty)$ as follows;

$$F^{(t)}_{k\ell}(s) = F_{k\ell}(t-s) \quad , \qquad 0 \leq s < \infty. \tag{13}$$

This function is called <u>the history up to time</u> t of the deformation gradient. The past history $F^{(t)}_{(r)k\ell}$ of the deformation gradient is just the <u>restriction</u> of $F^{(t)}_{k\ell}$ to the open interval $(0, \infty)$; i.e., $F^{(t)}_{(r)k\ell}(s)$ agrees with $F^{(t)}_{k\ell}(s)$ for all $s > 0$ but is left undefined for $s=0$. Obviously, a knowledge of the history $F^{(t)}_{k\ell}$ is equivalent to a knowledge of the <u>present value</u> $F_{k\ell}(t) = F^{(t)}_{k\ell}(0)$ <u>and</u> the past history $F^{(t)}_{(r)k\ell}$. Following Noll [1958, 1], we define a <u>simple material</u> to be a material for which the stress is determined when $F^{(t)}_{k\ell}$ is given. For such a material we can write

$$S_{ij} = \underset{=}{S}_{ij}(F^{(t)}_{k\ell}) \ . \tag{14}$$

---

*See, for example, rhe works of Green and Rivlin [1957, 1], Noll [1958, 1,] Coleman and Noll [1960, 1] [1961, 1], and Wang [1965, 7], which are summarized and extended in the exposition of Truesdell and Noll [1965, 6].

Here $\underline{S}_{ij}$ is a <u>functional</u> ; i.e., a function whose argument is a function, $F^{(t)}_{kl}$, and whose value is a tensor, S_{ij}. When we wish to emphasize that for a general simple material S_{ij} depends on both the present value and the past history of the deformation gradient, we write (14) in the form*

$$(15) \qquad S_{ij} = \underline{S}_{ij}\,(F^{(t)}_{(r)k\ell}\,;F_{k\ell}\,) ,$$

where, for short, we put $F_{kl} = F^{(t)}_{kl}\,(0)$. On comparing (3) and (15), we see that the elastic materials considered in the previous section are those simple materials for which the dependence of $\underline{S}_{ij}(F^{(t)}_{(r)k\ell};\, F_{k\ell}\,)$ on the past history $F^{(t)}_{(r)ij}$ is negligible. Here we do not assume that influence of $F^{(t)}_{(r)}$ on S_{ij} can be neglected; we assume merely that this influence is compatible with the principle of fading memory; i.e., with a smoothness postulate used by Coleman and Noll [1960, 1] [1961, 1] to render mathematical the intuitive idea that strains which occurred in the very distant past should have a smaller influence on the present stress than strains which occurred in the recent past.

Let us denote by $D_{F_{mn}}\,\underline{S}_{ij}(F^{(t)}_{k\ell}\,)$ the fourth-order tensor obtained by differentiating the stress with respect to the present value of deformation gradient holding fixed the past history :

$$(16) \qquad D_{F_{mn}}\,\underline{S}_{ij}(F^{(t)}_{k\ell}\,) = \frac{\partial}{\partial F_{mn}}\,\underline{S}_{ij}(F^{(t)}_{(r)k\ell}\,;\, F_{k\ell}).$$

---

*There is no summation over k and l in equations (15) and (16).

This tensor [1964, 1, 2] gives the moduli for instantaneous response to small strain impulses superimposed on $F^{(t)}_{k\ell}$ at time t.

Equation (16) defines a linear differential operator $D_{F_{mn}}$ mapping functionals into functionals ; this operator plays a central role in the theory of wave propagation [1965, 1-4] and in the thermodynamics of materials with memory [1964, 1, 2].

The theory of singular surfaces propagating in materials with memory is not an empty subject. Among the materials subsumed under the class of simple materials with fading memory are the materials of the theory of linear viscoelasticity. In that theory, Sips [1951, 1] , Lee and Kanter [1953, 2] , and Chu [1962, 1] have exhibited explicit solutions of the dynamical equations showing shock waves. It is an elementary exercise to construct from these solutions others showing acceleration waves. Further, Pipkin [1966, 1] has obtained exact solutions showing shock and acceleration waves for a special simple fluid with fading memory that gives rise to nonlinear field equations .

Recently [1965, 1-4], we have been able to extend to general non-linear materials with memory the classical propagation theorems given in the previous section. For example, we have the following extension [1965, 4] of the Fresnel-Hadamard and Ericksen - Truesdell theorems. Consider a wave of order 2 or greater traveling in the direction n_k in a general simple material with fading memory ; such a wave obeys the propagation condition (4)*, and the tensor $Q_{ij}(n_k)$, now called the instantaneous acoustic tensor, is given by the following remarkably simple generalization of (5):

$$(17) \qquad Q_{ij}(n_k) = F_{am} F_{b\ell} n_a n_b D_{F_{jm}} \underset{=}{S}_{i\ell}(F^{(t)}_{pq}) \quad .$$

For the validity of this theorem the past history of the material just in front of the wave may be arbitrary, subject only to certain natural tameness hypotheses .

Of course, we know that the stress in a general material with memory depends not only on the history of the deformation gradient but also on the history of a thermodynamic variable. Hence, (14) should be replaced by either

$$(18) \qquad S_{ij} = \bar{\underset{=}{S}}_{ij}(F^{(t)}_{k\ell}, \theta^{(t)})$$

or

$$(19) \qquad S_{ij} = \hat{\underset{=}{S}}_{ij}(F^{(t)}_{k\ell}, \eta^{(t)}),$$

where the function $\theta^{(t)}$ is the <u>history of the temperature</u> and the function $\eta^{(t)}$ the <u>history of the specific entropy</u>

$$(20) \qquad \theta^{(t)}(s) = \theta(t-s), \qquad \eta^{(t)}(s) = \eta(t-s) \qquad 0 \leq s < \infty .$$

---

*Propagation conditions of the type (4) are known for acceleration waves in several special materials; for elastic materials (2.18) was derived by Hadamard [1901, 1] and for the theory of linear viscoelasticity by Herrera and Gurtin [1965, 5]. We have recently seen a manuscript by Varley [1965, 7] in which he arrives at (4) for acceleration waves in materials of integral type.

When (18) is assumed, it is also reasonable to assume that the specific **Helmholtz** free energy ψ is given by a functional $\underline{\underline{p}}$ of the histories $F_{k\ell}^{(t)}$, and $\theta^{(t)}$,

(21) $$\psi = \underline{\underline{p}}\,(F_{k\ell}^{(t)}, \theta^{(t)}),$$

and that the heat flux vector q_i depends not only on the present temperature gradient $g_m = \frac{\partial}{\partial x_m}\theta$ but also on $F_{k1}^{(t)}$ and $\theta^{(t)}$,

(22) $$q_i = \underline{\underline{g}}_i(F_{k\ell}^{(t)}, \theta^{(t)} ; g_m)$$

When (19) is assumed, it is reasonable to postulate that the specific internal energy is given by a functional of $F_{k\ell}^{(t)}$ and $\eta^{(t)}$;

(23) $$\epsilon = \underline{\underline{e}}\,(F_{k\ell}^{(t)}, \eta^{(t)}).$$

Starting from assumptions somewhat more general than these, Coleman [1964, 1] has shown that the functionals $\bar{\underline{\underline{S}}}_{ij}$ and $\hat{\underline{\underline{S}}}_{ij}$ are compatible with the principle of fading memory and the second law of thermodynamics only if these functionals are determined by $\underline{\underline{p}}$ and $\underline{\underline{e}}$ through the relations,

(24a) $$\bar{\underline{\underline{S}}}_{ij}(F_{k\ell}^{(t)}, \theta^{(t)}) = D_{F_{ij}}\, \underline{\underline{p}}(F_{k\ell}^{(t)}, \theta^{(t)}),$$

(24b) $$\hat{\underline{\underline{S}}}_{ij}(F_{k\ell}^{(t)}, \eta^{(t)}) = D_{F_{ij}}\, \underline{\underline{e}}(F_{k\ell}^{(t)}, \eta^{(t)}),$$

where $D_{F_{ij}}$ is the operator defined in (16). These relations generalize to materials with memory the classical relations (10) and

and (11) for elastic materials.

Even when the history of a thermodynamic variable is brought in, the propagation condition (4) still holds for homothermal and homentropic acceleration waves [1965, 3, 4]. For homothermal waves the instantaneous acoustic thensor Q_{ij} is given by (17) with $S_{ij}(F^{(t)}_{pq})$ replaced by $\bar{S}_{ij}(F^{(t)}_{pq}, \theta^{(t)})$, and the function $\theta^{(t)}$, the history of the temperature up to the moment of arrival of the wave, is held fixed in the computation of the derivative (16). For homentropic acceleration waves $D_{F_{pq}}$ is applied to $\hat{S}_{ij}(F^{(t)}_{pq}, \eta^{(t)})$ and the function $\eta^{(t)}$, the history of the entropy at the wave, is held fixed in (16) and (17). These observations extend to materials with memory the first theorem of Duhem. The second theorem of Duhem also has a direct generalization [1965, 3, 4]: In a definite conductor of heat, every acceleration wave is homothermal ; in a non-conductor, every acceleration wave is homentropic. Here, by a definite conductor, we mean a material for which $-\dfrac{\partial q_i}{\partial g_m}$, computed using (22), is always a positive - definite tensor. The proof of the theorem is straightforward for a definite conductor, the proof for a non-conductor, i.e., a material with $q_i \equiv 0$, uses a generalization [1964, 1] of the relations (24).

We should like to bring to the attention of experimenters the following extension [1965, 4] of the classical symmetry condition (12) : *The relations (24) imply that, even in a material with memory, the instantaneous acoustic tensors for both homothermal and homentropic waves are always symmetric tensors in the sense* of equation (12) . This theorem appears to supply a method of testing the physical appropriateness of the relations (24). Fortunately, Truesdell [1961, 2] [1964, 3,] working in the theory of elastic materials, has found situations in which measurements of wave velocity can test the symmetry of the acoustic tensor. His analyses can be applied with small modifications to the theory of acceleration waves entering a general material with memory which previous to arrival of the wave had always been at rest in a fixed configuration.

REFERENCES

1901	1	J. Hadamard, Bull. Soc. Math. France 29, 50-60.
1951	1	R. Sips, J. Polymer Sci. 6, 285-293.
1953	1	J. L. Ericksen, J. Rational Mech. Anal. 2, 329-337.
	2.	E. H. Lee & L. Kanter, J. Appl. Phys. 24, 1115-1122.
1957	1	A. E. Green & R. S. Rivlin, Arch. Rational Mech. Anal. 1, 1-21.
1958	1	W. Noll, Arch. Rational Mech. Anal. 2, 197-226.
1960	1	B. D. Coleman & W. Noll 6, 355-370.
	2	C. Truesdell & R. A. Toupin, The classical field theories. In: <u>Flügge's Encyclopedia of Physics,</u> 3/1, 225-793.
1961	1	B.D. Coleman & W. Noll, Rev. Mod. Phys. 33, 239-249.
	2	C. Truesdell, Arch. Rational Mech. Anal. 8, 263-296.
1962	1	B.-T. Chu, March 1962, ARPA report from the Division of Engineering, Brown University.
1964	1	B. D. Coleman, Arch. Rational Mech. Anal. 17, 1-46.
	2	B. D. Coleman, Arch. Rational Mech. Anal. 17, 23`-254
	3	C. Truesdell, Internat. Sympos. Second-Order Effects, Haifa, 1962, pp. 187-199.
1965	1	B. D. Coleman, & M.E. Gurtin, I. Herrera, R., Arch. Rational Mech. Anal. 19, 1, 1-19.
	2	B.D. Coleman, & M.E. Gurtin, Arch. Rational Mech. Anal. 19, 4
	3	B.D. Coleman M.E. Gurtin, Arch. Rational Mech. Anal. 19.5

1965 4 B. D. Coleman & M. E. Gurtin, Arch. Rational Mech. Anal. **19**, 5 .

5 I. Herrera, R. & M.E. Gurtin, Quart. Appl. Math. **22**, 360-364.

6 C. Truesdell & W. Noll, The non-linear field theories of mechanics, In : Flügge's Encyclopedia of Physics, **3/3** , in press.

7 E. Varley, Arch. Rational Mech. Anal. **19** , 3, 215-225.

8 C.-C. Wang, Arch. Rational Mech. Anal. **18**, 117-126.

1966 1 A.C. Pipkin, Quart. Appl. Math. , in press.

CENTRO INTERNAZIONALE MATEMATICO ESTIVO

(C.I.M.E.)

L. DE VITO

SUI FONDAMENTI DELLA MECCANICA DEI SISTEMI CONTINUI (II)

Corso tenuto a Bressanone dal 31 maggio al 9 giugno 1965

SUI FONDAMENTI DELLA MECCANICA DEI SISTEMI CONTINUI (II)

di

Luciano De Vito

Università-Roma

In un recente lavoro, in collaborazione tra il dott. Innocente Mazzaroli e lo scrivente, [1] si è cominciato a sviluppare un ordine di idee, proposto dal Prof. Gaetano Fichera già alcuni anni fa - in uno dei Suoi corsi all'Istituto Nazionale di Alta Matematica - relativo all'introduzione di un punto di vista "globale" nella teoria della "Meccanica dei sistemi continui", in luogo del punto di vista "puntuale" che più spesso viene adottato. Secondo tale punto di vista "globale", le forze di volume e di superficie non saranno più rappresentate da integrali (in senso ordinario) di funzioni continue, bensì da funzioni di insieme affatto arbitrarie purchè numerabilmente additive. Nel sucitato lavoro si è pervenuti all'impostazione ed allo studio delle equazioni cardinali della "Statica dei sistemi continui", per il caso dei corpi indefinitamente estesi. Verrà ora qui esposto il caso dei corpi limitati (includendovi anche il caso dei corpi "fluidi").

Strumento essenziale per tale studio è la nozione di "vettore dotato di divergenza in senso debole (o in senso integrale)", che rientra, come caso particolarissimo, nella più generale nozione di " k-misura dotata di differenziale in senso debole" introdotta

---

(1) L. De Vito e I. Mazzaroli: "Sui fondamenti della Meccanica dei sistemi continui", Memorie Acc. Naz. Lincei, v. VII, 1965.

da G. Fichera [(2)] . Alla considerazione delle equazioni cardinali della "Statica" per corpi limitati - secondo l'attuale punto di vista - conviene quindi premettere lo studio dei vettori a divergenza debole e quello (strettamente connesso) delle funzioni a gradiente debole. A tale studio sono, appunto, dedicati i primi paragrafi .

1. Notazioni

In tutto quel che segue, ci riferiremo sempre allo spazio euclideo E^3 nel quale penseremo introdotto un arbitrario sistema di riferimento cartesiano ortogonale. Se con x (con ξ) indicheremo il punto generico di E^3 , con x^1, x^2, x^3, (con ξ^1, ξ^2, ξ^3) denoteremo le coordinate del detto punto, nel riferimento cartesiano introdotto. Denoteremo poi con I gli intervalli (superiormente aperti) di E^3 in relazione al fissato sistema cartesiano x^1, x^2, x^3, con I_{y^i} la proiezione ortogonale di I sul piano $x^i = 0$ (avendo indicato con y^i la coppia x^{i+1}, x^{i+2}; ove gli indici si intendono definiti a meno dell'aggiunzione di un multiplo di 3) , con (a, b) l'intervallo $a \leq t < b$ dell'asse reale. Con il termine funzione , intenderemo sempre una funzione scalare , reale, di x , oppure, sottintendendo "vettoriale", un vettore a tre componenti reali pensato come funzione del punto x ; invece di "funzione vettoriale" useremo spesso semplicemente il termine "vettore" . Se con u indichiamo un vettore , con u_1, u_2, u_3 indicheremo sempre le sue componenti rispetto al sistema cartesiano introdotto in E^3 . Con i simboli $\mu \equiv \mu(B)$, $\alpha \equiv \alpha(B)$ denoteremo sempre una misura (cioè una funzione definita sui boreliani B

(2) G. Fichera: "Spazi di k-misure e di forme differenziali". Proc. Intern. Symp. on Linear Spaces, Jerusalem 1960, Israel Ac. of Sciences & Humanities, Pergamon Press, 1961.

numerabilmente additiva) scalare o vettoriale ; in questo secondo caso sottintenderemo sempre che essa sia a tre componenti (le componenti cartesiane di μ, rispetto al sistema x^1, x^2, x^3 saranno indicate con μ_1, μ_2, μ_3). Con il termine <u>dominio</u> - che denoteremo con il simbolo D - intenderemo sempre un dominio "propriamente regolare" (nel senso di Fichera[3]) ; con ciò intendiamo che D è un dominio regolare, avente la frontiera $\mathcal{F}D$ composta di una unica superficie regolare semplice e chiusa, tale inoltre che sia possibile definire un versore $\lambda(x)$ per ogni $x \in \mathcal{F}D$, sempre "penetrante" nell'interno di D, continuo al variare di x su $\mathcal{F}D$, di classe 1 su ciascuna delle porzioni di superficie regolare delle quali si compone $\mathcal{F}D$[4] ; inoltre deve esistere un numero positivo $\rho_0 < 1$ tale che, per ogni fissato $\rho \in (0, \rho_0)$, l'insieme descritto dal punto $x = \xi + \rho\lambda(\xi)$ al variare di ξ su $\mathcal{F}D$ sia una superficie regolare semplice e chiusa, che costituisce la completa frontiera di un dominio regolare, che indichiamo con D_ρ ; infine, la trasformazione che ad ogni $\xi \in \mathcal{F}D$ e ad ogni $\rho \in (0, \rho_0)$ associa il punto $x = \xi + \rho\lambda(\xi)$ deve essere invertibile. Si potrà allora ricoprire $\mathcal{F}D$ con un numero finito di dominii T_K in guisa tale che ciascun punto di $\mathcal{F}D$ sia interno ad uno almeno di essi ed inoltre, per ogni K, l'insieme

---

(3) Cfr. G. Fichera: "Premesse ad una teoria generale dei problemi al contorno per le equazioni differenziali", Corsi Ist. Naz. Alta Mat., 1958, pag. 52.

(4) Le nozioni di "dominio regolare", "porzione di superficie regolare" etc. sono qui intese nel senso precisato in : Picone-Fichera, "Trattato di Analisi Matematica", Tumminelli editore, 1956, Roma.

T_K sia rappresentabile, con una trasformazione continua e "di classe 1 a pezzi", sul rettangolo R_K dello spazio cartesiano X^3, definito da $0 \leq t^i \leq 1$; $i=1,2$; $-1 \leq t^3 \leq 2$; possiamo supporre anche che, nella detta trasformazione, l'immagine di $T_K \cap D$ sia $Q \equiv \equiv (0 \leq t^i \leq 1; i=1,2; 0 \leq t^3 \leq 2)$ e l'immagine di $T_K \cap \mathcal{F}D$ sia $S \equiv \equiv (0 \leq t^i \leq 1; i=1,2; t^3=0)$. Per ogni h tale che $0 \leq h \leq 1$ e per ogni $j=1,2$, denoteremo con S_j^h la porzione di S definita da: $0 \leq t^i \leq 1$; $0 \leq t^j \leq 1-h$; $t^3=0$; $i \neq j$; $i,j=1,2$. Se u è una funzione definita sui punti di $\mathcal{F}D$, e quindi, in particolare, in punti di T_K, con $u_K(t^1,t^2)$ indicheremo la u pensata come funzione del punto (t^1,t^2) variabile in S. Se poi $u_K(t^1,t^2)$ è "di classe 1 a pezzi" su S, in corrispondenza ad ongi K, il vettore che, nella rappresentazione detta, per ogni K, coincide con il vettore di componenti $\frac{\partial u_K}{\partial t^1}$, $\frac{\partial u_K}{\partial t^2}$, sarà indicato con il simbolo $\operatorname{grad}_{\mathcal{F}D} u$.

Con ν indicheremo sempre la normale interna a $\mathcal{F}D$; talora useremo anche il simbolo $\nu_{\mathcal{F}D}$. Diremo poi che il dominio propriamente regolare D è di classe m se la sua frontiera è una superficie di classe m.

Con $C^m(D)$ si indicherà l'insieme delle funzioni di classe m in D; con $\mathring{C}^m(D)$ quello delle funzioni di $C^m(D)$ che hanno supporto contenuto in $D-\mathcal{F}D$; con $\mathcal{L}^p(D)$ l'insieme delle funzioni di modulo di potenza p-esima sommabile in D secondo Lebesgue, $1 \leq p \leq +\infty$(5), con $\mathcal{H}^p(D)$ l'insieme delle funzioni di $\mathcal{L}^p(D)$ che posseggono derivate prime (in senso generalizzato) in

(5) Se $p=+\infty$ e se f è una funzione lebseguiana in un insieme L lebesguiano, dicendo che $|f|^p$ è sommabile in L, intendiamo che $|f|$ è quasi ovunque limitata in L o, come anche si dice, pseudolimitata in L. Converremo poi che, per $p=+\infty$, il simbolo $\left(\int_L |f|^p dx\right)^{1/p}$ significhi pseudo estr. sup. $|f|$.

D, di modulo di potenza p-esima sommabile in D, con $\mathcal{H}^{p',p}(D)$ l'insieme delle funzioni di $\mathcal{L}^{p'}(D)$ che posseggono derivate parziali prime (in senso generalizzato) in D, di modulo di potenza p- esima sommabile in D, $1 \leq p' \leq +\infty$, $1 \leq p \leq +\infty$; con $\Lambda^{p',p}(\mathcal{F}D)$ l'insieme delle funzioni u definite su $\mathcal{F}D$, aventi modulo di potenza p'-esima sommabile su $\mathcal{F}D$ e, se $p > 1$, tali che, per ogni k, risultino sommabili in $(0,1)$ le funzioni di h:

$$\int_{S_1^h} \left| \frac{u_k(t^1+h,t^2)-u_k(t^1,t^2)}{h} \right|^p dt^1 dt^2 \quad , \int_{S_2^h} \left| \frac{u_k(t^1,t^2+h)-u_k(t^1,t^2)}{h} \right|^p dt^1 dt^2$$

$p \leq +\infty$; con $V(B)$ l'insieme delle misure reali (scalari o vettoriali) definite sulla famiglia di tutti i boreliani contenuti nel boreliano B di E^3; analogamente, se B è un boreliano di $\mathcal{F}D$, con $V(B)$ intenderemo la totalità delle misure reali (scalari o vettoriali) definite sui boreliani di $\mathcal{F}D$ contenuti in B. Porremo inoltre :

$$\|u\|_{\mathcal{L}^p(D)} = \left| \int_D |u|^p dx \right|^{1/p} \quad , \|u\|_{\mathcal{L}^p(\mathcal{F}D)} = \left| \int_{\mathcal{F}D} |u|^p d\sigma \right|^{1/p}$$

$$\|u\|_{\mathcal{H}^p(D)} = \|u\|_{\mathcal{L}^p(D)} + \|\operatorname{grad} u\|_{\mathcal{L}^p(D)} \quad , \quad \|u\|_{\mathcal{H}^{p',p}(D)} = \|u\|_{\mathcal{L}^{p'}(D)} +$$

$$+ \|\operatorname{grad} u\|_{\mathcal{L}^p(D)} \quad , \quad 1 \leq p \leq +\infty; \quad \|u\|_{\Lambda^{p',p}(\mathcal{F}D)} = \|u\|_{\mathcal{L}^{p'}(\mathcal{F}D)} +$$

$$+ \sum_k \left[\int_0^1 dh \int_{S_1^h} \left| \frac{u_k(t^1+h,t^2)-u_k(t^1,t^2)}{h} \right|^p dt^1 dt^2 + \right.$$

$$\left. + \int_0^1 dh \int_{S_2^h} \left| \frac{u_k(t^1,t^2+h)-u_k(t^1,t^2)}{h} \right|^p dt^1 dt^2 \right]^{1/p} , \quad 1 < p \leq +\infty;$$

$$\|u\|_{\Lambda^{p',1}(\mathcal{F}D)} = \|u\|_{\mathcal{L}^{p'}(\mathcal{F}D)} \quad , \quad \|u\|_{C^0(D)} = \max_{x \in D} |u(x)| .$$

Se μ è una misura appartenente a $V(B)$, porremo:

$$\| \mu \|_{V(B)} = V_{\mu}(B)$$

ove $V_{\mu}(B)$ è la variazione totale di μ su B[6].

Le norme testè definite introducono, come è noto, una struttura di spazio normato rispettivamente negli insiemi $\mathcal{L}^p(D)$, $\mathcal{L}^p(\mathcal{F}D)$ etc. Seguiteremo a indicare i detti spazi normati con i medesimi simboli $\mathcal{L}^p(D)$, $\mathcal{L}^p(\mathcal{F}D)$ etc.[7]. Lo spazio duale di uno spazio normato S sarà denotato con S^* e la norma in S^* con $\| \ \|_{S^*}$.

Converremo poi sempre di indicare con q, q', q'' etc. il duale rispettivamente di p, p', p'' etc. (cioè $\frac{1}{p} + \frac{1}{q} = 1$, $\frac{1}{p'} + \frac{1}{q'} = 1$, $\frac{1}{p''} + \frac{1}{q''} = 1$) con la convenzione che il duale di 1 è $+\infty$ e viceversa.

2. Divergenza debole di un vettore

Sia u un vettore di $\mathcal{L}^1(D)$. Si dice che u è dotato di divergenza debole su $D - \mathcal{F}D$ (rappresentata da una misura) se esiste $\mu \in V(D - \mathcal{F}D)$ tale che:

$$\int_D u \times \operatorname{grad} v \, dx = - \int_{D-\mathcal{F}D} v \, d\mu \tag{1}$$

per ogni $v \in \mathcal{C}^1(D)$, e la misura μ, univocamente determinata da questa condizione, chiamasi divergenza debole di u su $D - \mathcal{F}D$.

Si diche che u è dotato di divergenza debole su D (rappresentata da una misura) se esiste $\mu \in V(D)$ tale che:

---

(6) Cfr. loc. cit. in[1] p. 214.

(7) Gli spazi $\Lambda^{p',p}(\mathcal{F}D)$ sono stati introdotti da E. Gagliardo (per $p' = p$) cfr. E. Gagliardo, "Caratterizzazioni delle tracce sulla frontiera relative ad alcune classi di funzioni in n variabili", Rend. Sem. Mat. Univ. di Padova, 1957.

$$\int_D u \times \operatorname{grad} v \, dx = - \int_D v \, d\mu \tag{2}$$

per ogni $v \in C^1(D)$.

E' ovvio che :

I. Se $u \in \mathcal{L}^1(D)$ è dotato di divergenza debole $\mu \in V(D)$ su D, e di divergenza debole $\tilde{\mu} \in V(D-\mathcal{F}D)$ su $D-\mathcal{F}D$, risulta $\tilde{\mu}(B) = \mu(B)$ per ogni boreliano $B \subset D-\mathcal{F}D$.

II. Se $u \in \mathcal{L}^1(D)$ è dotato di divergenza debole $\mu \in V(D)$ su D allora u è anche dotato di divergenza debole $\tilde{\mu} \in V(D-\mathcal{F}D)$ su $D-\mathcal{F}D$ e riesce : $\tilde{\mu}(B) = \mu(B)$ per ogni boreliano $B \subset D-\mathcal{F}D$.

In generale, invece, non è vero che, dall'essere $u \in \mathcal{L}^1(D)$ dotato di divergenza debole su $D-\mathcal{F}D$, segua che u è dotato di divergenza debole su D. Si ha però che ;

III. Se $u \in \mathcal{L}^1(D)$ è dotato di divergenza debole $\mu \in V(D-\mathcal{F}D)$ su $D-\mathcal{F}D$, esiste un vettore $u_0 \in \mathcal{L}^1(D)$ dotato di divergenza debole su $D-\mathcal{F}D$ identicamente nulla, tale che $u+u_0$ sia dotato di divergenza debole $\in V(D)$ su D.

Sia infatti $\alpha \in V(\mathcal{F}D)$ tale che $\alpha(\mathcal{F}D) = -\mu(D-\mathcal{F}D)$ e si consideri la $\tilde{\mu} \in V(D)$ così definita : $\tilde{\mu}(B) = \mu[B \cap (D-\mathcal{F}D)] + \alpha(B \cap \mathcal{F}D)$.
Si ha : $\tilde{\mu}(D) = 0$. Esiste allora $\tilde{u} \in \mathcal{L}^1(D)$ che ha per divergenza debole su D la $\tilde{\mu}$ (come segue da un principio esistenziale di G. Fichera[8] e dalla diseguaglianza di Poincarè :

$$\inf_{\text{cost.}} \| v + \text{cost} \|_{C^0(D)} \leq K \| \operatorname{grad} v \|_{\mathcal{L}^p(D)} \;, \quad p>3 \;)$$

---

[8] Cfr. G. Fichera: "Alcuni recenti sviluppi della teoria dei problemi al contorno per le equazioni alle derivate parziali", Atti Convegno Internaz. sulle equazioni a deriv. parziali , Trieste , 1954 .

Il vettore $\tilde{u}$ ha per divergenza debole su $D-\mathcal{F}D$ la μ e quindi differisce da u per un vettore $u_0 \in \mathcal{L}^1(D)$ che ha divergenza debole su $D-\mathcal{F}D$ nulla.

Se u è dotato di divergenza debole μ su D o su $D-\mathcal{F}D$, e se la misura μ è assolutamente continua sui boreliani contenuti in $D-\mathcal{F}D$, con "densità" $f \in \mathcal{L}^1(D)$, la f dicesi la <u>divergenza generalizzata</u> (locale) di u e si indicherà con $\operatorname{div} u$; si dirà anche, in tal caso, che u ha divergenza debole rappresentata, nei punti di $D-\mathcal{F}D$, dalla funzione $f = \operatorname{div} u$. Dicendo che $u \in \mathcal{L}^1(D)$ ha come divergenza debole su $D-\mathcal{F}D$ (su D) la funzione $f \in \mathcal{L}^{q'}(D)$ intendiamo che ha come divergenza debole su $D-\mathcal{F}D$ (su D) una misura assolutamente continua con densità $f \in \mathcal{L}^{q'}(D)$.

3. Gradiente debole

Sia u una funzione di $\mathcal{L}^1(D)$. Si dice che u <u>è dotata di gradiente debole su</u> $D-\mathcal{F}D$ (rappresentato da una misura vettoriale) se esiste $\mu \in V(D-\mathcal{F}D)$ tale che:

$$(3) \qquad \int_D u \operatorname{div} v \, dx = - \int_{D-\mathcal{F}D} v \, d\mu$$

per ogni $v \in \overset{\circ}{C}{}^1(D)$ e μ è il <u>gradiente debole su</u> $D-\mathcal{F}D$ <u>di</u> u.

Si dice che u <u>è dotata di gradiente debole su</u> D (rappresentata da una misura) se esiste $\mu \in V(D)$ tale che:

$$(4) \qquad \int_D u \operatorname{div} v \, dx = - \int_D v \, d\mu$$

per ogni $v \in C^1(D)$; μ <u>è il gradiente debole di</u> u <u>su</u> D.

E' ovvio che:

IV. <u>Se</u> $u \in \mathcal{L}^1(D)$ <u>è dotata di gradiente debole</u> $\mu \in V(D)$ <u>su</u> D

e di gradiente debole $\tilde{\mu} \in D-\mathcal{F}D$ su $D-\mathcal{F}D$, risulta $\mu(B)=\tilde{\mu}(B)$ per ogni boreliano $B \subset D-\mathcal{F}D$.

V. Se $u \in \mathcal{L}^1(D)$ è dotata di gradiente debole $\mu \in V(D)$ su D, allora u è dotata anche di gradiente debole $\tilde{\mu} \in V(D-\mathcal{F}D)$ su $D-\mathcal{F}D$ e riesce: $\tilde{\mu}(B)=\mu(B)$ per ogni boreliano $B \subset D-\mathcal{F}D$.

Sussiste anche il seguente teorema :

VI . Se $u \in \mathcal{L}^1(D)$ è dotata di gradiente debole $\mu \in V(D-\mathcal{F}D)$ su $D-\mathcal{F}D$, allora u è anche dotata di gradiente debole $\tilde{\mu} \in V(D)$ su D e riesce: $\tilde{\mu}(B)=\mu(B)$ per ogni boreliano $B \subset D-\mathcal{F}D$.

La dimostrazione di questo teorema seguirà dai risultati dei successivi paragrafi 7 e 8 .

4. Gradiente debole su $D-\mathcal{F}D$.

Si ha intanto, come è noto, che :

VII. Condizione necessaria e sufficiente perchè $\mu \in V(D-\mathcal{F}D)$ sia il gradiente debole su $D-\mathcal{F}D$ di una funzione $u \in \mathcal{L}^1(D)$ è che riesca: $\int_{D-\mathcal{F}D} v \, d\mu = 0$ per ogni vettore $v \in \overset{\circ}{C}{}^1(D)$ che abbia divergenza nulla[9] .

Se u è dotata di gradiente debole $\mu \in V(D-\mathcal{F}D)$ su $D-\mathcal{F}D$ e se, $\mu(B)=\int_B f \, dx$, allora, come è noto, u è assolutamente continua secondo Tonelli in D e le componenti di f sono le derivate parziali prime , in senso generalizzato, di u [10] . Con le notazioni del paragrafo 1 , si avrà allora : $u \in \mathcal{H}^1(D)$. Scriveremo anche $f = \operatorname{grad} u$ (intendendo il gradiente in senso generalizzato). Se inoltre

---

[9] Cfr. L. S. Schwartz: "Théorie des distributions", Act. Sci. Industr. Parigi, 1957, t. I. p. 59 . .

[10] Cfr. G. Fichera, loc. cit . in[3] , cap. III .

si ha : $u \in \mathcal{L}^{p'}(D)$, $f \in \mathcal{L}^{p}(D)$ sarà $u \in \mathcal{H}^{p',p}(D)$. Come è noto, dallo essere $u \in \mathcal{H}^{1,p}(D)$ segue $u \in \mathcal{H}^{p',p}(D)$ e risulta :

$$(5)_{p',p} \qquad \inf_{cost.} \| u + cost \|_{\mathcal{L}^{p'}(D)} \leq K_{p',p}(D) \| \operatorname{grad} u \|_{\mathcal{L}^{p}(D)} ,$$

$$1 \leq p' \leq \frac{3p}{3-p} , 1 \leq p < 3 ; \; 1 \leq p' < +\infty , p = 3 ; \; 1 \leq p' \leq +\infty , 3 < p \leq +\infty$$

per ogni $u \in \mathcal{H}^{p',p}(D)$[11] ; inoltre, se $3 < p \leq +\infty$, si ha, di più, $u \in C^{0}(D)$. Se poi $u \in \mathcal{H}^{p',p}(D)$, allora u risulta dotata di gradiente debole su $D - \mathcal{F}D$ rappresentato dalla misura assolutamente continua $\mu(B) = \int_B \operatorname{grad} u \, dx$[10].

VIII. Se $u \in \mathcal{L}^{1}(D)$ è dotata di gradiente debole $\mu \in V(D - \mathcal{F}D)$ su D si ha : $u \in \mathcal{L}^{p'}(D)$, $1 \leq p' \leq \frac{3}{2}$ e riesce :

$$(6) \qquad \inf_{cost.} \| u + cost \|_{\mathcal{L}^{p'}(D)} \leq K \, V_{\mu}(D - \mathcal{F}D) , \quad 1 \leq p' \leq \frac{3}{2}$$

Dato che u ha per gradiente debole su $D - \mathcal{F}D$ una misura $\mu \in V(D - \mathcal{F}D)$, risulta $u \in \mathcal{L}^{p'}(D)$ per ogni $p' < \frac{3}{2}$[12]. Sia $\mathcal{K}_{\varepsilon}(x - \xi)$ un nucleo regolarizzatore nel senso di Friedrichs [13] e poniamo

$$(7) \qquad u_n(x) = \int_D \mathcal{K}_{1/n}(x - \xi) u(\xi) \, d\xi .$$

Se D' è un dominio propriamente regolare $\subset D - \mathcal{F}D$ riuscirà :

$$(8) \qquad \lim_{n \to \infty} \| u_n - u \|_{\mathcal{L}^{p'}(D')} = 0 .$$

Per ongi $x \in D'$, non appena n è abbastanza grande, si avrà :

---

(11) Cfr. S. L. Sobolev: "Su un teorema di analisi funzionale", Math. Sbornik , 1938 e E. Gagliardo: "Proprietà di alcune classi di funzioni in piu variabili", Ricerche di Mat., 1958 (e bibliografia qui citata). Le limitazioni per p, p', come è noto , sono le migliori possibili.

(12) Cfr. loc. cit. in (9) .

(13) Cfr. loc. cit. in (10) .

$$\operatorname{grad} u_m(x) = \int_D \operatorname{grad}_x \mathcal{K}_{1/m}(x-\xi)\, u(\xi)\, d\xi =$$

$$= -\int_D \operatorname{grad}_\xi \mathcal{K}_{1/m}(x-\xi)\, u(\xi)\, d\xi = \int_{D-\mathcal{F}D} \mathcal{K}_{1/m}(x-\xi)\, d_\xi \mu$$

donde, con un facile calcolo :

(9) $$\| \operatorname{grad} u_m \|_{\mathcal{L}^1(D')} \leq V_\mu(D-\mathcal{F}D)^{(14)}, \quad m > \frac{1}{\overline{\mathcal{F}D\,\mathcal{F}D'}}.$$

Detta $\mathcal{V}_{q'}(D')$ la varietà lineare dei vettori di $\mathcal{L}^\infty(D')$ che hanno divergenza debole su D' contenuta in $\mathcal{L}^{q'}(D')$, per ogni $v \in \mathcal{V}_{q'}(D')$ si ha:

$$\int_{D'} v \times \operatorname{grad} u_m\, dx = -\int_{D'} u_m \operatorname{div} v\, dx$$

Da qui e da (8) si trae che esiste finito il limite :

(9') $$\lim_{m\to\infty} \int_{D'} v \times \operatorname{grad} u_m\, dx = -\int_{D'} u \operatorname{div} v\, dx \equiv F_{D'}(v)$$

per ogni $v \in \mathcal{V}_{q'}(D')$. $F_{D'}$ è funzionale lineare e continuo sulla varietà $\mathcal{V}_{q'}(D')$ ove pensiamo di aver introdotto la norma : $\| \ \|_{\mathcal{L}^\infty(D')}$, dato che, in forza di (9), riesce :

(10) $$|F_{D'}(v)| \leq \|\mu\|_{V(D-\mathcal{F}D)} \|v\|_{\mathcal{L}^\infty(D')}, \quad v \in \mathcal{V}_{q'}(D').$$

Allora $F_{D'}$ può essere prolungato in un funzionale lineare e continuo su tutto $\mathcal{L}^\infty(D')$ (che seguiteremo a denotare con il medesimo simbolo $F_{D'}$) tale che :

(11) $$\|F_{D'}\|_{[\mathcal{L}^\infty(D')]^*} \leq \|\mu\|_{V(D-\mathcal{F}D)}$$

e, per ogni $v \in \mathcal{V}_{q'}(D')$, riesce : $\int_{D'} u \operatorname{div} v\, dx = -F_{D'}(v)$, talchè $F_{D'}$ risulta ortogonale alla varietà $\mathcal{V}_{D'}$ di tutti i vettori di $\mathcal{L}^\infty(D')$ che hanno divergenza debole su D' identicamente

---

(14) Questa proprietà delle u_m è stata osservata da J. Serrin: "On the Differentiability of functions of several variables", Arch. Rat. Mech. Analysis", 1961 .

nulla. Consideriamo ora l'equazione :

(12) $$\int_{D'} \tilde{u} \operatorname{div} v \, dx = -F_{D'}(v) \quad , \quad v \in \mathcal{V}_3(D')$$

nell'incognita $\tilde{u} \in \mathcal{L}^{3/2}(D)$. Dato che $F_{D'} \perp \mathcal{V}_{D'}$, per il citato principio esistenzale di G. Fichera, la (12) avrà soluzione, in forza della seguente formula di maggiorazione :

(13) $$\mathcal{J}_{D'} \| v + v_0 \|_{\mathcal{L}^\infty(D')} \leq K(D') \underset{v \in \mathcal{V}_3(D')}{} \| \operatorname{div} v \|_{\mathcal{L}^3(D')}$$

la quale è duale, nel senso di Fichera[15], della formula di maggiorazione $(5)_{\frac{3}{2},1}$ ove si assuma $K(D') = K_{\frac{3}{2},1}(D')$. Tale soluzione sarà, ovviamente, unica a meno dell'aggiunzione di una costante. Per il principio di dualità di G. Fichera e per (11) si avrà :

(14) $$\inf_{\text{cost.}} \| \tilde{u} + \text{cost} \|_{\mathcal{L}^{3/2}(D')} \leq K(D') \| \mu \|_{V(D - \mathcal{F}D)} .$$

Se ora, come "funzione di prova", in (12), assumiamo una $v \in \overset{\circ}{C}{}^1(D)$, avremo :

$$\int_{D'} \tilde{u} \operatorname{div} v \, dx = -\int_{D'} v \, d\mu$$

Ne viene che u differisce da $\tilde{u}$ in D' per una costante e quindi riesce : $u \in \mathcal{L}^{3/2}(D')$: inoltre, per (14), si ha :

(15) $$\inf_{\text{cost.}} \| u + \text{cost} \|_{\mathcal{L}^{3/2}(D')} \leq K(D') \| \mu \|_{V(D - \mathcal{F}D)} .$$

Possiamo ora assumere, come D', uno qualsiasi dei domini D_ς dei quali alla definizione di dominio propriamente regolare, supponendo $0 < \varsigma < \frac{1}{2}\varsigma_0$. Sia $v \in C^1(D_\varsigma)$. Per ogni punto $x \in D - D_\varsigma$ poniamo: $v(x) = v[\xi + t\lambda(\xi)] = v[\xi + (2\varsigma - t)\lambda(\xi)]$. La v, così prolungata a tutto D, risulta ivi uniformemente lipschitziana; inoltre esistono due costanti

(15) Cfr. G. Fichera, loc. cit. in (3), p. 38 .

A_1, A_2, dipendenti solo da D, tali che

$$A_1 \left| \operatorname{grad} v(x)_{x=\xi+t\lambda(\xi)} \right| \leq \left| \operatorname{grad} v(x)_{x=\xi+(2\rho-t)\lambda(\xi)} \right| \leq A_2 \left| \operatorname{grad} v(x)_{x=\xi+t\lambda(\xi)} \right|.$$

Esiste allora una costante L (dipendente solo da D) tale che:

$$\int_{D-D_\rho} |\operatorname{grad} v|\, dx \leq L \int_{D_\rho - D_{2\rho}} |\operatorname{grad} v|\, dx, \quad 0<\rho<\frac{\rho_0}{2}.$$

Ne viene:

$$\inf_{\text{cost.}} \| v + \text{cost.} \|_{\mathcal{L}^{3/2}(D')} \leq K_{3/2,1}(D)(L+1)\, \| \operatorname{grad} v \|_{\mathcal{L}^1(D')}$$

per ogni $v \in C^1(D')$. Se ne deduce che, come costante $K(D') \equiv K_{3/2,1}(D')$, ogni volta che $D' \equiv D_\rho$, $0<\rho<\frac{\rho_0}{2}$, si può assumere il numero (dipendente solo da D) $K_{3/2,1}(D)(L+1)$. Da (15) si trae allora:

$$\inf_{\text{cost}} \| u + \text{cost} \|_{\mathcal{L}^{3/2}(D')} \leq K_{3/2,1}(D)(L+1)\, \| \mu \|_{V(D-\mathcal{F}D)}$$

e da qui scende subito $u \in \mathcal{L}^{3/2}(D)$. La maggiorazione (6) è poi immediata conseguenza - in virtù del citato principio di G. Fichera - del fatto che, per ogni $\mu \in V(D-\mathcal{F}D)$ ortogonale a ciascuna $v \in \overset{\circ}{C}{}^1(D)$ con $\operatorname{div} v = 0$, esiste $u \in \mathcal{L}^{p'}(D)$, $1 \leq p' \leq 3/2$, che ha μ come gradiente debole su $D-\mathcal{F}D$.

4. Gradiente debole su D

IX. <u>Condizione necessaria e sufficiente perchè $\mu \in V(D)$ sia il gradiente debole su D di una funzione $u \in \mathcal{L}^1(D)$ è che riesca:</u>

$$\int_D v\, d\mu = 0 \tag{16}$$

<u>per ogni $v \in C^1(D)$ che abbia divergenza nulla</u>.

La necessità è ovvia. Mostriamo la sufficienza. Se indichiamo con I un intervallo chiuso, contenente D nel proprio interno, e se prolunghiamo μ in una $\tilde{\mu} \in V(I)$ ponendo, per ogni $B \subset I$:

$\tilde{\mu}(B)=\mu(B\cap D)$, la condizione (16) implica, per il teor. VII, l'esistenza di $\tilde{u}\in\mathcal{L}^1(\bar{I})$ che ha per gradiente debole su $\bar{I}-\mathcal{F}\bar{I}$ la $\tilde{\mu}$. Sia ora v un qualsiasi vettore di $C^1(\bar{I})$ identicamente nullo in D; dall'essere, per ogni siffatto v: $\int_{\bar{I}}\tilde{u}\,\mathrm{div}\,v\,dx=0$ si deduce $\tilde{u}=c$ (c costante) in $\bar{I}-D$. Allora $u=\tilde{u}-c$ soddisfa (4) per ogni $v\in C^1(D)$ e quindi ha per gradiente debole su D la μ.

X. <u>Se $\mu\in V(D)$ verifica la (16) per ogni $v\in C^1(D)$ avente divergenza nulla, esiste una ed una sola soluzione $u\in\mathcal{L}^1(D)$ della equazione (4). Tale soluzione appartiene a $\mathcal{L}^{3/2}(D)$ e verifica la diseguaglianza:</u>

(17) $$\|u\|_{\mathcal{L}^{3/2}(D)}\leq K\,\|\mu\|_{V(D)}$$

<u>con K costante indipendente da u</u>.

Dalla dimostrazione precedente appare che ogni soluzione $u\in\mathcal{L}^1(D)$ di (4) differisce per una costante da una $\tilde{u}$ che ha per gradiente debole su $D-\mathcal{F}D$ la μ. Per il teor. VIII si ha: $\tilde{u}\in\mathcal{L}^{3/2}(D)$ e quindi è $u\in\mathcal{L}^{3/2}(D)$. L'unicità di u è evidente. Il sussistere di (17) è conseguenza del principio esistenziale, già ricordato, di G. Fichera.

Supponiamo ora che la $\mu\in V(D)$, verificante (16), sia assolutamente continua con densità $f\in\mathcal{L}^p(D)$. Allora, poichè la soluzione u di (4) (unica) ha per gradiente debole su $D-\mathcal{F}D$ la μ, per i già citati risultati di Sobolev si ha: $u\in\mathcal{H}^{p',p}(D)$ ove p' e p verificano le limitazioni già indicate per la maggiorazione $(5)_{p',p}$. Si ha pertanto il teorema:

XI. <u>Esiste una costante $H_{p',p}(D)$, dipendente solo dai numeri p', p e da D, tale che, per ogni $u\in\mathcal{L}^{p'}(D)$ avente per</u>

gradiente debole su D la misura $\mu(B)=\int_B f\,dx$, $f\in\mathcal{L}^p(D)$, riesce :

$$(18)_{p',p} \qquad \|u\|_{\mathcal{L}^{p'}(D)} \leq H_{p',p}(D)\,\|f\|_{\mathcal{L}^p(D)}$$

$$1\leq p'\leq \frac{3p}{3-p},\ 1\leq p<3\ ;\ 1\leq p'<+\infty,\ p=3\ ;\ 1\leq p'\leq+\infty,\ 3<p\leq+\infty.$$

In particolare, dalla $(18)_{p',p}$ segue :

$$(19)_{p',p} \qquad \|v\|_{\mathcal{L}^{p'}(D)} \leq H_{p',p}(D)\,\|\operatorname{grad} v\|_{\mathcal{L}^p(D)}\ ,\ v\in \overset{\circ}{C}{}^1(D)$$

$$1\leq p'\leq \frac{3p}{3-p},\ 1\leq p<3\ ;\ 1\leq p'<+\infty,\ p=3\ ;\ 1\leq p'\leq+\infty,\ 3<p\leq+\infty.$$

6. Teoremi d'esistenza e formule di maggiorazione per l'operatore divergenza debole su D.

XII. Condizione necessaria e sufficiente perchè la misura $\mu\in V(D)$ sia la divergenza debole su D di un vettore $u\in\mathcal{L}^1(D)$ è che riesca $\mu(D)=0$.

E' immediata conseguenza dell'applicazione del principio esistenziale di G. Fichera, già citato, alla equazione (2) e della formula di maggiorazione

$$(20)_p \qquad \inf_{\text{cost.}}\|v+\text{cost.}\|_{C^0(D)} \leq K_{\infty,p}(D)\,\|\operatorname{grad} v\|_{\mathcal{L}^p(D)}\ ,\quad p>3,\ v\in C^1(D),$$

che è inclusa nella $(5)_{\infty,p}$.

XIII . Se $\mu\in V(D)$ è tale che $\mu(D)=0$, esistono infiniti vettori $u\in\mathcal{L}^1(D)$ che hanno per divergenza debole su D la misura μ. Fissato comunque $q<3/2$ tra di essi ve n'è sempre almeno uno appartenente a $\mathcal{L}^q(D)$. In generale non ve n'è alcuno che appartenga a $\mathcal{L}^{\bar q}(D)$ con $\bar q\geq 3/2$. Esiste una costante $K_q(D)$, $1\leq q<3/2$, tale che, in corrispondenza ad un qualsiasi vettore $u\in\mathcal{L}^q(D)$ avente μ per divergenza debole su D , risulta :

$$(21)_q \qquad \inf_{u_0\in\mathcal{U}_q}\|u+u_0\|_{\mathcal{L}^q(D)} \leq K_q(D)\,\|\mu\|_{V(D)}\ ,\ 1\leq q<3/2$$

ove con $\mathcal{U}_q$ si intende l'insieme dei vettori $u_0 \in \mathcal{L}^q(D)$ che hanno divergenza debole su D identicamente nulla.

Dall'applicazione del citato principio esistenziale di G. Fichera e dalla $(20)_p$ si trae l'esistenza di una soluzione u dell'equazione (2) (termine noto μ verificante la condizione di compatibilità $\mu(D)=0$) appartenente a $\mathcal{L}^q(D)$ con q duale di p, $3<p$. La $(21)_q$ sussiste in quanto duale (nel senso di Fichera) della $(20)_p$. L'esistenza di infinite autosoluzioni è evidente. Se, per un fissato $\bar{q} \geq 3/2$, in corrispondenza ad ongi $\mu \in V(D)$ con $\mu(D)=0$, esistesse una soluzione u di (2) contenuta in $\mathcal{L}^{\bar{q}}(D)$, in virtù del citato principio esistenziale di G. Fichera dovrebbe sussistere la formula di maggiorazione (necessaria) $(20)_{\bar{p}}$ con $\bar{p} \leq 3$. Ma è noto che $(20)_{\bar{p}}$, con $\bar{p} \leq 3$, non è vera.

XIV. Se $\mu \in V(D)$ è tale che $\mu(D)=0$ e se μ è assolutamente continua con densità $f \in \mathcal{L}^{q'}(D)$, $1 \leq q' \leq +\infty$, esistono infiniti vettori $u \in \mathcal{L}^q(D)$ che hanno per divergenza debole su D la f e riesce:

$$(22)_{q,q'} \qquad \inf_{u_0 \in \mathcal{U}_q} \| u+u_0 \|_{\mathcal{L}^q(D)} \leq K_{q,q'}(D) \, \| \operatorname{div} u \|_{\mathcal{L}^{q'}(D)}, \quad \operatorname{div} u = f$$

$$1 \leq q < 3/2, \; 1 \leq q' \leq +\infty; \; q=3/2, \; 1 < q' \leq +\infty; \; 3/2 < q \leq +\infty, \; \frac{3q}{3+q} \leq q' \leq +\infty,$$

ove si convenga che, per $q=+\infty$, sia $3q/(3+q)=3$. $\mathcal{U}_q$ ha il significato specificato nell'enunciato del teorema precedente. Le limitazioni per q e q' sono le migliori possibili.

La tesi segue facilmente dall'applicazione del principio di esistenza e di dualità di G. Fichera alla equazione (2) e dalle diseguaglianze $(5)_{p',p}$ delle quali le $(22)_{q,q'}$ sono le duali. Se la limitazione $3/2 < q \leq +\infty$, $\frac{3q}{3+q} \leq q' \leq +\infty$ fosse migliorabile (rispetto a q'), per il citato principio esistenziale di G. Fichera dovrebbe sussistere la formula di maggiorazione $\inf_{\text{cost.}} \|v+\text{cost}\|_{\mathcal{L}^{p'}(D)} \leq K \| \text{grad}\, v \|_{\mathcal{L}^{p}(D)}$ per $1 \leq p < 3$ e per qualche $p' > \frac{3p}{3-p}$, il che invece non è possibile (come fu esplicitamente notato da Sobolev nel luogo citato). Resta quindi solo da esaminare il caso $q'=1$, $q=3/2$. Supponiamo allora, per assurdo, che, per ogni $f \in \mathcal{L}^1(D)$ e tale che $\int_D f\, dx = 0$, esista $v \in \mathcal{L}^{3/2}(D)$ avente per divergenza debole su D la f. Detta $\mathcal{V}$ la totalità dei vettori di $\mathcal{L}^{3/2}(D)$ aventi divergenza debole su D contenuta in $\mathcal{L}^1(D)$, dovrebbe allora aversi :

$$\inf_{v_0 \in \mathcal{U}_{3/2}} \| v + v_0 \|_{\mathcal{L}^{3/2}(D)} \leq K \| \operatorname{div} v \|_{\mathcal{L}^1(D)}, \quad v \in \mathcal{V}.$$

In virtù del citato principio esistenziale di G. Fichera, l'equazione $\int_D u \operatorname{div} v\, dx = -\int_D v \times f\, dx$, $v \in \mathcal{V}$, con termine noto $f \in \mathcal{L}^3(D)$ ed incognita $u \in \mathcal{L}^{\infty}(D)$ ammette soluzione in corrispondenza ad ogni f ortogonale ad $\mathcal{U}_{3/2}$ e riesce :

$$\inf_{\text{cost.}} \| u + \text{cost} \|_{\mathcal{L}^{\infty}(D)} \leq K \| f \|_{\mathcal{L}^3(D)}$$

in particolare, questa diseguaglianza dovrebbe sussistere per $u \in C^1(D)$ e ciò , come è noto , è assurdo.

7. Teoremi d'esistenza e formule di maggiorazione per l'operatore divergenza debole su $D - \mathcal{F}D$.

XV . Assegnato comunque $\mu \in V(D-\mathcal{F}D)$ esistono infiniti vettori $u \in \mathcal{L}^1(D)$ che hanno divergenza debole su $D-\mathcal{F}D$ coincidente con μ . Fissato comunque $q < 3/2$ tra di essi ve n'è sempre almeno uno appartenente a $\mathcal{L}^q(D)$. In generale, non ve n'è alcuno che appartenga ad $\mathcal{L}^q(D)$ con $q \geq 3/2$. Esiste una costante $H_q(D)$ tale che, in corrispondenza ad un qualsiasi vettore $u \in \mathcal{L}^q(D)$ avente μ come divergenza debole su $D-\mathcal{F}D$, risulta :

$$(23)_q \qquad \inf_{u_0 \in \tilde{\mathcal{U}}_q} \| u + u_0 \|_{\mathcal{L}^q(D)} \leq H_q(D) \| \mu \|_{V(D-\mathcal{F}D)} , \quad 1 \leq q < 3/2$$

ove con $\tilde{\mathcal{U}}_q$ si intende l'insieme dei vettori di $\mathcal{L}^q(D)$ che hanno per divergenza debole su $D-\mathcal{F}D$ la misura identicamente nulla.

Segue immediatamente dall'applicazione del principio esistenziale di G. Fichera alla equazione (1) e dalla formula di maggiorazione :

$$(24)_p \qquad \| v \|_{C^0(D)} \leq H_{\infty,p}(D) \| \operatorname{grad} v \|_{\mathcal{L}^p(D)} , \quad p > 3, \quad v \in C^0(D) ,$$

che è inclusa nella $(19)_{\infty,p}$.

XVI . Se $\mu \in V(D-\mathcal{F}D)$ è una misura assolutamente continua con densità $f \in \mathcal{L}^{q'}(D)$ esistono infiniti vettori $u \in \mathcal{L}^q(D)$ che hanno per divergenza debole μ su $D-\mathcal{F}D$ e riesce :

$$(24)_{q,q'} \qquad \inf_{u_0 \in \tilde{\mathcal{U}}_q} \| u + u_0 \|_{\mathcal{L}^q(D)} \leq H_{q,q'}(D) \| \operatorname{div} u \|_{\mathcal{L}^{q'}(D)}$$

$$1 \leq q < 3/2 , \; 1 \leq q' \leq +\infty \; ; \; q = 3/2 , \; 1 < q' \leq +\infty \; ; \; 3/2 < q \leq +\infty , \; \frac{3q}{3+q} \leq q' \leq +\infty$$

(ove si convenga che, per $q = \infty$, si ha $\frac{3q}{3+q} = 3$) ; $\tilde{\mathcal{U}}_q$ ha il significato specificato nel teorema precedente. Le limitazioni per q e per q' sono le migliori possibili .

Segue dal teor. XVI, in base alla seguente osservazione. Posto $\tilde{f} = f - (\operatorname{mis} D)^{-1} \int_D f \, dx$, si ha $\tilde{f} \in \mathcal{L}^{q'}(D)$ e $\int_D \tilde{f} \, dx = 0$. Allora, per il teor. XIV esiste un vettore $\tilde{u} \in \mathcal{L}^q(D)$ che ha per divergenza debole su D la misura $\tilde{\mu}(B) = \int_B \tilde{f} \, dx$. Sia $u' \in C^\infty(D)$ tale che $\operatorname{div} u' = (\operatorname{mis} D)^{-1} \int_D f \, dx$. Allora, il vettore $u = \tilde{u} + u' \in \mathcal{L}^q(D)$ ha per divergenza debole su $D-\mathcal{F}D$ la misura μ . L'impossibilità di migliorare le limitazioni per q e per q' segue da quella - nota - di migliorare le limitazioni per p e p' in relazione alla $(19)_{p',p}$ con un ragionamento perfettamente analogo ad un fatto nella dimostrazione del teor. XIV .

8. Traccia sul contorno per funzioni dotate di gradiente debole su $D-\mathcal{F}D$.

Sussiste il seguente teorema :

XVII . Se $u \in \mathcal{L}^1(D)$ ha gradiente debole su $D-\mathcal{F}D$ rappresentato dalla misura μ , esiste una ed una sola funzione $u^* \in \mathcal{L}^1(\mathcal{F}D)$ che gode delle seguenti proprietà :

1) riesce :

$$\lim_{\rho \to 0^+} \int_{\mathcal{F}D} |u[x+\rho\lambda(x)] - u^*(x)| \, d\sigma_x = 0, \quad \lim_{\rho \to 0^+} u[x+\rho\lambda(x)] = u^*(x) \quad x \text{ q.o.} \in \mathcal{F}D \tag{25}$$

ove i limiti devono pensarsi fatti prescindendo da un insieme di misura nulla per ρ ;

2) sussiste la formula di maggiorazione

$$\|u\|_{\mathcal{L}^1(\mathcal{F}D_\rho)} \leq K \left[\|u^*\|_{\mathcal{L}^1(\mathcal{F}D)} + \|\mu\|_{V(D-\mathcal{F}D)} \right] \tag{26}$$

ove K è una costante dipendente solo da D ed ove ρ varia in $(0, \rho_0)$ privato di un conveniente insieme di misura nulla ;

3) si ha :

$$\|u^*\|_{\mathcal{L}^1(\mathcal{F}D)} \leq H \left[\|u\|_{\mathcal{L}^1(D)} + \|\mu\|_{V(D-\mathcal{F}D)} \right] \tag{27}$$

ove H è una costante dipendente solo da D .

4) per ogni $v \in C^1(D)$ risulta :

$$\int_D u \, \mathrm{div}\, v \, dx = - \int_{\mathcal{F}D} u^* \cdot v \times \nu \, d\sigma - \int_{D-\mathcal{F}D} v \times d\mu .$$

Dimostriamo la (25) . Basterà far vedere che, per ogni porzione di superficie regolare Γ contenuta in $\mathcal{F}D$, riesce :

$$\lim_{\rho \to 0^+} \int_\Gamma |u[x+\rho\lambda(x)] - u^*(x)| \, d\sigma = 0, \quad \lim_{\rho \to 0} u[x+\rho\lambda(x)] = u^*(x) \quad \text{q.o. in } \Gamma .$$

Basterà anche limitarsi a considerare il caso in cui Γ sia contenuta in un piano. Possiamo quindi supporre che Γ sia contenuta nel piano $x^i=0$ e sia costituita dalla chiusura dell'intervallo $I^o_{y^i}$ di tale piano (definito dalle limitazioni $a^{i+1} \leq x^{i+1} < b^{i+1}$, $a^{i+2} \leq x^{i+2} < b^{i+2}$). Supporremo anche che D contenga l'intervallo $I^o = I^o_{y^i} \times I^o_{x^i}$ con $I^o_{x^i} \equiv (a^i \leq x^i < 0)$. Assumeremo come versore $\lambda(x)$ per $x \in \Gamma$ il versore normale interno. Con le notazioni introdotte in 1, si tratta allora di mostrare che esiste una $u^*(y^i) \equiv u^*(x^{i+1}, x^{i+2}) \in \mathcal{L}^1(\overline{I^o_{y^i}})$ tale che:

$$(28) \qquad \lim_{\varepsilon \to 0^-} \int_{I^o_{y^i}} |u(\varepsilon, y^i) - u^*(y^i)|\, dy^i = 0 \,, \quad \lim_{\varepsilon \to 0^-} u(\varepsilon, y^i) = u^*(y^i) \quad \text{q. o. in } I^o_{y^i}$$

ove i limiti si intendono fatti prescindendo da un conveniente insieme di misura nulla per ε. Indicate con p_{μ_i} e q_{μ_i} le variazioni positiva e negativa di μ_i e con $\frac{\partial \alpha}{\partial y^i}[I_{x^i}; y^i]$ la derivata parziale di α rispetto a y^i (nel senso di Lebesgue) [16], per ogni intervallo I_{y^i} del piano $x^i=0$ contenuto in $I^o_{y^i}$ e per ogni intervallo I_{x^i} dell'asse reale contenuto, con la sua chiusura, in $(a^i, 0)$, riesce:

$$\mu_i(I_{x^i} \times I_{y^i}) = \int_{I_{y^i}} \frac{\partial \mu_i}{\partial y^i}(I_{x^i}; y^i)\, dy^i \,, \quad p_{\mu_i}(I_{x^i} \times I_{y^i}) = \int_{I_{y^i}} \frac{\partial p_{\mu_i}}{\partial y^i}(I_{x^i}; y^i)\, dy^i \,,$$

$$q_{\mu_i}(I_{x^i} \times I_{y^i}) = \int_{I_{y^i}} \frac{\partial q_{\mu_i}}{\partial y^i}(I_{x^i}; y^i)\, dy^i \quad [17].$$

Posto $I^{(\varepsilon)}_{x^i} \equiv (a^i, \varepsilon)$, con $a^i < \varepsilon < 0$, si ha:

$$\lim_{\varepsilon \to 0^-} \int_{I_{y^i}} \frac{\partial p_{\mu_i}}{\partial y^i}(I^{(\varepsilon)}_{x^i}; y^i)\, dy^i = p_{\mu_i}(I^o_{x^i} \times I_{y^i})$$

$$\lim_{\varepsilon \to 0^-} \int_{I_{y^i}} \frac{\partial q_{\mu_i}}{\partial y^i}(I^{(\varepsilon)}_{x^i}; y^i)\, dy^i = q_{\mu_i}(I^o_{x^i} \times I_{y^i}).$$

---

(16) Cfr. loc. cit. in (1) pp. 215 e segg..

(17) Cfr. loc. cit. in (1) teor. IX.5 e teor. III. 2 .

Dal teorema di Beppo Levi si trae allora che la funzione di y^i: $\frac{\partial P_{\mu_i}}{\partial y^i}(I_{x^i}^{(\varepsilon)}; y^i)$ è sommabile in I_{y^i} uniformemente rispetto ad ε, che esiste q.o. il limite in I_{y^i}: $\lim_{\varepsilon \to 0^-} \frac{\partial P_{\mu_i}}{\partial y^i}(I_{x^i}^{(\varepsilon)}; y^i)$, che tale limite è sommabile in I_{y^i} e riesce:

$$\lim_{\varepsilon \to 0^-} \int_{I_{y^i}} \frac{\partial P_{\mu_i}}{\partial y^i}(I_{x^i}^{(\varepsilon)}; y^i) dy^i = \int_{I_{y^i}} \lim_{\varepsilon \to 0^-} \frac{\partial P_{\mu_i}}{\partial y^i}(I_{x^i}^{(\varepsilon)}; y^i) dy^i.$$

Analogo ragionamento si può fare per Q_{μ_i}. Se ne conclude che esiste q.o. il limite

$$\lim_{\varepsilon \to 0^-} \left[\frac{\partial P_{\mu_i}}{\partial y^i}(I_{x^i}^{(\varepsilon)}; y^i) - \frac{\partial Q_{\mu_i}}{\partial y^i}(I_{x^i}^{(\varepsilon)}; y^i) \right] = \lim_{\varepsilon \to 0^-} \frac{\partial \mu_i}{\partial y^i}(I_{x^i}^{(\varepsilon)}; y^i),$$

tale limite è sommabile in $I_{y^i}^0$ e, uniformemente al variare di $I_{y^i} \subset I_{y^i}^0$, risulta:

$$\mu_i(I_{x^i}^0 \times I_{y^i}) = \lim_{\varepsilon \to 0^+} \mu_i(I_{x^i}^{(\varepsilon)} \times I_{y^i}) \equiv \lim_{\varepsilon \to 0^-} \int_{I_{y^i}} \frac{\partial \mu_i}{\partial y^i}(I_{x^i}^{(\varepsilon)}; y^i) dy^i =$$

$$= \int_{I_{y^i}} \lim_{\varepsilon \to 0^-} \frac{\partial \mu_i}{\partial y^i}(I_{x^i}^{(\varepsilon)}; y^i) dy^i.$$

Ma si ha:

$$u(\varepsilon, y^i) - u(a^i; y^i) = \frac{\partial \mu_i}{\partial y^i}(I_{x^i}^{(\varepsilon)}; y^i)$$

per tutte le coppie di punti (ε, y^i), (a^i, y^i) che appartengono al complementare di un opportuno insieme di misura nulla[18]. Ne viene che, scelto a^i fuori di un certo insieme di misura nulla, esiste finito il limite (q.o. in I_{y^i})

$$\lim_{\varepsilon \to 0^-} u(\varepsilon, y^i) = u(a^i, y^i) + \lim_{\varepsilon \to 0^-} \frac{\partial \mu_i}{\partial y^i}(I_{x^i}^{(\varepsilon)}; y^i) = u^*(y^i)$$

ove, nel calcolo del limite, si prescinda da un conveniente insieme di misura nulla per ε. Il limite $u^*(y^i)$ risulta sommabile su $I_{y^i}^0$ e, uniformemente rispetto a $I_{y^i} \subset I_{y^i}^0$ riesce

---

[18] Cfr. Loc. cit. in (1), teor. V.5 .

$$\lim_{\varepsilon\to 0}\int_{I_{y^i}} u(\varepsilon, y^i)\,dy^i = \int_{I_{y^i}} u(a^i; y^i)\,dy^i + \int_{I_{y^i}} \lim_{\varepsilon\to 0} \frac{\partial\mu_i}{\partial y^i}(I^{(\varepsilon)}_{x^i}; y^i)\,dy^i =$$

$$= \int_{I_{y^i}} u(a^i; y^i)\,dy^i + \mu_i(I^\circ_{x^i}\times I_{y^i}) = \int_{I_{y^i}} u^*(y^i)\,dy^i ,$$

ove i limiti si intendano fatti prescindendo da un conveniente insieme di misura nulla per ε. Dalla uniformità di tale relazione di limite rispetto a I_{y^i}, si trae immediatamente la dimostrazione di (25).

Dimostrazione di (26). Basterà limitarsi a provare che:

$$(29)\qquad \int_{I_{y^i}} |u(a^i; y^i)|\,dy^i \leq \int_{I_{y^i}} |u^*(y^i)|\,dy^i + V_{\mu_i}(I^\circ_{x^i}\times I_{y^i})$$

prescindendo, per ε, da un conveniente insieme di misura nulla.

Dalla precedente dimostrazione si deduce, tra l'altro, che, pur di prescindere da un conveniente insieme di misura nulla, riesce:

$$|u(a^i; y^i)| \leq |u(\varepsilon, y^i)| + \left|\frac{\partial\mu_i}{\partial y^i}(I^{(\varepsilon)}_{x^i}; y^i)\right| \leq$$

$$\leq |u(\varepsilon, y^i)| + V_{\frac{\partial\mu_i}{\partial y^i}}(I^{(\varepsilon)}_{x^i}; y^i)$$

donde, integrando rispetto a y^i su I_{y^i}:

$$\int_{I_{y^i}} |u(a^i; y^i)|\,dy^i \leq \int_{I_{y^i}} |u(\varepsilon, y^i)|\,dy^i + \int_{I_{y^i}} V_{\frac{\partial\mu_i}{\partial y^i}}(I^{(\varepsilon)}_{x^i}; y^i)\,dy^i =$$

$$= \int_{I_{y^i}} |u(\varepsilon, y^i)|\,dy^i + V_{\mu_i}(I^{(\varepsilon)}_{x^i}\times I_{y^i}) \quad (19).$$

Da qui, passando al limite per $\varepsilon\to 0$, si trae l'asserto.

Dimostrazione di (27). Basterà far vedere che risulta:

$$(30)\qquad \int_{I_{y^i}} |u^*(y^i)|\,dy^i \leq H\left(\int_{I^\circ_{x^i}\times I_{y^i}} |u(x^i; y^i)|\,dx + V_{\mu_i}(I^\circ_{x^i}\times I_{y^i})\right)$$

con H costante indipendente da u. Dalla dimostrazione prece-

(19) Cfr. loc.cit. in(1), teor. III.2 e teor. IX. 5 .

dente si trae :

$$\int_{I_{y^i}} |u(\xi,y^i)|dy^i \leq \int_{I_{y^i}} |u(t,y^i)|dy^i + V_{\mu_i}(I^{\circ}_{x^i} \times I_{y^i}),$$

ξ e t q.o., donde , passando al limite, per $\xi \to 0$:

$$\int_{I_{y^i}} |u^*(y^i)|dy^i \leq \int_{I_{y^i}} |u(t,y^i)|dy^i + V_{\mu_i}(I^{\circ}_{x^i} \times I_{y^i}).$$

Da qui, integrando rispetto t , si deduce la (30) .

La proprietà 4) si ottiene subito , passando al limite, per $\rho \to 0$, nella relazione

$$\int_{D_\rho} u \operatorname{div} v \, dx + \int_{\mathcal{F}D_\rho} u \,(v \times \nu_{\mathcal{F}D_\rho})\, d\sigma = - \int_{D_\rho} v \, d\mu , \quad v \in C^1(D), \quad \rho \text{ q.o. } \in (0,\rho_0),$$

e ricordando la proprietà 1) .

L'indipendenza di u^* dalla particolare scelta del versore λ di cui alla definizione di dominio propriamente regolare, è di verifica immediata.

La funzione u^* , di cui si è provata l'esistenza e l'unicità, chiamasi <u>traccia</u> di u su $\mathcal{F}D$ [20].

Per la funzione traccia sussistono anche i seguenti teoremi.

XVIII. <u>Se $u \in \mathcal{L}^1(D)$ ha per gradiente debole su $D - \mathcal{F}D$ la misura μ , la u appartiene a $\mathcal{L}^{3/2}(D)$ e, indicata con u^* la sua traccia su $\mathcal{F}D$, riesce :</u>

(31) $$\inf_{\text{cost.}} \left[\| u + \text{cost} \|_{\mathcal{L}^{3/2}(D)} + \| u^* + \text{cost} \|_{\mathcal{L}^1(\mathcal{F}D)} \right] \leq K \, \| \mu \|_{V(D - \mathcal{F}D)}$$

<u>ove K è una costante dipendente solo da D .</u>

---

(20) Tale nozione di traccia è stata data da G. Fichera nel 1949(cfr. G. Fichera, "Sull'esistenza e sul calcolo delle soluzioni dei problemi al contorno relativi all'equilibrio di un corpo elastico", Ann. Sc. Norm. Sup. Pisa, 1950, Pubblicazione dell'INAC n. 248, 1949) e ripresa, più tardi, da Stampacchia (cfr. G. Stampacchia, "Problemi al contorno per equazioni di tipo ellettico a derivate parziali e questioni di calcolo delle variazioni connesse", Ann. di Mat. pura e appl., 1952)

La (31) si deduce immediatamente applicando il già citato principio esistenziale e di dualità di G. Fichera all'equazione

$$(32) \qquad \int_D u \,\mathrm{div}\, v \, dx + \int_{\mathcal{F}D} u\, v \times \nu \, d\sigma = - \int_{D-\mathcal{F}D} v \, d\mu \quad , v \in C^1(D)$$

ove il termine noto è $\mu \in V(D-\mathcal{F}D)$ e l'incognita e $u \in \mathcal{L}^{3/2}(D) \cap \mathcal{L}^1(\mathcal{F}D)$ (lo spazio $\mathcal{L}^{3/2}(D) \cap \mathcal{L}^1(\mathcal{F}D)$ è normalizzato mediante la norma $\| \ \| = \| \ \|_{\mathcal{L}^{3/2}(D)} + \| \ \|_{\mathcal{L}^1(\mathcal{F}D)}$.

XIX. Se $u \in \mathcal{L}^{p'}(D)$ ha per gradiente debole su $D-\mathcal{F}D$ la $\mu \in V(D-\mathcal{F}D)$, e se μ è assolutamente continua con densità $f \in \mathcal{L}^p(D)$, la traccia u^* di u su $\mathcal{F}D$ appartiene a $\mathcal{L}^{p''}(\mathcal{F}D)$ e riesce:

$$(33)_{p,p',p''} \qquad \inf_{\text{cost}} \left(\| u + \text{cost} \|_{\mathcal{L}^{p'}(D)} + \| u^* + \text{cost} \|_{\mathcal{L}^{p''}(\mathcal{F}D)} \right) \leq K_{p,p',p''}(D) \, \| \text{grad}\, u \|_{\mathcal{L}^p(D)}$$

ove p, p', p'' soddisfano le seguenti limitazioni: $\qquad f = \text{grad}\, u$

$$1 \leq p < 3 \quad , 1 \leq p' \leq \frac{3p}{3-p} \quad , \quad 1 \leq p'' \leq \frac{2p}{3-p}$$

$$p = 3 \quad , 1 \leq p' < +\infty \quad , \quad 1 \leq p'' < +\infty$$

$$3 < p \leq +\infty \quad , 1 \leq p' \leq +\infty \quad , \quad 1 \leq p'' \leq +\infty .$$

Tali limitazioni non possono essere migliorate.

Si consideri, anche questa volta, l'equazione (32) ove, però, ora, il termine noto μ è una misura assolutamente continua con densità $f \in \mathcal{L}^p(D)$; l'incognita u è una funzione dello spazio $\mathcal{L}^{p'}(D) \cap \mathcal{L}^{p''}(\mathcal{F}D)$ normalizzato con la norma: $\| \ \|_{\mathcal{L}^{p'}(D)} + \| \ \|_{\mathcal{L}^{p''}(\mathcal{F}D)}$. Se ora si tiene presente che una $u \in \mathcal{H}^{p',p}(D)$ ha la traccia appartenente a $\mathcal{L}^{p''}(\mathcal{F}D)$ con p, p', p'' verificanti le limitazioni sopra dette, come segue dai noti risultati di Sobolev [21] (e tali limitazioni

---

[21] Cfr. L. S. Sobolev ed E. Gagliardo loc. cit. in [11].

non sono migliorabili, come osservato da Sobolev), l'applicazione del citato principio di dualità di G. Fichera all'equazione (32), consente di dimostrare subito il teorema .

9. Traccia sul contorno per funzioni dotate di gradiente debole su D

Per il teor. V e per quanto detto nel n. precedente, si ha che ogni funzione dotata di gradiente debole su D possiede traccia u^* su $\mathcal{F}D$ nel senso precisato al numero precedente e quindi sussistono, per tale traccia, tutte le proprietà indicate nei teorr. XVII e XVIII. Si ha inoltre :

XX. Se $u \in \mathcal{L}^1(D)$ ha come gradiente debole su D la misura $\mu \in V(D)$, la u appartiene a $\mathcal{L}^{3/2}(D)$ e, indicata con u^* la sua traccia su $\mathcal{F}D$, riesce:

(34) $$\inf_{\text{cost.}} \left(\| u + \text{cost} \|_{\mathcal{L}^{3/2}(D)} + V_\tau(\mathcal{F}D) \right) \leq K \, V_\mu(D)$$

$$\tau(B) = \int_B (u^* + \text{cost}) \nu \, d\sigma - \mu(B)$$

con K indipendente da u .

Si consideri l'equazione

(35) $$\int_D u \, \text{div}\, v \, dx + \int_{\mathcal{F}D} v \, d\mu^* = - \int_D v \, d\mu , \quad v \in C^1(D)$$

ove il termine noto μ appartiene a $V(D)$, e l'incognita è contenuta nello spazio delle coppie (u, μ^*) con $u \in \mathcal{L}^{3/2}(D), \mu^* \in V(\mathcal{F}D)$ normalizzato con la norma $\| \ \| = \|u\|_{\mathcal{L}^{3/2}(D)} + \|\mu^*\|_{V(\mathcal{F}D)}$. L'applicazione del principio di dualità di G. Fichera all'equazione (35) (tenendo presente la proprietà 4) delle tracce) consente di provare immediatamente l'asserto.

XXI. Se $u \in \mathcal{L}^1(D)$ ha per gradiente debole su D la misura

$\mu \in V(D)$ riesce :

(36) $$\lim_{\varrho \to 0} \int_{\Gamma_{\mathcal{F}D_\varrho}} u \, d\sigma = \int_{\Gamma'_{\mathcal{F}D}} \nu x \, d\mu$$

ove $\Gamma_{\mathcal{F}D}$ è una porzione di superficie regolare contenuta in $\mathcal{F}D$ e $\Gamma_{\mathcal{F}D_\varrho}$ è la porzione di $\mathcal{F}D_\varrho$ che corrisponde a $\Gamma_{\mathcal{F}D}$ nella corrispondenza posta tra i punti di $\mathcal{F}D$ e quelli di $\mathcal{F}D_\varrho$ dalla equazione $x = \xi + \varrho\lambda(\xi)$, $\xi \in \mathcal{F}D_\varrho$, $x \in \mathcal{F}D$. Tale relazione di limite è uniforme al variare di $\Gamma_{\mathcal{F}D}$ su $\mathcal{F}D$; inoltre il limite deve intendersi eseguito prescindendo da un insieme di misura nulla per la variabile ϱ.

Se ci poniamo nel caso particolare considerato nella dimostrazione del teor. XVII (il che, d'altra parte, non è restrittivo rispetto al caso generale), facendo uso delle notazioni là introdotte, si vede subito che, per quasi tutti gli ε negativi e abbastanza vicini a zero, riesce :

$$\int_{I_{y^i}} u(\varepsilon, y^i) \, dy^i = -\mu_i(\overline{J^{(\varepsilon)}_{x^i}} \times I_{y^i}), \quad J^{(\varepsilon)}_{x^i} \equiv (\varepsilon \leq x^i < 0)$$

donde, passando al limite, per $\varepsilon \to 0$, si trae :

$$\lim_{\varepsilon \to 0} \int_{I_{y^i}} u(\varepsilon, y^i) \, dy^i = -\mu_i(I_{y^i}).$$

L'uniformità, rispetto a I_{y^i}, di questa relazione di limite, segue dai risultati del teor. XVII.

Da qui, in particolare, si trae che :

XXII . Se $u \in \mathcal{L}^1(D)$ ha per gradiente debole su D la misura $\mu \in V(D)$ indicata con u^* la traccia di u su $\mathcal{F}D$, risulta :

$$\int_{B \cap \mathcal{F}D} u^* \, d\sigma = \int_{B \cap \mathcal{F}D} \nu x \, d\mu$$

per ogni boreliano $B \subset D$, e si ha quindi che la misura così definita sui boreliani $B_{\mathcal{F}D}$ di $\mathcal{F}D$: $\mu^*(B_{\mathcal{F}D}) = \int_{B_{\mathcal{F}D}} \nu x \, d\mu$ è superficialmente assolutamente continua. Risulta inoltre :

(37) $$\lim_{\rho\to 0} u[x+\rho\lambda(x)] = \nu x \frac{d\mu}{d\sigma} \quad \text{in } \mathcal{L}^1(\mathcal{F}D) \text{ e q.o. su } \mathcal{F}D$$

avendo indicato con $\frac{d\mu}{d\sigma}$ la derivata bidimensionale (nel senso di Vitali) della misura $\mu(B_{\mathcal{F}D})$, definita sui boreliani $B_{\mathcal{F}D}$ di $\mathcal{F}D$, rispetto all'ordinaria misura di Lebesgue sulla superficie $\mathcal{F}D$. Nel calcolo del limite (37) si intende di prescindere da un insieme opportuno di misura nulla per la variabile ρ.

XXIII. Se $u \in \mathcal{L}^1(D)$ ha gradiente debole su D rappresentato da una misura $\mu \in V(D)$ e se $\mu(B) = \int_B f\, dx$ per ogni boreliano $B \subset D$, la traccia di u su $\mathcal{F}D$ è nulla.

XXIV. Condizione necessaria e sufficiente perchè $u \in \mathcal{L}^{p'}(D)$ abbia gradiente debole su D rappresentato da una misura assolutamente continua con densità $f \in \mathcal{L}^p(D)$, è che $u \in \mathcal{H}^{p',p}(D)$ e che la traccia di u su $\mathcal{F}D$ sia nulla.

XXV. Se $u \in \mathcal{H}^{p',p}(D)$ ha traccia nulla su $\mathcal{F}D$, riesce :

(18)$'_{p',p}$ $$\|u\|_{\mathcal{L}^{p'}(D)} \leq H_{p',p}(D) \, \|\operatorname{grad} u\|_{\mathcal{L}^p(D)}$$

$$1 \leq p' \leq \frac{3p}{3-p}, \; 1 \leq p < 3 \; ; \; 1 \leq p' < +\infty, \; p = 3 \; ; \; 1 \leq p' \leq +\infty, \; 3 < p \leq +\infty$$

e tali limitazioni per p e per p' sono le migliori possibili.

10. Proprietà delle tracce delle funzioni dotate di gradiente debole.

XXVI. Se $u^*(x) \in \Lambda^{p',p}(\mathcal{F}D)$, $1 \leq p' \leq +\infty$, $1 \leq p \leq +\infty$, esiste una $u \in \mathcal{H}^{p',p}(D)$ che ha come traccia su $\mathcal{F}D$ la u^* e riesce :

$$(38)_{p',p} \qquad \inf_{u_0 \in \mathcal{U}^0_{p',p}} \|u+u_0\|_{\mathcal{H}^{p',p}(D)} \leq R_{p',p}(D)\, \|u^*\|_{\Lambda^{p',p}(\mathcal{F}D)}$$

ove $R_{p',p}(D)$ è una costante dipendente solo da p, p' e da D ed ove $\mathcal{U}^0_{p',p}$ è l'insieme delle $u \in \mathcal{H}^{p',p}(D)$ che hanno traccia nulla su $\mathcal{F}D$ (cioè in forza del teor. XXIV, $\mathcal{U}^0_{p',p}$ è l'insieme delle $u \in \mathcal{L}^{p'}(D)$ che hanno gradiente debole su D rappresentato da una misura assolutamente continua con densità $\in \mathcal{L}^p(D)$. Se $u^* \in C^0(\mathcal{F}D) \cap \Lambda^{\infty,p}(\mathcal{F}D)$, $1 \leq p \leq +\infty$, esiste una $u \in C^0(D) \cap \mathcal{H}^p(D)$ che ha come traccia u^* su $\mathcal{F}D$ e riesce

$$(39)_p \qquad \inf_{u_0 \in \mathcal{U}^0_p} \left(\|u+u_0\|_{C^0(D)} + \|\operatorname{grad}(u+u_0)\|_{\mathcal{L}^p(D)}\right) \leq R_p(D)\, \|u^*\|_{\Lambda^{\infty,p}(\mathcal{F}D)}$$

ove $\mathcal{U}^0_p$ è l'insieme delle funzioni di $C^0(D) \cap \mathcal{H}^p(D)$ che hanno traccia nulla su $\mathcal{F}D$

XXVII. Se $u \in \mathcal{H}^{p',p}(D)$ con $1 \leq p' \leq p \leq +\infty$, la sua traccia u^* appartiene a $\Lambda^{p',p}(\mathcal{F}D)$ e risulta:

$$(40)_{p',p} \qquad \|u^*\|_{\Lambda^{p',p}(\mathcal{F}D)} \leq R'_{p',p}(D)\, \|u\|_{\mathcal{H}^{p',p}(D)}$$

con $R'_{p',p}(D)$ costante dipendente solo da p', p, D.

I Teoremi XXVI e XXVII, nel caso $p' = p$, sono stati dati da E. Gagliardo (cfr. loc. cit. in [7]); l'estensione qui indicata si consegue, sostanzialmente, con le stesse dimostrazioni di Gagliardo[22].

Il teor. XXVII non si estende al caso $p' > p$. Se infatti sussistesse la $(40)_{p',p}$ per $p' > p$, sarebbe anche $\|u^*\|_{\mathcal{L}^{p'}(\mathcal{F}D)} \leq R\, \|u\|_{\mathcal{H}^{p',p}(D)}$ e questa diseguaglianza è assurda; per constatarlo basta far ricorso, $p' > p$ nel caso del piano alla successione $\{u_n(x)\}$

---

(22) Questi teoremi di Gagliardo sono stati successivamente generalizzati da varii Autori (cfr. S. V. Uspenskii Doklady Akad. Nauk. 1960).

$$u_n(x)\begin{cases} = n^{-\beta}(1-x^2 n)^n (x^1+1/n)^{-\alpha/p'} &, 0\leq x^1\leq 1, 0\leq x^2\leq \frac{1}{n} \\ = 0 &, 0\leq x^1\leq 1, \frac{1}{n}\leq x^2\leq 1\end{cases}$$

con $\beta = 2\frac{p-1}{p}$, $2p'\frac{p-1}{p}+1<\alpha<\frac{p'}{p}$, $\alpha\neq 1$, $\alpha<2p'\frac{p-1}{p}+3$

dopo aver scelto p e p' in guisa tale che : $1\leq p\leq\frac{3}{2}$, $2p'\frac{p-1}{p}+$ $+1<\frac{p'}{p}$, $p<p'$.

11. Traccia per i vettori dotati di divergenza debole su $D-\mathcal{F}D$.

XXVIII. Se $u\in\mathcal{L}^q(D)$, con $1\leq q\leq+\infty$, possiede divergenza debole su $D-\mathcal{F}D$ rappresentata da una misura $\mu\in V(D-\mathcal{F}D)$, esiste uno ed un solo elemento $\varphi\in[C^0(\mathcal{F}D)\cap\Lambda^{\infty,p}(\mathcal{F}D)]^*$ [23] tale che :

$$(41)\qquad \int_D u \operatorname{grad} v\, dx + \int_{D-\mathcal{F}D} v\, d\mu = -\langle\varphi, v\rangle \quad , v\in C^1(D)$$

ove $\langle\varphi, v\rangle$ indica il valore del funzionale φ (lineare e continuo su $C^0(\mathcal{F}D)\cap\Lambda^{\infty,p}(\mathcal{F}D)$) calcolato su v ; riesce :

$$(42)\qquad \|\varphi\|_{[C^0(\mathcal{F}D)\cap\Lambda^{\infty,p}(\mathcal{F}D)]^*}\leq R_p(D)\left[\|u\|_{\mathcal{L}^q(D)}+\|\mu\|_{V(D-\mathcal{F}D)}\right]$$

Si consideri la (41) come un'equazione ove l'incognita è l'elemento $\varphi\in[C^0(\mathcal{F}D)\cap\Lambda^{\infty,p}(\mathcal{F}D)]^*$ ed il termine noto è costituito dalla coppia (u,μ) con $u\in\mathcal{L}^q(D), 1\leq q\leq+\infty$, $\mu\in V(D-\mathcal{F}D)$, (lo spazio ove varia il termine noto si pensa normalizzato con la norma $\|\ \| = \|u\|_{\mathcal{L}^q(D)}+\|\mu\|_{V(D-\mathcal{F}D)}$). Applicando il principio esistenziale e di dualità di G. Fichera alla detta equazione, si ottiene la tesi.

XXIX. Se $u\in\mathcal{L}^q(D)$, con $1\leq q\leq+\infty$, è dotato di divergenza debole su $D-\mathcal{F}D$ contenuta in $\mathcal{L}^{q'}(D)$, $1\leq q'\leq+\infty$, esiste uno ed

---

(23) L'insieme $C^0(\mathcal{F}D)\cap\Lambda^{\infty,p}(\mathcal{F}D)$ è qui pensato come spazio normato con la norma di $\Lambda^{\infty,p}(\mathcal{F}D)$.

un solo elemento $\varphi \in [\Lambda^{p',p}(\partial D)]^*$ tale che :

(43) $$\int_D u \times \mathrm{grad}\, v\, dx + \int_D v\, \mathrm{div}\, u\, dx = -\langle \varphi, v \rangle \quad , v \in C^1(D)$$

e riesce :

(44) $$\|\varphi\|_{[\Lambda^{p',p}(\partial D)]^*} \leq R_{p',p}(D) \left(\|u\|_{\mathcal{L}^q(D)} + \|\mathrm{div}\, u\|_{\mathcal{L}^{q'}(D)} \right).$$

Il simbolo $\langle \varphi, v \rangle$ ha significato analogo a quello precisato nel teorema precedente. La dimostrazione è analoga a quella del teorema precendente.

Naturalmente, se esiste una funzione $\varphi \in \mathcal{L}^1(\partial D)$ (oppore una misura $\varphi \in V(\partial D)$) per cui riesca:

(45) $$\int_D u \times \mathrm{grad}\, v\, dx + \int_{D-\partial D} v\, d\mu = -\int_{\partial D} \varphi v\, d\sigma , \ v \in C^1(D)$$

(oppure $\int_D u \times \mathrm{grad}\, v\, dx + \int_{D-\partial D} v\, d\mu = -\int_{\partial D} v\, d\varphi , v \in C^1(D)$) , sarà $\langle \varphi, v \rangle = \int_{\partial D} \varphi v\, d\sigma$ (oppure $\langle \varphi, v \rangle = \int_{\partial D} v\, d\varphi$) e quindi, in tal caso, il funzionale $\varphi \in [\Lambda^{p',p}(\partial D)]^*$ si potrà identificare con la funzione $\varphi \in \mathcal{L}^1(\partial D)$ (o con la misura $\varphi \in V(\partial D)$) che, d'altra parte, è univocamente determinata dalla (45)[24]. Questa circostanza giustifica la denominazione di <u>traccia</u> di u su ∂D (più precisamente sarebbe la traccia di $u \times \nu$) che si dà al funzionale φ . E' però da notare che esistono dei vettori - e si possono facilmente costruire - che hanno divergenza debole su $D - \partial D$ anche molto regolare (ad es. identicamente nulla) e per i quali la traccia φ non è nè una funzione nè una misura .

Caso particolare della (44) è la seguente :

---

[24] Naturalmente, la rappresentazione $\langle \varphi, v \rangle = \int_{\partial D} \varphi v\, d\sigma$, anche nel suddetto caso, sussiste solo per $v \in C^1(D)$ mentre, in generale, essa non potrà sussistere per una qualsiasi $v \in \Lambda^{p',p}(\partial D)$; basta pensare al caso $p' < +\infty$. Saranno indicate, in seguito, altre rappresentazioni per il funzionale φ .

$$\|v \times \nu\|_{[\Lambda^{p',p}(\partial D)]^*} \leq R_{p',p}(D) \left(\|v\|_{\mathcal{L}^q(D)} + \|\operatorname{div} v\|_{\mathcal{L}^{q'}(D)}\right), \quad v \in C^1(D) \tag{46}$$

$$1 \leq p' \leq +\infty, \quad 1 \leq p \leq +\infty .$$

XXX. Se il vettore $u \in \mathcal{L}^q(D)$ con $1 < q \leq +\infty$, è dotato di divergenza debole su $D - \partial D$ contenuta in $\mathcal{L}^{q'}(D)$, con $q \leq q' \leq +\infty$, indicata con $\varphi \in [\Lambda^{p',p}(\partial D)]^*$ la sua traccia su ∂D, riesce:

$$(47)_{p',p} \qquad \inf_{u_0 \in W_{q,q'}} \left(\|u + u_0\|_{\mathcal{L}^q(D)} + \|\operatorname{div}(u+u_0)\|_{\mathcal{L}^{q'}(D)}\right) \leq R'_{p',p}(D) \, \|\varphi\|_{[\Lambda^{p',p}(\partial D)]^*},$$

ove $W_{q,q'}$ è l'insieme dei vettori di $\mathcal{L}^q(D)$ che hanno divergenza debole su D contenuta in $\mathcal{L}^{q'}(D)$. Il teorema non è vero per $1 < q' < q$.

Se si considera l'equazione ;

$$\int_D u \times \operatorname{grad} v \, dx + \int_D v f \, dx = - \langle \varphi, v \rangle, \quad v \in C^1(D). \tag{48}$$

ove il termine noto φ appartiene a $[\Lambda^{p',p}(\partial D)]^*$ e l'incognita è la coppia (u, f) ; con $u \in \mathcal{L}^q(D), f \in \mathcal{L}^{q'}(D)$, la tesi segue dall'applicazione del principio, già citato, di esistenza e di dualità di G. Fichera alla (48), tenendo conto della $(40)_{p',p}$.

In particolare, per ogni $v \in C^1(D)$ riesce :

$$(49)_{q,q'} \qquad \inf_{w_0 \in W_{q,q'}} \left(\|v + w_0\|_{\mathcal{L}^q(D)} + \|\operatorname{div}(v+w_0)\|_{\mathcal{L}^{q'}(D)}\right) \leq R'_{p',p}(D) \, \|v \times \nu\|_{[\Lambda^{p',p}(\partial D)]^*},$$

ed anche :

$$1 < q \leq q' \leq +\infty,$$

$$(50)_{q,q'} \qquad \inf_{w_0 \in W_{q,q'}} \left(\|v + w_0\|_{\mathcal{L}^q(D)} + \|\operatorname{div}(v+w_0)\|_{\mathcal{L}^{q'}(D)}\right) \leq R_{p',p}(D) \, \|v \times \nu\|_{\mathcal{L}^{q'}(\partial D)},$$

$$1 \leq q \leq q' \leq +\infty, \quad 1 < q'.$$

Riguardo quest'ultima, basta osservare che, ovviamente, riesce: $\|v \times \nu\|_{[\Lambda^{p',p}(\partial D)]^*} \leq \|v \times \nu\|_{\mathcal{L}^{q'}(\partial D)}$; si può inoltre assumere anche $q = 1$ dato che la $(50)_{1,q'}$ segue da $(50)_{q,q'}$ con $1<q$.

Se poi il vettore $u\in\mathcal{L}^q(D)$ ha divergenza debole su $D-\mathcal{F}D$ contenuta in $\mathcal{L}^{q'}(D)$, ed ha traccia su $\mathcal{F}D$ rappresentata da una funzione φ di $\mathcal{L}^{q''}(\mathcal{F}D)$ con $q''\geq q'$, con lo stesso ragionamento testè fatto si vede che riesce:[25]

$$(51)_{q,q'} \qquad \inf_{u_0\in W_{q,q'}}\left(\|u+u_0\|_{\mathcal{L}^q(D)}+\|\operatorname{div}(u+u_0)\|_{\mathcal{L}^{q'}(D)}\right)\leq R'_{p',p}(D)\,\|\varphi\|_{\mathcal{L}^{q''}(\mathcal{F}D)}$$

$$1\leq q\leq q'\leq+\infty,\ q'>1.$$

Una rappresentazione dei funzionali appartenenti a $[\Lambda^{p',p}(\mathcal{F}D)]^*$, che sussiste quali si siano p, p', è fornita dal seguente teorema:

XXXI. Sia D di classe 1. Se $\varphi\in[\Lambda^{p',p}(\mathcal{F}D)]^*$ con $1\leq p\leq +\infty$, $1\leq p'\leq+\infty$, esistono due misure α e α' appartenenti a $V(\mathcal{F}D)$, la prima vettoriale a due componenti e la seconda scalare, tali che:

$$\langle\varphi,v\rangle=\int_{\mathcal{F}D} v\,d\alpha'+\int_{\mathcal{F}D}\operatorname{grad}_{\mathcal{F}D}v\times d\alpha$$

per ogni $v\in C^1(\mathcal{F}D)$. Se $p<+\infty$ (se $p'<+\infty$) si può assumere α (si può assumere α') assolutamente continua con densità in $\mathcal{L}^q(\mathcal{F}D)$ (in $\mathcal{L}^{q'}(\mathcal{F}D)$). Se $p=1$, qualunque sia φ può assumersi $\alpha\equiv 0$, ed il caso $p=1$ è l'unico che goda di detta proprietà

E' conseguenza di teoremi noti sulla rappresentazione dei funzionali lineari e continui negli spazi $\mathcal{H}^1$, ma può essere ridimostrato subito facendo ricorso al principio esistenziale di G. Fichera. Basta infatti applicare tale principio all'equazione

$$\int_{\mathcal{F}D} v\,d\alpha'+\int_{\mathcal{F}D}\operatorname{grad}_{\mathcal{F}D}v\times d\alpha=\langle\varphi,v\rangle \qquad ,\ v\in C^1(\mathcal{F}D)$$

---

[25] In tutte le formule (47) (49) (50) (51) l'insieme $W_{q,q'}$ può essere sostituito con $\overset{0}{C}{}^1(D)$ come apparirà dai teoremi del n. 16.

ove $\varphi\in[\Lambda^{p',p}(\mathcal{F}D)]^*$ é il termine noto, e l'incognita è costituita dalla coppia (α',α) di misure di $V(\mathcal{F}D)$, tenendo conto della formula di maggiorazione

$$\|v\|_{\Lambda^{p',p}(\mathcal{F}D)}\leq K_{p',p}(\mathcal{F}D)\left(\|v\|_{\mathcal{L}^{p'}(\mathcal{F}D)}+\|\operatorname{grad}_{\mathcal{F}D} v\|_{\mathcal{L}^{p}(\mathcal{F}D)}\right) \qquad (26)$$

$$v\in C^1(\mathcal{F}D),\ 1\leq p'\leq+\infty,\ 1<p\leq+\infty$$

e, per $p=1$ ricordando che $\|\ \|_{\Lambda^{p',1}(\mathcal{F}D)}=\|\ \|_{\mathcal{L}^{p'}(\mathcal{F}D)}$.

Da qui si deduce il seguente teorema.

XXXII. Il vettore $u\in\mathcal{L}^q(D)$, $1\leq q\leq+\infty$, possegga divergenza debole $\mu\in V(D-\mathcal{F}D)$ su $D-\mathcal{F}D$. Sia Σ un insieme aperto di $\mathcal{F}D$ contenuto in un piano che, per semplicità, supponiamo coincidente con il piano $x^i=0$. Ammettiamo che la normale interna spiccata nei punti di Σ abbia l'orientazione dell'asse x^i. Esistono allora una famiglia $\{I_{y^i}\}$, elementare ovunque densa in Σ, di intervalli del piano[27] $x^i=0$, contenuti in Σ, una misura α'

---

26) Tale formula di maggiorazione, banale per $p=+\infty$, nel caso $p<+\infty$ é una immediata conseguenza della diseguaglianza:

$$\int_0^{1-h}\left|\frac{1}{h}\int_t^{t+h}f(\tau)\,d\tau\right|^p dt\leq\int_0^1|f(t)|^p\,dt \quad ,\ 1<p<+\infty$$

dovuta a C. B. Morrey; cfr. "Functions of several variables and absolute continuity" Duke Math. J. v. 6, pp. 187-215, 1940.

27) Seguendo G. Fichera, diremo che una famiglia di insiemi è "elementare" se è chiusa rispetto al prodotto e se inoltre la differenza tra due suoi qualsiansi insiemi può porsi come unione di un numero finito di insiemi della famiglia stessa a due a due disgiunti. Dicendo che una famiglia di insiemi di Σ è "ovunque densa", intendiamo che la famiglia totalmente additiva minima che la contiene è quella dei boreliani di Σ.

definita sui boreliani di Σ ed un vettore Ψ (a due componenti) $\in \mathcal{L}^q(\Sigma)$ tali che:

$$\lim_{\varepsilon\to 0^+} \int_{I_{y^i}} \nu(0,y^i)\times u(\varepsilon,y^i)\,dy^i = \alpha^i(I_{y^i}) + \int_{\mathcal{F} I_{y^i}} \Psi\times\nu^*\,d\sigma\,,\quad I_{y^i}\in\{I_{y^i}\}$$

ove $\nu(x)$ è la normale interna a $\mathcal{F}D$ nei punti x di Σ ed ove ν^* indica la normale interna a $\mathcal{F} I_{y^i}$ sul piano $x^i = 0$. Se $q = +\infty$ si può assumere $\Psi \equiv 0$. Se μ è assolutamente continua con densità in $\mathcal{L}^{q'}(D)$, α^i è pure assolutamente continua con densità in $\mathcal{L}^{q'}(\Sigma)$, $1 < q' \leq +\infty$. Il limite, nella precedente relazione, si intende fatto prescindendo da un conveniente insieme di misura nulla per ε.

Sia I_{y^i} un intervallo di Σ definito dalle limitazioni $a^{i+1} \leq x^{i+1} < b^{i+1}$, $a^{i+2} \leq x^{i+2} < b^{i+2}$. Sia $\varepsilon > 0$. Si consideri la funzione $v \equiv v^{(\varepsilon)}(x) = \varphi_\varepsilon(x^i;\eta^i)\cdot\varphi_\varepsilon(x^{i+1};a^{i+1},b^{i+1})\cdot\varphi_\varepsilon(x^{i+2};a^{i+2},b^{i+2})$ ove si è posto:

$$\varphi_\varepsilon(t;\eta)\begin{cases} = 1 & 0 \leq t < \eta \\ = \varepsilon^{-1}(x-\eta-\varepsilon) & \eta \leq t < \eta+\varepsilon \\ = 0 & \eta+\varepsilon \leq t \quad \text{oppure} \quad t < 0 \end{cases}$$

$$\varphi_\varepsilon(t;a,b)\begin{cases} = 0 & t < a-\varepsilon \quad \text{oppure} \quad t \geq b+\varepsilon \\ = \varepsilon^{-1}(t-a+\varepsilon) & a-\varepsilon \leq t < a \\ = 1 & a \leq t < b \\ = -\varepsilon^{-1}(t-b-\varepsilon) & b \leq t < b+\varepsilon\,. \end{cases}$$

Se ora si tiene presente che, in base al teorema precedente, riesce:

$$\int_D u\times \operatorname{grad} v^{(\varepsilon)}\,dx = -\int_{D-\mathcal{F}D} v^{(\varepsilon)}\,d\mu - \int_{\mathcal{F}D} v^{(\varepsilon)}\,d\alpha' - \int_{\mathcal{F}D} \operatorname{grad}_{\mathcal{F}D} v^{(\varepsilon)}\times d\alpha$$

e si passa al limite, in questa relazione, per $\varepsilon \to 0$, prescindendo da un conveniente insieme di misura nulla per ε , pur di scegliere l'intervallo I_{y^i} in una conveniente famiglia elementare ovunque densa in Σ , si ottiene la tesi.

12. Traccia per i vettori dotati di divergenza debole su D

Poichè i vettori di $\mathcal{L}^1(D)$ che hanno divergenza debole su D hanno anche divergenza debole su $D-\mathcal{F}D$, la nozione di traccia e le proprietà di tale traccia considerate nel numero precedente, seguitano a sussistere anche per i vettori dotati di divergenza debole su D . Di più , questa volta si può dire che :

XXXIII. Se il vettore $u \in \mathcal{L}^1(D)$ ha divergenza debole $\mu \in V(D)$ su D , la sua traccia su $\mathcal{F}D$ è rappresentata da una misura e precisamente dalla μ stessa , pensata come elemento di $V(\mathcal{F}D)$.

Il teorema è ovvio .

Da teor. XXXII segue subito che :

XXXIV. Il vettore $u \in \mathcal{L}^1(D)$ possegga divergenza debole su D rappresentata dalla misura $\mu \in V(D)$. Sia Σ un insieme aperto di $\mathcal{F}D$, contenuto in un piano, che, per semplicità , supponiamo coincidere con il piano $x^i = 0$. Ammettiamo che la normale interna spiccata nei punti di Σ abbia l'orientazione dell'asse x^i . Esistono allora una famiglia $\{I_{y^i}\}$ elementare, ovunque densa su Σ , di intervalli del piano $x^i = 0$, contenuti in Σ , tali che :

$$\lim_{\varepsilon \to 0^+} \int_{I_{y^i}} \nu(0,y^i) \times u(\varepsilon, y^i)\, dy^i = \mu(I_{y^i}) \quad , \; I_{y^i} \in \{I_{y^i}\} \tag{52}$$

ove il limite si intende fatto prescindendo da un opportuno insieme di misura nulla per ε .

Naturalmente, la relazione di limite (52) , in generale, non è uniforme rispetto a I_{y^i} ; se lo fosse, la $\mu(I_{y^i})$ dovrebbe essere assolutamente continua su $\{I_{y^i}\}$, laddove $\mu(I_{y^i})$ può anche essere totalmente singolare. E' da notare che, anche se $\mu(I_{y^i})$ fosse assolutamente continua su $\{I_{y^i}\}$, non si potrebbe, in generale, concludere che la (52) è uniforme rispetto a I_{y^i} . In effetti, se il suddetto limite fosse uniforme, allora la funzione $\nu(0,y^i) \times u(\xi,y^i)$ risulterebbe convergente in $\mathcal{L}^1(\Sigma)$, per $\varepsilon \to 0$, laddove si può mostrare, con esempi, che, in generale, ciò non si verifica. Riferiamoci, per semplicità, al caso del piano. Poniamo :

$$u_2(x^1,x^2) \begin{cases} = 0 & 0 \leq x^1 \leq 1 \ , \ \frac{1}{2} \leq x^2 \leq 1 \\ = -2n(2n+1)\left(x^2 - \frac{1}{2n}\right) & \frac{2k}{2n} \leq x^1 < \frac{2k+1}{2n} \ , \ \frac{1}{2n+1} \leq x^2 \leq \frac{1}{2n} \\ = 2n(2n+1)\left(x^2 - \frac{1}{2n}\right) & \frac{2k+1}{2n} \leq x^1 < \frac{2k+2}{2n} \ , \ \frac{1}{2n+1} \leq x^2 \leq \frac{1}{2n} \\ = (2n+1)(2n+2)\left(x^2 - \frac{1}{2n+2}\right) & \frac{2k}{2n} \leq x^1 < \frac{2k+1}{2n} \ , \ \frac{1}{2n+2} \leq x^2 \leq \frac{1}{2n+1} \\ = -(2n+1)(2n+2)\left(x^2 - \frac{1}{2n+2}\right) & \frac{2k+1}{2n} \leq x^1 < \frac{2k+2}{2n} \ , \ \frac{1}{2n+2} \leq x^2 \leq \frac{1}{2n+1} \end{cases}$$

$$k = 0, 1, \ldots, n-1$$

$$u_1(x^1,x^2) = -\int_0^{x^1} \frac{\partial u_2(t,x^2)}{\partial t}\, dt, \qquad u = u_1, u_2 \ . \quad \text{Si ha } u \in \mathcal{L}^{\infty}(Q),$$

essendo Q il quadrato : $0 \leq x^1 \leq 1$, $0 \leq x^2 \leq 1$. Sia ora v una qualsiasi funzione di $C^1(Q)$. Per $0 < \varepsilon < 1/2$, con integrazione per parti si ha :

$$\int_0^1 dx^1 \int_\varepsilon^1 u \times \mathrm{grad}\, v \, dx^2 = -\int_0^1 v(x^1,\varepsilon)\, u_2(x^1,\varepsilon)\, dx^1 .$$

Con un calcolo, si verifica che, quale si sia $v \in C^1(Q)$, risulta: $\lim_{\varepsilon \to 0^+} \int_0^1 v(x^1,\varepsilon)\, u_2(x^1,\varepsilon)\, dx^1 = 0$. Ne viene : $\int_Q u \times \mathrm{grad}\, v\, dx = 0$. Pertanto il vettore u ha divergenza debole su Q identicamente nulla (é chiaro come le definizioni qui date di divergenza debole si trasportino al caso del piano). Ne viene che la traccia di u sul lato $x^2 = 0$ di Q é identicamente nulla. E' però evidente che $\nu(x^1,0)\, u(x^1,\varepsilon) = u_2(x^1,\varepsilon)$ non ammette limite per $\varepsilon \to 0$, neppure se si prescinde da un conveniente insieme di misura nulla. Inoltre, per ogni intervallo $I_{x^1} \subset (0,1)$ si ha :

$$\min \lim_{\varepsilon \to 0} \int_{I_{x^1}} |u_2(x^1,\varepsilon)|\, dx^1 = 0 , \quad \max \lim_{\varepsilon \to 0} \int_{I_{x^1}} |u_2(x^1,\varepsilon)|\, dx^1 = 0.$$

Con questo esempio si è visto, dunque, che per la traccia di un vettore dotato di divergenza debole non sussistono, in generale, le analoghe della (25) e della (26) (relative alla traccia di una funzione dotata di gradiente). E' facile vedere che non sussiste neppure la analoga della (27). Basta, per questo, nel caso bidimensionale, considerare la successione di vettori : $u^{(m)} \equiv (u_1^{(m)}, u_2^{(m)})$ con $u_1^{(m)} = |x^1|^m \mathrm{sen}(m^2 x^2)$, $u_2^{(m)} = m^{-1} |x^1|^{m-1} \cos(m^2 x^2)$. Si ha $u^{(m)} \in C^1(Q)$ ove $Q \equiv (0 \leq x^1 \leq 1,\ 0 \leq x^2 \leq 1)$. Inoltre riesce $\mathrm{div}\, u^{(m)} = 0$, $\lim_{m \to \infty} \| u^{(m)} \|_{\mathcal{L}^q(Q)} = 0$ quale si sia q, $1 \leq q < +\infty$. Se sussistesse la analoga di (27) e cioè : $\int_{\partial Q} |u \times \nu|\, d\sigma \leq H\left(\|u\|_{\mathcal{L}^q(Q)} + V_u(Q - \partial Q) \right)$ nel caso attuale si avrebbe : $\int_0^1 |u_1^{(m)}(1,x^2)|\, dx^2 \leq H \, \| u^{(m)} \|_{\mathcal{L}^q(Q)}$ donde $\lim_{m \to \infty} \int_0^1 |\mathrm{sen}(m^2 x^2)|\, dx^2 = 0$, il che è falso.

XXXV. <u>Se</u> $\mu \in V(D)$ <u>è tale che</u> $\mu(D) = 0$ <u>e se riesce</u> :

$$\mu(B) = \int_B f\,dx + \int_{B \cap \mathcal{F}D} \varphi\, d\sigma \,, \quad f \in \mathcal{L}^{q'}(D) \,, \quad \varphi \in \mathcal{L}^{q''}(\mathcal{F}D) \,,$$

esistono infiniti vettori $u \in \mathcal{L}^q(D)$ che hanno per divergenza debole μ su D e per traccia su $\mathcal{F}D$ la φ, e riesce:

$$(53)_{q,q',q''} \qquad \inf_{u_0 \in \mathcal{U}_q} \| u + u_0 \|_{\mathcal{L}^q(D)} \leq K_{p,p',p''}(D) \left[\| f \|_{\mathcal{L}^{q'}(D)} + \| \varphi \|_{\mathcal{L}^{q''}(\mathcal{F}D)} \right]$$

$$1 \leq q < \tfrac{3}{2} \,, \quad 1 \leq q' \leq +\infty \,, \quad 1 \leq q'' \leq +\infty$$

$$q = \tfrac{3}{2} \,, \quad 1 < q' \leq +\infty \,, \quad 1 < q'' \leq +\infty$$

$$\tfrac{3}{2} < q \leq +\infty \,, \quad \tfrac{3q}{3+q} \leq q' \leq +\infty \,, \quad \tfrac{2q}{3} \leq q'' \leq +\infty$$

ove $\mathcal{U}_q$ è l'insieme dei vettori di $\mathcal{L}^q(D)$ che hanno divergenza debole su D identicamente nulla. Le limitazioni q, q', q'' sono le migliori possibili.

Il teorema d'esistenza e la formula di maggiorazione si deducono rispettivamente dal principio esistenziale e di dualità di G. Fichera, facendo ricorso al teor. XIX ed alla formula di maggiorazione $(33)_{p,p',p''}$ della quale la $(53)_{q,q',q''}$ è la duale (nel senso di Fichera). E' poi facile vedere, in virtù della già osservata non migliorabilità delle limitazioni per p, p', p'' relative alla $(33)_{p,p',p''}$, che l'unico caso da esaminare, ai fini di una eventuale migliorabilità delle limitazioni per q, q', q'', è il seguente: $q = \frac{3}{2}$, $1 < q' \leq +\infty$, $q'' = 1$. Consideriamo dapprima la situazione: $q = \frac{3}{2}$, $1 < q' < +\infty$, $q'' = 1$. Se la $(53)_{q,q',q''}$ sussistesse in riferimento a questa scelta di q, q, q'', indicata con $\mathcal{V}$ la totalità dei vettori $v \in \mathcal{L}^{3/2}(D)$ dotati di divergenza debole su D della forma

$\mu(B)=\int_B f\,dx+\int_{B\cap \mathcal{F}D}\varphi\,d\sigma$ con $f\equiv \mathrm{div}\,v\in\mathcal{L}^{q'}(D)$, $\varphi\equiv v\times\nu\in\mathcal{L}^{1}(\mathcal{F}D)$, in virtù del citato principio esistenziale di G. Fichera si potrebbe enunciare la seguente proposizione : l'equazione

$$\int_D u\,\mathrm{div}\,v\,dx+\int_{\mathcal{F}D}u^* v\times\nu\,d\sigma=-\int_D v\,g\,dx\ ,\quad v\in\mathcal{V},$$

con termine noto $g\in\mathcal{L}^3(D)$ ed incognita (u,u^*), $u\in\mathcal{L}^{p'}(D)$, $u^*\in\mathcal{L}^{\infty}(\mathcal{F}D)$ ammette soluzione in corrispondenza ad ogni $g\in\mathcal{L}^3(D)$ che sia ortogonale ad $\mathcal{U}_{3/2}$ e riesce, per ogni soluzione u : $\inf_{\text{cost.}}\left(\|u+\text{cost.}\|_{\mathcal{L}^{p'}(D)}+\right.$ $\left.+\|u^*+\text{cost.}\|_{\mathcal{L}^{\infty}(\mathcal{F}D)}\right)\leq K(D)\|g\|_{\mathcal{L}^3(D)}$. In particolare, quindi, per ogni $u\in C^1(D)$ si ha :

$$\inf_{\text{cost.}}\left(\|u+\text{cost.}\|_{\mathcal{L}^{p'}(D)}+\|u+\text{cost.}\|_{C^0(\mathcal{F}D)}\right)\leq K(D)\,\|\mathrm{grad}\,u\|_{\mathcal{L}^3(D)}\ .$$

D'altra parte, questa diseguaglianza è falsa perchè implica che ogni $u\in\mathcal{L}^{p'}(D)$ avente gradiente in $\mathcal{L}^3(D)$ è continua su $\mathcal{F}D$. Nel caso $q=\frac{3}{2}$, $q'=+\infty$, $q''=1$, con un ragionamento analogo a quello ora esposto, si perviene alla diseguaglianza $\inf_{\text{cost.}}\left(\|u+\text{cost.}\|_{\mathcal{L}^3(D)}+\|u+\text{cost.}\|_{\mathcal{L}^{\infty}(\mathcal{F}D)}\right)\leq K(D)\|\mathrm{grad}\,u\|_{\mathcal{L}^{3}}$ la quale è falsa, come subito si vede.

Dal teor. XXXV si deduce subito che :

XXXVI. <u>Se $u\in\mathcal{L}^q(D)$ ha divergenza debole su D rappresentata da una misura assolutamente continua su $D-\mathcal{F}D$, di densità $f\in\mathcal{L}^{q'}(D)$, modificando u con l'aggiunzione di un conveniente vettore di $\mathcal{L}^q(D)$ a divergenza debole su $D-\mathcal{F}D$ identicamente nulla, si può fare in modo che u abbia su $\mathcal{F}D$ una traccia $\varphi\in\mathcal{L}^{q''}(\mathcal{F}D)$ comunque prefissata purchè risulti $\int_{\mathcal{F}D}\varphi\,d\sigma=-\int_D f\,dx$ ed inoltre q,q',q'' soddisfino le limitazioni di cui alla $(53)_{q,q',q''}$.</u>

<u>Se $u\in\mathcal{L}^q(D)$, con $1\leq q<3/2$, ha divergenza debole $\mu\in V(D)$ su D, modificando u con l'aggiunzione di un vettore di $\mathcal{L}^q(D)$ a divergenza debole su $D-\mathcal{F}D$ identicamente nulla, si può sempre fare</u>

in modo che u abbia su $\mathcal{F}D$ una traccia φ comunque prefissata, contenuta in $\mathcal{L}^{q''}(\mathcal{F}D)$ con $1 \leq q'' \leq +\infty$ e purchè sia $\int_{\mathcal{F}D} \varphi \, d\sigma = -\mu(D-\mathcal{F}D)$

13. Altri teoremi d'esistenza e formule di maggiorazione per l'operatore di divergenza debole

XXXVII. Esiste una costante $K_{p',p}(D)$ tale che, per ogni $v \in \mathring{C}^1(D)$ (per ogni $v \in C^1(D)$) riesce:

$$(54)_{p',p} \qquad \inf_{v_0 \in V_0} \|v+v_0\|_{\mathcal{L}^q(D)} \leq K_{p',p}(D) \|\mathrm{div}\, v\|_{\mathcal{L}^{q'}(D)}, \quad \begin{array}{ll} 1 \leq q < 3/2, & 1 \leq q' \leq +\infty \\ q = 3/2, & 1 < q' \leq +\infty \\ 3/2 < q \leq +\infty, & \frac{3q}{3+q} \leq q' \leq +\infty, \end{array}$$

ove V_0 è l'insieme dei vettori di $\mathring{C}^1(D)$ (di $C^1(D)$) che hanno divergenza nulla. Le limitazioni per q e per q' non sono migliorabili.

E' conseguenza della parte necessaria del principio esistenziale di G. Fichera, in relazione ai teoremi di esistenza XIII, XIV, XV, XVI.

XXXVIII. Se $f \in \mathcal{L}^{q'}(D)$, con $3 \leq q' \leq +\infty$, esiste un vettore $u \in C^0(D)$ che ha per divergenza debole su $D-\mathcal{F}D$ la funzione f e riesce

$$(55) \qquad \inf_{u_0 \in U_0} \|u+u_0\|_{C^0(D)} \leq K(D) \|f\|_{\mathcal{L}^{q'}(D)}$$

ove con U_0 si è indicato l'insieme dei vettori di $C^0(D)$ che hanno divergenza debole su $D-\mathcal{F}D$ identicamente nulla. Se $q' < 3$, in generale non esiste alcun vettore $u \in C^0(D)$ che abbia per divergenza debole su $D-\mathcal{F}D$ la f.

Sia $3 \leq q' < +\infty$. Sia $\{f_n\}$ una successione di polinomi convergente ad f in $\mathcal{L}^{q'}(D)$ e sia u_n un vettore di $C^1(D)$ che abbia divergenza eguale ad f_n nei punti di D. In forza di $(54)_{p',1}$, la successione $\{u_n\}$ risulta convergente in $C^0(D)/\overline{V}_0$ ove $\overline{V}_0$ è la chiusura, in $C^0(D)$, della varietà V_0 dei vettori di $\overset{\circ}{C}{}^1(D)$ che hanno divergenza nulla (tale spazio quoziente si intende normalizzato con la norma $\|v\| = \inf_{v_0 \in \overline{V}_0} \max_D |v(x) + v_0(x)|$). Esiste allora un elemento $u \in C^0(D)$ al quale $\{u_n\}$ converge in $C^0(D)/\overline{V}_0$. Passando al limite, per $n \to \infty$, nella relazione $\int_D u_n \times \operatorname{grad} v \, dx = -\int_D f_n v \, dx$, si ottiene

$$\int_D u \times \operatorname{grad} v \, dx = -\int_D f v \, dx \ , \quad v \in \overset{\circ}{C}{}^1(D) .$$

La (55) è conseguenza del principio di dualità di G. Fichera. Il caso $q' = +\infty$ si riduce immediatamente al caso di prima. Se il teorema fosse vero per $q' < 3$, in virtù del principio esistenziale di G. Fichera (parte necessaria), dovrebbe sussistere la $(54)_{p',1}$ con $p' > 3/2$ e ciò è assurdo.

Analogamente si vede che:

XXXIX. Se $f \in \mathcal{L}^{q'}(D)$ con $3 \leq q' \leq +\infty$, $\int_D f \, dx = 0$ esiste un vettore $u \in C^0(D)$ che ha per divergenza debole su D la f (cioè la misura ass. cont. di densità f), e quindi avente traccia nulla su $\mathcal{F}D$, e sussiste la (55) ove però V_0 è, questa volta, l'insieme dei vettori di $C^0(D)$ che hanno divergenza debole, su D, identicamente nulla.

XL. Per ogni vettore $v \in C^1(D)$ riesce:

$$(56)_{q,q',q''} \quad \inf_{v_0 \in V_0} \|v + v_0\|_{\mathcal{L}^q(D)} \leq K_{p,p',p''}(D) \left[\|\operatorname{div} v\|_{\mathcal{L}^{q'}(D)} + \|v \times \nu\|_{\mathcal{L}^{q''}(\mathcal{F}D)} \right]$$

$1\leq q<3/2$, $1\leq q'\leq +\infty$, $1\leq q''\leq+\infty$; $q=3/2$, $1<q'\leq+\infty$, $1<q''\leq+\infty$;

$\frac{3}{2}<q<+\infty$, $\frac{3q}{3+q}\leq q'\leq+\infty$, $\frac{2q}{3}\leq q''\leq+\infty$; $q=+\infty$, $3\leq q'\leq+\infty$, $q''=+\infty$

ove $\mathcal{V}_0$ è l'insieme dei vettori $\in \overset{\circ}{C}{}^1(D)$ che hanno divergenza nulla. Le limitazioni per q, q', q'' sono le migliori possibile.

Segue dalla parte necessaria del principio esistenziale di G. Fichera e dal teor. XIX.

XLI. Siano D (propriamente regolare) di classe 2, $f\in\mathcal{L}^{q'}(D)$, con $3\leq q'\leq+\infty$, e $\varphi\in C^0(\mathcal{F}D)$, con $\int_D f\,dx+\int_{\mathcal{F}D}\varphi\,d\sigma=0$. Esiste un $u\in C^0(D)$ che ha per divergenza debole su $D-\mathcal{F}D$ la f e tale che $u\times\nu=\varphi$ su $\mathcal{F}D$. Riesce :

$$\inf_{u_0\in\mathcal{U}_0}\|u+u_0\|_{C^0(D)}\leq K\left(\|f\|_{\mathcal{L}^{q'}(D)}+\|\varphi\|_{C^0(\mathcal{F}D)}\right) \tag{57}$$

ove K è una costante dipendente solo che da D e q' ed ove $\mathcal{U}_0$ è l'insieme dei vettori di $C^0(D)$ che hanno divergenza su D identicamente nulla. La limitazione per q' non può essere migliorata.

Sia $3\leq q'<+\infty$. Sia $\{f_n\}$ una successione di funzioni di $C^\infty(D)$ che converge verso f in $\mathcal{L}^{q'}(D)$ e tale che $\int_D f_n dx=\int_D f\,dx$. Sia $\{\varphi_n\}$ una successione di funzioni di $C^2(\mathcal{F}D)$ che converge a φ in $C^0(\mathcal{F}D)$ e tale che $\int_{\mathcal{F}D}\varphi_n d\sigma=\int_{\mathcal{F}D}\varphi\,d\sigma$. Esiste, per ogni n , un vettore $w_n\in C^1(D)$ tale che $\operatorname{div} w_n=f_n$ in D , $w_n\times\nu=\varphi_n$ su $\mathcal{F}D$. In forza di $(56)_{\infty,q',\infty}$ si ha che $\{w_n\}$ converge, in $C^0(D)/\overline{\mathcal{V}}_0$, verso una $u\in C^0(D)$ (ove $\overline{\mathcal{V}}_0$ è la chiusura, in $C^0(D)$, dello insieme $\mathcal{V}_0$ dei vettori di $\overset{\circ}{C}{}^1(D)$ che hanno divergenza nulla) . Se ora si passa al limite, per $n\to\infty$, nella identità

$\int_D u_u \times \operatorname{grad} v \, dx = -\int_D v f_u \, dx - \int_{\mathcal{F}D} v \varphi_u \, d\sigma$, si ottiene la tesi, nel caso in esame $3 \leq q' < +\infty$. La (57) è conseguenza del principio di dualità di G. Fichera. Il caso $q' = +\infty$ si riduce immediatamente al caso di prima. La non migliorabilità delle imitazioni per q' segue da quella, già osservata, relativa al teor. XXXVIII.

Sia ora D_1 un dominio propriamente regolare $\subset D - \mathcal{F}D$ tale che $D_2 \equiv D - (D_1 - \mathcal{F}D_1)$ sia, esso pure, un dominio propriamente regolare. Sussiste allora il seguente teorema.

XLII. Se $\mu \in V(D)$ è una misura assolutamente continua sui boreliani contenuti in $D_1 - \mathcal{F}D_1$, con densità $\in \mathcal{L}^q(D_1)$, $1 \leq q \leq +\infty$ e se $\mu(D) = 0$, esiste un vettore $u \in \mathcal{L}^{q_1}(D_1) \cap \mathcal{L}^{q_2}(D_2)$ con $1 \leq q_1 < 3/2$, $1 \leq q \leq +\infty$; $q_1 = \frac{3}{2}$, $1 < q \leq +\infty$; $3/2 < q_1 \leq +\infty$, $\frac{3q_1}{3+q_1} \leq q \leq +\infty$; $1 \leq q_2 < 3/2$, che ha per divergenza debole la misura μ, e riesce:

(58) $$\inf_{u_0 \in U_0} \left[\|u + u_0\|_{\mathcal{L}^{q_1}(D_1)} + \|u + u_0\|_{\mathcal{L}^{q_2}(D_2)} \right] \leq L_{q, q_1, q_2}(D_1, D_2) \left[\left\| \frac{d\mu}{dx} \right\|_{\mathcal{L}^q(D_1)}, \|\mu\|_{V(D_2)} \right]$$

ove U_0 è l'insieme dei vettori di $\mathcal{L}^{q_1}(D_1) \cap \mathcal{L}^{q_2}(D_2)$ che hanno divergenza debole su D identicamente nulla.

Sia $\mathcal{V}$ lo spazio delle $v \in C^1(D)$ in cui si sia introdotta la norma : $\|v\|_{\mathcal{V}} = \|v\|_{\mathcal{L}^{p_1}(D_1)} + \|v\|_{C^0(D_2)}$.
Il duale $\mathcal{V}^*$ di $\mathcal{V}$, se $p < +\infty$, sarà l'insieme delle misure $\alpha \in V(D)$ assolutamente continue sui boreliani contenuti in $D_1 - \mathcal{F}D_1$ e con densità in $\mathcal{L}^q(D_1)$, ove la norma è così definita :

$$\|\alpha\|_{\mathcal{V}^*} = \max \left[\left\| \frac{d\alpha}{dx} \right\|_{\mathcal{L}^q(D_1)}, \|\alpha\|_{V(D_2)} \right].$$

Sia poi $\mathcal{B}_{p_1, p_2}$ lo spazio delle funzioni f di $\mathcal{L}^{p_1}(D_1) \cap \mathcal{L}^{p_2}(D_2)$ ove la norma è $\|f\|_{\mathcal{B}_{p_1,p_2}} = \|f\|_{\mathcal{L}^{p_1}(D_1)} + \|f\|_{\mathcal{L}^{p_2}(D_2)}$.

Il duale $\mathcal{B}^*_{p_1,p_2}$, se $p_1<+\infty$, $p_2<+\infty$, sarà lo spazio delle funzioni w di $\mathcal{L}^{p_1}(D_1)\cap\mathcal{L}^{p_2}(D_2)$ ove la norma é :

$$\|w\|_{\mathcal{B}^*_{p_1,p_2}} = \max\left(\|w\|_{\mathcal{L}^{q_1}(D_1)},\ \|w\|_{\mathcal{L}^{q_2}(D_2)}\right).$$

La tesi del teorema segue facilmente dalla applicazione del principio esistenziale e di dualità di G. Fichera alla equazione:

$$\int_D u \times \operatorname{grad} v\, dx = -\int_D v\, d\alpha \qquad , v\in C^1(D)$$

con termine noto $\alpha\in V^*$ e incognita $u\in\mathcal{B}^*_{p_1,p_2}$.

Da quí segue, in particolare, che sussistono le seguenti formule di maggiorazione

$$\text{(59)}\quad \inf_{\text{cost.}}\left[\|v+\text{cost.}\|_{\mathcal{L}^{q}(D_1)}+\|v+\text{cost}\|_{\mathcal{L}^{q}(D_2)}\right]\leqslant K_{q,q_1,q_2}(D_1,D_2)\cdot$$

$$\cdot\left[\|\operatorname{grad} v\|_{\mathcal{L}^{p_1}(D_1)}+\|\operatorname{grad} v\|_{\mathcal{L}^{p_2}(D_2)}\right]$$

$3<p_2\leqslant+\infty$; $1\leqslant p_1\leqslant+\infty$, $1\leqslant p\leqslant 3/2$; $\frac{3p}{3+p}\leqslant p_1\leqslant+\infty$, $\frac{3}{2}<p<+\infty$; $3<p_1\leqslant+\infty$, $p=+\infty$;

$$\text{(60)}\quad \inf_{v_0\in V_0}\left[\|v+v_0\|_{\mathcal{L}^{q_1}(D_1)}+\|v+v_0\|_{\mathcal{L}^{q_2}(D_2)}\right]\leqslant K_{q,q_1,q_2}(D_1,D_2)\cdot$$

$$\cdot\left[\|\operatorname{div} v\|_{\mathcal{L}^{q}(D_1)}+\|\operatorname{div} v\|_{\mathcal{L}^{1}(D_2)}\right]$$

$1\leqslant q_2\leqslant 3/2$; $1<q_1\leqslant+\infty$, $3\leqslant q\leqslant+\infty$; $1\leqslant 3q/3-q$; $1<q<3$; $1\leqslant q_1<3/2$, $q=1$.

Queste limitazioni per p,p_1,p_2, q,q_1,q_2 non sono migliorabili.

14. Gradiente "forte"

Si dice che $u\in\mathcal{L}^{p'}(D)$ è dotato di gradiente "forte" $f\in\mathcal{L}^{p}(D)$ su $D-\mathcal{A}D$ se esiste una successione $\{u_n\}$ di funzioni di $C^1(D)$ tale che

$$\lim_{n\to\infty} \|u_n - u\|_{\mathcal{L}^{p'}(D)} = \lim_{n\to\infty} \|\operatorname{grad} u_n - f\|_{\mathcal{L}^p(D)} = 0, \quad 1 \leq p \leq +\infty, \; 1 \leq p' \leq +\infty.$$

E' noto il seguente teorema:

XLIII. La funzione $u \in \mathcal{L}^{p'}(D)$ é dotata di gradiente forte $f \in \mathcal{L}^p(D)$ su $D - \mathcal{F}D$ se e solo se u ha come gradiente debole su $D - \mathcal{F}D$ il vettore f, $1 \leq p \leq +\infty$, $1 \leq p' \leq +\infty$, ove si conviene di assumere come $\mathcal{L}^\infty(D)$ lo spazio $C^0(D)$ [28].

Si dice che $u \in \mathcal{L}^{p'}(D)$ é dotato di gradiente "forte" $f \in \mathcal{L}^p(D)$ su D se esiste una successione $\{u_n\}$ di funzioni di $\mathcal{E}^1(D)$ tale che $\lim_{n\to\infty} \|u_n - u\|_{\mathcal{L}^{p'}(D)} = \lim_{n\to\infty} \|\operatorname{grad} u_n - f\|_{\mathcal{L}^p(D)} = 0$.

XLIV. La funzione $u \in \mathcal{L}^{p'}(D)$ é dotata di gradiente forte $f \in \mathcal{L}^p(D)$ su D se e solo se u ha come gradiente debole su D il vettore f, $1 \leq p' \leq +\infty$, $1 \leq p \leq +\infty$ ove si conviene di assumere come $\mathcal{L}^\infty(D)$ lo spazio $C^0(D)$.

La dimostrazione può condursi in modo analogo a quello seguito da Gagliardo nella citata dimostrazione del teorema precedente (si veda anche la dimostrazione del teor. L).

Diremo che $u \in \mathcal{L}^{p'}(D)$ ha gradiente "forte" su D rappresentato dalla misura

$$\mu(B) = \int_B f\,dx + \int_{B \cap \mathcal{F}D} \varphi\,d\sigma, \quad f \in \mathcal{L}^p(D), \; \varphi \in \mathcal{L}^{p''}(\mathcal{F}D)$$

---

[28] Cfr. E. Gagliardo loc. cit. in [11] ove il teorema é dimostrato in ipotesi molto generali e per dominii limitati ed a frontiera "localmente lipschitziana" tra i quali rientrano i dominii propriamente regolari che qui vengono considerati. Cfr. anche V. M. Babich: "Sul problema del prolungamento delle funzioni" Uspekhi Matematicheskikh Nauk, v. 54, 1953, relativamente a dominii di classe 1 .

se esiste una successione $\{u_n\}$ di funzioni di $C^1(D)$ tale che:

$$(61) \quad \lim_{n\to\infty} \|u_n - u\|_{\mathcal{L}^{p'}(D)} = 0, \ \lim_{n\to\infty} \|\operatorname{grad} u_n - f\|_{\mathcal{L}^{p}(D)} = 0, \ \lim_{n\to\infty} \|u_n - \varphi\|_{\mathcal{L}^{p''}(\mathcal{F}D)} = 0$$

$$1 \le p \le +\infty, \ 1 \le p' \le +\infty, \ 1 \le p'' \le \infty .$$

XLV. Siano $1 \le p < +\infty, 1 \le p' \le +\infty, 1 \le p'' \le +\infty$ e supponiamo che D sia di classe 2. La funzione $u \in \mathcal{L}^{p'}(D)$ è dotata di gradiente forte su D rappresentato dalla misura $\mu(B) = \int_B f\,dx + \int_{B\cap\mathcal{F}D} \varphi\, d\sigma$, $f \in \mathcal{L}^p(D), \varphi \in \mathcal{L}^{p''}(\mathcal{F}D)$ se e solo se u ha come gradiente debole su D la μ ; con $\mathcal{L}^\infty$ qui si intende lo spazio C^0 .

Se u é dotato di gradiente forte nel senso detto sopra è anche dotato di gradiente debole, rappresentato dalla medesima misura che ne rappresenta il gradiente forte. Mostriamo il viceversa. Consideriamo dapprima il caso $1 \le p < +\infty, 1 \le p' < +\infty, 1 \le p'' < +\infty$. Sia $\{\varphi_n\}$ una successione di funzioni di $C^2(\mathcal{F}D)$ tale che :

$$(62) \quad \lim_{n\to\infty} \left(\|\varphi_n - \varphi\|_{\mathcal{L}^{p''}(\mathcal{F}D)} + \|\varphi_n - \varphi\|_{\Lambda^{p',p}(\mathcal{F}D)} \right) = 0 .$$

Alla costruzione di una tale successione si può provvedere, ad esempio, con il metodo degli operatori regolarizzatori secondo Friedrichs . Sia $v_n \in C^1(D)$ tale che $v_n = \varphi_n$ su $\mathcal{F}D$. Per $(38)_{p',p}$ riesce:

$$(63) \quad \inf_{w \in \overset{\circ}{U}_{p',p}} \| u - v_n + w \|_{\mathcal{H}^{p',p}(D)} \le R_{p',p}(D) \|\varphi_n - \varphi\|_{\Lambda^{p',p}(\mathcal{F}D)} ,$$

ove $\overset{\circ}{U}_{p',p}$ è l'insieme delle funzioni di $\mathcal{H}^{p',p}(D)$ che hanno traccia nulla su $\mathcal{F}D$; da (62) (63) segue :

$$\lim_{n\to\infty} \inf_{w \in \overset{\circ}{U}_{p',p}} \| u - v_n + w \|_{\mathcal{H}^{p',p}(D)} = 0 .$$

Dato che $\overset{\circ}{\mathcal{C}}^1(D)$ é una base per $\overset{\circ}{\mathcal{U}}^1(D)$ con la metrica di $\mathcal{H}^{p',1}(D)$ (cfr. teor. XLIV) esiste $w_n \in \overset{\circ}{\mathcal{C}}^1(D)$ tale che $\lim_{n\to\infty} \| u - v_n + w_n \|_{\mathcal{H}^{p',p}(D)} = 0$. Posto $u_n = v_n - w_n$, da qui e da (62) viene che $\{u_n\}$ verifica le (61).

Se poi é $p = +\infty$, dovrà necessariamente essere $u \in \mathcal{C}^1(D)$ e in tal caso l'asserto è banale.

15. Divergenza "forte" su $D - \mathcal{F}D$.

Diremo che il vettore $u \in \mathcal{L}^q(D)$, $1 \leq q \leq +\infty$, ha come divergenza "forte" su $D - \mathcal{F}D$ la funzione $f \in \mathcal{L}^{q'}(D)$, $1 \leq q' \leq +\infty$, se esiste una successione $\{u_n\}$ di vettori di $\mathcal{C}^1(D)$ tale che:

(64) $$\lim_{n\to\infty} \left(\| u_n - u \|_{\mathcal{L}^q(D)} + \| \operatorname{div} u_n - f \|_{\mathcal{L}^{q'}(D)} \right) = 0 .$$

E' evidente che:

XLVI. Se $u \in \mathcal{L}^q(D)$ ha per divergenza forte su $D - \mathcal{F}D$ la funzione $f \in \mathcal{L}^{q'}(D)$, allora u ha per divergenza debole su $D - \mathcal{F}D$ la f, $1 \leq q \leq +\infty$, $1 \leq q' \leq +\infty$.

Mostriamo che sussiste il teorema inverso:

XLVII. Sia $1 \leq q \leq +\infty$, $1 \leq q' \leq +\infty$. Se $u \in \mathcal{L}^q(D)$ ha per divergenza debole su $D - \mathcal{F}D$ la funzione $f \in \mathcal{L}^{q'}(D)$, allora u ha per divergenza forte su $D - \mathcal{F}D$ la f, ove si assuma, come $\mathcal{L}^\infty$, lo spazio $\mathcal{C}^0$.

Consideriamo dapprima il caso $1 \leq q' \leq q < +\infty$. La dimostrazione, in tal caso, é del tutto analoga a quella, citata, di Gagliardo relativa al teor. XLIII, cioé al gradiente.

Siano $S_1, S_2, \dots, S_m$ sfere aperte che ricoprano D. Se

$\bar{S}_K \subset D - \mathcal{F}D$, é facile costruire una successione $\{u_n^{(K)}(x)\}$ di vettori di $C^1(D)$ tale che:

$$\lim_{n\to\infty}\left(\| u_n^{(K)} - u \|_{\mathcal{L}^q(\bar{S}_K)} + \| \operatorname{div} u_n^{(K)} - f \|_{\mathcal{L}^{q'}(\bar{S}_K)}\right) = 0 .$$

Si può, ad esempio, ricorrere al metodo degli operatori regolarizzatori di Friedrichs. Sia ora K tale che $\bar{S}_K \cap \mathcal{F}D \neq \emptyset$: porremo allora $J_K = \bar{S}_K \cap D$. Per le ipotesi fatte su $\mathcal{F}D$, è sempre possibile pensare di aver eseguito il ricoprimento di D in modo tale che esista un numero positivo δ_0 verificante la seguente condizione: per ogni numero positivo $\delta < \delta_0$ esiste un vettore ω di modulo eguale a δ siffatto che l'insieme J_K^δ, ottenuto traslando J_K del vettore ω, sia interno a D [29]. Per $x \in J_K$ e per ogni funzione v definita in D, poniamo $v_K^\delta(x) = v(x+\omega)$. In corrispondenza ad ogni $\varepsilon > 0$ si può costruire un $\delta_\varepsilon < \delta_0$ tale che:

$$\| u_K^\delta - u_K \|_{\mathcal{L}^q(J_K)} + \| f_K^\delta - f \|_{\mathcal{L}^{q'}(J_K)} < \varepsilon \quad , \quad \delta < \delta_\varepsilon , \tag{65}$$

come segue subito dal teor. di Vitali sul passaggio al limite sotto segno di integrale. Poiché $J_K^\delta \subset D - \mathcal{F}D$, esiste, come s'é osservato, $\tilde{u} \in C^1(D)$ tale che:

$$\| \tilde{u} - u \|_{\mathcal{L}^q(J_K^\delta)} + \| \operatorname{div} \tilde{u} - f \|_{\mathcal{L}^{q'}(J_K^\delta)} < \varepsilon .$$

Ma si ha:

$$\| \tilde{u} - u \|_{\mathcal{L}^q(J_K^\delta)} = \| \tilde{u}_K^\delta - u_K^\delta \|_{\mathcal{L}^q(J_K)} , \; \| \operatorname{div} \tilde{u} - f \|_{\mathcal{L}^{q'}(J_K^\delta)} = \| \operatorname{div} \tilde{u}_K^\delta - f_K^\delta \|_{\mathcal{L}^{q'}(J_K)} ,$$

donde:

$$\| \tilde{u}_K^\delta - u_K^\delta \|_{\mathcal{L}^q(J_K)} + \| \operatorname{div} \tilde{u}_K^\delta - f_K^\delta \|_{\mathcal{L}^{q'}(J_K)} < \varepsilon .$$

---

(29) Si può, anzi, vedere che D gode di tale proprietà anche se non è propriamente regolare, purché abbia la frontiera "localmente lipschitziana".

i e da (65) viene :

$$\| \tilde{u}_K^{\delta} - u \|_{\mathcal{L}^q(J_K)} + \| div\, \tilde{u}_K^{\delta} - f \|_{\mathcal{L}^{q'}(J_K)} < 2\varepsilon .$$

è $\tilde{u}_K^{\delta} \in C^1(J_K)$, resta provato che, in corrispondenza ad ogni K, esi-
ıa successione $\{u_m^{(K)}(x)\}$ di funzioni di $C^1(D \cap \bar{S}_K)$ tale che :

$$(\| u_m^{(K)}(x) - u(x) \|_{\mathcal{L}^q(D \cap \bar{S}_K)} + \| div\, u_n^{(K)}(x) - f(x) \|_{\mathcal{L}^{q'}(D \cap \bar{S}_K)} = 0 .$$

Per completare la dimostrazione, nell'attuale ipotesi, basterà onsiderare una "partizione dell'unità" : $\sum_1^m{}_K \alpha^{(K)}(x) \equiv 1$

va alle sfere aperte S_K e porre $u_n(x) = \sum_1^m{}_K \alpha^{(K)}(x)\, u_n^{(K)}(x)$. Riesce
nente :

$$\lim_{n \to \infty} \| u_n - u \|_{\mathcal{L}^q(D)} = 0 .$$

inoltre :

$$div\, u_n = \sum_1^m{}_K u_m^{(K)} \times grad\, \alpha^{(K)} + \sum_1^m{}_K \alpha^{(K)}\, div\, u_n^{(K)} ,$$

:

$$\lim_{n \to \infty} \| div\, u_n - f \|_{\mathcal{L}^{q'}(D)} = 0 .$$

indi, in tal caso la tesi è acquisita .

Il caso $1 \leq q' \leq q \leq +\infty$ si tratta in modo del tutto analogo.

Prima di passare a considerare gli altri casi, dobbiamo dare un na di completezza.

XLVIII. Sia $1 \leq q < +\infty$ (sia $q = +\infty$). L'insieme V_0 dei vettori $v \in C^1(D)$ anno divergenza nulla in D , è denso, con la metrica di $\mathcal{L}^q(D)$ (di), nell'insieme dei vettori $\in \mathcal{L}^q(D)$ (dei vettori $\in C^0(D)$) che hanno ;enza debole su $D - \mathcal{F}D$ identicamente nulla.

Consideriamo dapprima il caso $1 < q < +\infty$. Sia $f \in \mathcal{L}^p(D)$ e supponia-
e f sia ortogonale a V_0 . Esiste allora una funzione $u \in \mathcal{L}^p(D)$
ıa f come gradiente debole su D . Ne viene l'esistenza di una suc-
ıne $\{u_m\}$ di funzioni di $\overset{0}{C}{}^1(D)$ tali che $\lim_{m \to \infty} \| u_m - u \|_{\mathcal{H}^p(D)} = 0$ (teor.
. Allora, se $w \in \mathcal{L}^q(D)$ ha divergenza debole su $D - \mathcal{F}D$ identicamen-
la, per ogni n si avrà : $\int_D v \times grad\, u_n\, dx = 0$ e quindi,

passando al limite per $n\to\infty$: $\int_D w f\, dx = 0$. Da qui si trae subito l'asserto nel caso $1 < q < +\infty$. Consideriamo ora il caso $q = 1$. Sia $f \in \mathcal{L}^\infty(D)$ e supponiamo che f sia ortogonale a $\mathcal{V}_0$. Esiste allora, come è noto, una funzione $u \in \mathcal{L}^\infty(D)$ (anzi $\in C^0(D)$) che ha per gradiente debole su D il vettore f ,e quindi riesce:

(66) $$\int_D u \operatorname{div} v\, dx = -\int_D v f\, dx \qquad , v \in C^1(D).$$

Sia ora w un vettore di $\mathcal{L}^1(D)$ avente divergenza debole su $D-\mathcal{F}D$ identicamente nulla. Per la prima parte del teor. XLVII già dimostrata (cioè quella relativa al caso $1 \leq q' \leq q \leq +\infty$), esiste una successione $\{v_n\}$ di vettori di $C^1(D)$ tale che:

$$\lim_{n\to\infty} \left(\| v_n - w \|_{\mathcal{L}^1(D)} + \| \operatorname{div} v_n \|_{\mathcal{L}^1(D)} \right) = 0 .$$

Dalla (66) si trae, per ogni n :

$$\int_D u \operatorname{div} v_n\, dx = -\int_D v_n f\, dx$$

e passando al limite per $n\to\infty$, in questa relazione , si ha : $\int_D w f\, dx = 0$
Resta , dunque , anche in tal caso, provato l'asserto.

Esaminiamo , da ultimo il caso $q = +\infty$. Sia data una misura $\alpha \in V(D)$ la quale sia ortogonale a $\mathcal{V}_0$. Esiste una funzione $u \in \mathcal{L}^1(D)$ tale che :

(67) $$\int_D u \operatorname{div} v\, dx = -\int_D v\, d\alpha \qquad , v \in C^1(D) .$$

Sia $w \in C^0(D)$ un vettore dotato di divergenza debole su $D-\mathcal{F}D$ identicamente nulla. Per la prima parte del teor. XLVII già dimostrata (caso $q = q' = +\infty$), esiste una successione $\{v_n\}$ di vettori di $C^1(D)$ tale che :

$$\lim_{n\to\infty} \left(\| v_n - v \|_{C^0(D)} + \| \operatorname{div} v_n \|_{C^0(D)} \right) = 0$$

lla (67) si trae , per ogni n : $\int_D u \operatorname{div} v_n \, dx = -\int_D v_n \, d\alpha$

passando al limite per $n \to \infty$:

$$\int_D w \, d\alpha = 0,$$

ide, anche in tal caso, l'asserto.

Possiamo ora riprendere e completare la dimostrazione del r. XLVII.

Consideriamo il caso $1 \leq q < q' < +\infty$. Dato che $f \in \mathcal{L}^{q'}(D)$, esi- un vettore $\tilde{u} \in \mathcal{L}^{q'}(D)$ che ha f come divergenza debole su $D - \mathcal{F}D$. Il vettore $u_0 = u - \tilde{u}$, per l'ipotesi $q < q'$, appartiene a $\mathcal{L}^q(D)$ ed ha divergenza debole su $D - \mathcal{F}D$ identicamente nulla. rtanto, in forza del teor. XLVIII, esiste una successione $\{u'_n\}$ di tori di $C^1(D)$ che converge a u_0 in $\mathcal{L}^q(D)$. Con il cedimento indicato nella prima parte della dimostrazione del teor. VII si può costruire una successione $\{\tilde{u}_n\}$ di vettori di $C^1(D)$ e che:

$$\lim_{n \to \infty} \left(\| \tilde{u}_n - \tilde{u} \|_{\mathcal{L}^{q'}(D)} + \| \operatorname{div} \tilde{u}_n - f \|_{\mathcal{L}^{q'}(D)} \right) = 0 .$$

ora, posto $u_n = \tilde{u}_n + u'_n$ si ha :

$$\lim_{n \to \infty} \left(\| u_n - u \|_{\mathcal{L}^q(D)} + \| \operatorname{div} u_n - f \|_{\mathcal{L}^{q'}(D)} \right) = 0$$

de viene l'asserto. Il caso $1 \leq q < q' = +\infty$ si tratta in modo perfetta- nte analogo a questo.

16. Divergenza "forte" su D .

Diremo che il vettore $u \in \mathcal{L}^q(D)$, $1 \leq q < +\infty$, ha per divergenza rte" su D la funzione $f \in \mathcal{L}^{q'}(D)$, $1 \leq q' \leq +\infty$ se esiste una successione $\{u_m\}$ di vettori di $\overset{\circ}{C}{}^1(D)$ tale che :

(68) $$\lim_{m\to\infty}\left(\|u_m-u\|_{\mathcal{L}^q(D)}+\|\operatorname{div}u_m-f\|_{\mathcal{L}^{q'}(D)}\right)=0.$$

E' evidente che :

XLIX. Se $u\in\mathcal{L}^q(D)$ ha per divergenza forte su D la funzione $f\in\mathcal{L}^{q'}(D)$, allora u ha per divergenza debole su D la f.

Mostriamo ora il teorema inverso.

L. Sia $1\leq q\leq+\infty$, $1\leq q'\leq+\infty$. Se $u\in\mathcal{L}^q(D)$ ha per divergenza debole su D la funzione $f\in\mathcal{L}^{q'}(D)$, allora u ha per divergenza forte su D la f, ove si assuma come $\mathcal{L}^\infty(D)$ lo spazio $C^0(D)$.

Consideriamo dapprima il caso $1\leq q'\leq q<+\infty$. Siano $S_1, S_2, \ldots, S_m$ dei campi a forma di "cilindro circolare retto definito" che ricoprano D. Se $\overline{S}_k\subset D-\mathcal{F}D$, è facile costruire una successione $\{u_n^{(k)}\}$ di vettori di $\overset{\circ}{C}{}^1(D)$ tale che:

$$\lim_{n\to\infty}\left(\|u_n^{(k)}-u\|_{\mathcal{L}^q(\overline{S}_k)}+\|\operatorname{div}u_n^{(k)}-f\|_{\mathcal{L}^{q'}(\overline{S}_k)}\right)=0,$$

ricorrendo, per esempio, agli operatori regolarizzatori di Friedrichs. Sia ora k tale che $\overline{S}_k\cap\mathcal{F}D\neq\emptyset$. Porremo $J_k=\overline{S}_k\cap D$. Per l'ipotesi che D sia propriamente regolare, si può sempre assumere il ricoprimento di D in guisa tale che, se $\overline{S}_k\cap\mathcal{F}D\neq\emptyset$, siano soddisfatte le seguenti condizioni:

1) esiste un punto $x_0\in\overline{S}_k\cap\mathcal{F}D$ siffatto che, introdotto un opportuno sistema di coordinate cartesiane ortogonali ξ^1, ξ^2, ξ^3 con origine in x_0, l'insieme J_k risulti definito dalle seguenti limitazioni:

$$|\xi^1|^2+|\xi^2|^2 \leq |R_K|^2 \quad , \quad \Psi_K(\xi^1,\xi^2) \leq \xi^3 \leq \ell_K$$

ove: R_K é il raggio del cilindro S_K ; $\Psi_K(\xi^1,\xi^2)$ è una funzione definita nel cerchio (chiuso) di centro l'origine e raggio R_K, ivi continua e di classe 1 a pezzi, tale che $\xi^3=\Psi_K(\xi^1,\xi^2)$, $|\xi^1|^2+|\xi^2|^2 \leq |R_K|^2$ sia la rappresentazione parametrica di $S_K \cap \partial D$ (parametri ξ^1,ξ^2); ℓ_K é un numero positivo maggiore di

$$\rho_K = \max_{|\xi^1|^2+|\xi^2|^2 \leq R_K^2} |\Psi_K(\xi^1,\xi^2)|,$$

2) esiste un numero positivo δ_0 tale che: $\delta_0 < R_K$, $2\delta_0 < \ell_K - \rho_K$ e tale inoltre che l'insieme definito da

$$|\xi^1|^2+|\xi^2|^2 \leq |R_K|^2 \quad , \quad \Psi_K(\xi^1,\xi^2)-2\delta \leq \xi^3 < \Psi_K(\xi^1,\xi^2)$$

è contenuto nel complementare di D per ogni $\delta \in (0,\delta_0)$.

Denotiamo con J_K^δ l'insieme definito da :

$$|\xi^1|^2+|\xi^2|^2 \leq (R_K-\delta)^2 \quad , \quad \Psi_K(\xi^1,\xi^2) \leq \xi^3 \leq \ell_K-\delta \, ,$$

con I_K^δ quello definito da :

$$|\xi^1|^2+|\xi^2|^2 \leq R_K^2 \quad , \quad \Psi_K(\xi^1,\xi^2)-\delta \leq \xi^3 \leq \ell_K$$

e con $\tilde{I}_K^\delta$ quello definito da:

$$|\xi^1|^2+|\xi^2|^2 \leq R_K^2 \quad , \quad \Psi_K(\xi^1,\xi^2)+\delta \leq \xi^3 \leq \ell_K \, .$$

Pensiamo ora di prolungare u ed f fuori di D con il valore zero. E' chiaro che la u , così prolungata , è dotata di divergenza debole su $I_K^\delta - \partial I_K^\delta$ e tale divergenza debole coincide con la f prolungata nel modo detto. Poniamo $u(x)=u_K(\xi)$ e $f(x)=f_K(\xi)$ per $x \in I_K^\delta$ ove $\xi \equiv (\xi^1,\xi^2,\xi^3)$ rappresenta il punto x nel nuovo riferimento . Poniamo inoltre, per $\xi \in I_k^\delta$:

$$u_k^\delta(\xi) \equiv u_k^\delta(\xi^1, \xi^2, \xi^3) = u_k(\xi^1, \xi^2, \xi^3 - \delta) \;,\; f_k^\delta(\xi) \equiv f_k^\delta(\xi^1, \xi^2, \xi^3) = f_k(\xi^1, \xi^2, \xi^3 - \delta).$$

Fissato $\varepsilon > 0$, esiste $\delta_\varepsilon > 0 (\varepsilon < \delta_0)$ tale che, per $\delta < \delta_\varepsilon$ riesce:

(69) $$\| u_k^\delta(\xi) - u_k(\xi) \|_{\mathcal{L}^q(J_k^\delta)} + \| f_k^\delta(\xi) - f_k(\xi) \|_{\mathcal{L}^{q'}(J_k^\delta)} < \varepsilon$$

(come si è sopra osservato). Il vettore $u_k^\delta(\xi)$ ha per divergenza debole, su $I_k^\delta - \mathcal{F} I_k^\delta$, la funzione f_k^δ . Poichè è $J_k^\delta \subset I_k^\delta - \mathcal{F} I_k^\delta$, si può costruire un vettore $v_k^\delta \in C^1(I_k^\delta)$ tale che :

(70) $$\| v_k^\delta(\xi) - u_k^\delta(\xi) \|_{\mathcal{L}^q(J_k^\delta)} + \| \operatorname{div} v_k^\delta(\xi) - f_k^\delta(\xi) \|_{\mathcal{L}^{q'}(J_k^\delta)} < \varepsilon .$$

Si può anzi supporre che v_k^δ possa prolungarsi a tutto lo spazio in modo da avere supporto contenuto in $D - \mathcal{F} D$. Se, infatti, ad esempio, si assume

$$v_k^\delta(x) = \int_{I_k^\delta} \mathcal{K}_\eta (x-y)\, u_k^\delta(y)\, dy$$

ove $\mathcal{K}_\eta(x-y)$ è un nucleo regolarizzatore nel senso di Friedrichs (nullo per $|x-y| \geq \eta$), é noto che, per η abbastanza piccolo, risultano soddisfatte le (70) ; se poi si osserva che il supporto di u_k^δ é contenuto in $\tilde{I}_k^\delta$, si vede che risulta :

$$v_k^\delta (x) = \int_{\tilde{I}_k^\delta} \mathcal{K}_\eta (x-y)\, u_k^\delta(y)\, dy$$

e quindi il supporto di $v_k^\delta(x)$ risulta contenuto in $(I_k^\delta \cap D) - \mathcal{F} D$ non appena $\eta < \overline{\tilde{I}_k^\delta\ \mathcal{F} D}$. Pertanto v_k^δ potrà prolungarsi a tutto lo spazio in modo che sia soddisfatta la condizione : supporto di v_k^δ contenuto in $D - \mathcal{F} D$. Da (69) (70) segue, d'altra parte,

$$\| v_k^\delta - u \|_{\mathcal{L}^q(J_k^\delta)} + \| \operatorname{div} v_k^\delta - f \|_{\mathcal{L}^{q'}(J_k^\delta)} < 2\varepsilon .$$

Per completare la dimostrazione basterà ora ricorrere ad una "partizione dell'unità "esattamente come si è fatto nella dimostrazione della prima parte del teor.XLVII sfruttando l'ipotesi $q' \leq q$.

Il caso $1 \leq q' \leq q \leq +\infty$ si tratta in modo perfettamente analogo.

Per includere gli altri casi (cioè $1 \leq q < q' \leq +\infty$) non c'è che da ripetere la dimostrazione fatta in relazione al teor. XLVII sfruttando il seguente teorema di completezza , analogo al teor. XLVIII :

LI. Sia $1 \leq q < +\infty$ (sia $q = +\infty$). L'insieme $\mathcal{V}_0$ dei vettori di $\overset{o}{C}{}^1(D)$ che hanno divergenza identicamente nulla in D è denso, con la metrica di $\mathcal{L}^q(D)$ (di $C^0(D)$) nell'insieme dei vettori di $\mathcal{L}^q(D)$ (di $C^0(D)$) che hanno divergenza debole su D identicamente nulla.

La dimostrazione di questo teorema è perfettamente analoga a quella del teor. XLVIII.

Diremo che $u \in \mathcal{L}^q(D)$ ha divergenza "forte" su D rappresentata dalla misura

(71) $$\mu(B) = \int_B f\,dx + \int_{B \cap \mathcal{F}D} \varphi\, d\sigma, \quad f \in \mathcal{L}^{q'}(D), \ \varphi \in \mathcal{L}^{q''}(\mathcal{F}D)$$

se esiste una successione $\{u_n\}$ di vettori di $C^1(D)$ tale che:

(72) $$\lim_{n\to\infty} \| u_n - u \|_{\mathcal{L}^q(D)} = 0, \ \lim_{n\to\infty} \| \operatorname{div} u_n - f \|_{\mathcal{L}^{q'}(D)} = 0, \ \lim_{n\to\infty} \| u_n \times \nu - \varphi \|_{\mathcal{L}^{q''}(\mathcal{F}D)} = 0$$

$$1 \leq q \leq +\infty, \ 1 \leq q' \leq +\infty, \ 1 \leq q'' \leq +\infty.$$

E' evidente che:

LII. Se $u \in \mathcal{L}^q(D)$ ha divergenza forte su D rappresentata dalla misura (71) allora u ha come divergenza debole su D

la misura μ .

Mostriamo, viceversa, che:

LIII. Se D è di classe 2 e se $u \in \mathcal{L}^q(D)$ ha divergenza debole su D rappresentata dalla misura (71), allora u ha come divergenza forte su D la misura μ ; $1 \leq q < 3/2$, $1 \leq \leq q' \leq +\infty$, $1 \leq q'' \leq +\infty$; $q = 3/2$, $1 \leq q' \leq +\infty$, $1 < q'' \leq +\infty$; $\frac{3}{2} < q < +\infty$, $1 \leq q' \leq +\infty$, $\frac{2}{3} q \leq q'' \leq +\infty$; $q = +\infty$, $1 \leq q' \leq +\infty$, $q'' = +\infty$; $q = \frac{3}{2}$, $q' = 1$, $q'' = 1$, ove si convenga di assumere come $\mathcal{L}^\infty(D)$ lo spazio $C^0(D)$ e analogamente, come $\mathcal{L}^\infty(\mathcal{F}D)$ lo spazio $C^0(\mathcal{F}D)$.

Esamineremo, intanto, i primi tre casi . Facciamo, dapprima, l'ipotesi :

(73) $$\int_D f\, dx = \int_{\mathcal{F}D} \varphi\, d\sigma = 0 .$$

Esiste allora $u' \in \mathcal{L}^q(D)$ che ha per divergenza debole su D la misura $\mu(B) = \int_{B \cap \mathcal{F}D} \varphi\, d\sigma$ (teor. XXXV) e riesce , per $(53)_{p,p',p''}$:

(74) $$\inf_{u_0 \in \mathcal{U}_q} \| u' + u_0 \|_{\mathcal{L}^q(D)} \leq K_{p,p',p''}(D) \| \varphi \|_{\mathcal{L}^{q''}(\mathcal{F}D)}$$

ove $\mathcal{U}_q$ è l'insieme dei vettori di $\mathcal{L}^q(D)$ che hanno divergenza debole su D identicamente nulla. Sia $\{\varphi_n\}$ una successione di funzioni di $C^2(\mathcal{F}D)$ tali che :

$$\int_{\mathcal{F}D} \varphi_n\, d\sigma = 0 , \quad \lim_{n \to \infty} \| \varphi_n - \varphi \|_{\mathcal{L}^{q''}(\mathcal{F}D)} = 0 .$$

Esiste $u_n \in C^1(D)$ tale che: $\operatorname{div} u_n = 0$ in D , $u_n \times \nu = \varphi_n$ su $\mathcal{F}D$; dalla precedente diseguaglianza si trae allora :

$$\lim_{n \to \infty} \inf_{u_0 \in \mathcal{U}_q} \| u' - u_n + u_0 \|_{\mathcal{L}^q(D)} = 0 .$$

Per il teor. LI esiste quindi un vettore $w_n \in \overset{\circ}{C}^1(D)$, con $\operatorname{div} w_n = 0$ tale che

$$\lim_{n\to\infty} \| u' - u_n + w_n \|_{\mathcal{L}^q(D)} = 0 , \quad \lim_{n\to\infty} \| \varphi - (u_n - w_n) \times \nu \|_{\mathcal{L}^{q''}(\mathcal{F}D)} = 0 .$$

Posto $u'' = u - u'$, si ha $u'' \in \mathcal{L}^q(D)$; inoltre u'' avrà divergenza debole su D data dalla misura $\mu(B) = \int_B f\, dx$. Ne viene (teor. L) l'esistenza di $v_n \in \overset{\circ}{C}^1(D)$ tale che:

$$\lim_{n\to\infty} \| v_n - u'' \|_{\mathcal{L}^q(D)} = 0 , \quad \lim_{n\to\infty} \| \operatorname{div} v_n - f \|_{\mathcal{L}^{q'}(D)} = 0 .$$

Dunque, $\{u_n + v_n - w_n\}$ è la successione di cui alla tesi, nelle attuali ipotesi.

Se la (73) non è soddisfatta, poniamo $c = \frac{1}{3 \operatorname{mis} D} \int_D f\, dx$,

$$\tilde{u}(x) = u(x) - cx , \quad \tilde{\varphi}(x) = \varphi(x) - c\nu \times x , \quad \tilde{f}(x) = f(x) - 3c$$

ove x è il vettore di componenti x^1, x^2, x^3. Si ha:

$$\int_D \tilde{u} \times \operatorname{grad} v\, dx = - \int_D \tilde{f}\, v\, dx - \int_{\mathcal{F}D} v \tilde{\varphi}\, d\sigma$$

e $\int_D \tilde{f}\, dx = 0$ donde $\int_{\mathcal{F}D} \tilde{\varphi}\, d\sigma = 0$. In tal modo ci si è ricondotti al caso precedente.

Esaminiamo ora il caso $q = +\infty, 1 \leq q' \leq +\infty, q'' = +\infty$. Si può ripetere la dimostrazione ora esposta, con la sola avvertenza di far uso del teor. XLI (in luogo del teor. XXXV) per riguardo all'esistenza di u', e della formula di maggiorazione (57) (in luogo della $(53)_{p,p',p''}$) per conseguire la (74), nella quale, naturalmente, l'insieme U_q andrà sostituito con quello dei vettori di $C^0(D)$ che hanno divergenza su D identicamente nulla.

Esaminiamo, da ultimo, il caso : $q = \frac{3}{2}$, $q' = q'' = 1$. Possiamo sempre supporre, come prima si è visto : $\int_{\mathcal{F}D} \varphi\, d\sigma = 0$. Sia $\tilde{u}$ un vettore di $\mathcal{H}^1(D)$ tale che $\tilde{u} \times \nu = \varphi$ su $\mathcal{F}D$ (cfr. teor.

XXVI) ; ed anzi sarà $\tilde{u} \in \mathcal{L}^{3/2}(D)$. Il vettore $u-\tilde{u}$ appartiene quindi a $\mathcal{L}^{3/2}(D)$, ed è dotato di divergenza debole su D rappresentata dalla misura $\mu(B)=\int_B (f-\operatorname{div}\tilde{u})dx$. Esiste quindi $u'_n \in \overset{\circ}{C}{}^1(D)$ tale che:

$$\lim_{n\to\infty} \| u'_n - u + \tilde{u} \|_{\mathcal{L}^{3/2}(D)} = 0, \quad \lim_{n\to\infty} \| \operatorname{div} u'_n - f + \operatorname{div}\tilde{u} \|_{\mathcal{L}^1(D)} = 0$$

(teor. L). Siano $\tilde{u}_i$ le componenti di $\tilde{u}$. Si ha $\tilde{u}_i \in \mathcal{L}^1(\mathcal{F}D)$. Sia $\tilde{u}_i^{(n)} \in C^1(D)$ e tale che $\lim_{n\to\infty} \| \tilde{u}_i^{(n)} - \tilde{u}_i \|_{\mathcal{L}^1(\mathcal{F}D)} = 0$. Risulta (per $(38)_{1,1}$ e per il teor. L)

$$\inf_{u_0 \in \overset{\circ}{C}{}^1(D)} \| \tilde{u}_i - \tilde{u}_i^{(n)} - u_0 \|_{\mathcal{H}^{1,1}(D)} \leq R_{1,1}(D) \| \tilde{u}_i - \tilde{u}_i^{(n)} \|_{\mathcal{L}^1(\mathcal{F}D)}.$$

Esiste quindi $w_i^{(n)} \in \overset{\circ}{C}{}^1(D)$ tale che:

$$\lim_{n\to\infty} \| \tilde{u}_i - \tilde{u}_i^{(n)} - w_i^{(n)} \|_{\mathcal{H}^{1,1}(D)} = 0.$$

Se ne deduce l'esistenza di una successione di costanti $\{c_i^{(n)}\}$ tale che

$$\lim_{n\to\infty} \| \tilde{u}_i - \tilde{u}_i^{(n)} - w_i^{(n)} - c_i^{(n)} \|_{\mathcal{L}^{3/2}(D)} = 0 ,$$

$$\lim_{n\to\infty} \| \operatorname{grad}(\tilde{u}_i - \tilde{u}_i^{(n)} - w_i^{(n)} - c_i^{(n)}) \|_{\mathcal{L}^1(D)} = 0 .$$

Dalla $(40)_{1,1}$ si trae: $\lim_{n\to\infty} \| \tilde{u}_i - \tilde{u}_i^{(n)} - w_i^{(n)} - c_i^{(n)} \|_{\mathcal{L}^1(\mathcal{F}D)} = 0$.
Indicato con u''_n il vettore che ha per i-esima componente $\tilde{u}_i^{(n)} + w_i^{(n)} + c_i^{(n)}$, si ha allora che $\{u'_n + u''_n\}$ è la successione di cui alla tesi nel caso $q = \frac{3}{2}$, $q' = q'' = 1$.

Altre caratterizzazioni di tipo "forte" per i vettori dotati di divergenza debole su D (con divergenza rappresentata da una misura della forma (71)) sono date dai seguenti teoremi.

LIV. <u>Sia D di classe 2. Condizione necessaria e sufficiente perchè $u \in \mathcal{L}^q(D)$ abbia divergenza debole su D rappresentata da una misura della forma (71), è che esista una successione $\{u_n\}$ con</u>

$u_n \in C^1(D)$ tale che:

$$\lim_{n\to\infty} \|u_n - u\|_{\mathcal{L}^{\min(q,q'')}(D)} = 0, \quad \lim_{n\to\infty} \|u_n \times \nu - \varphi\|_{\mathcal{L}^{q''}(\mathcal{F}D)} = 0, \quad \lim_{n\to\infty} \|\operatorname{div} u_n - f\|_{\mathcal{L}^{q'}(D)} = 0$$

$1 \le q \le +\infty, \ 1 \le q' \le +\infty, \ 1 \le q'' \le +\infty$

convenendo di assumere come $\mathcal{L}^\infty$ lo spazio C^0 (sia in relazione a D che a $\mathcal{F}D$)

La condizione è ovviamente sufficiente. Mostriamone la necessità. Come si è visto nella dimostrazione precedente, possiamo supporre che sia $\int_D f\,dx = \int_{\mathcal{F}D} \varphi\, d\sigma = 0$. Sia $\varphi_n \in C^2(\mathcal{F}D)$ tale che: $\int_{\mathcal{F}D} \varphi_n\, d\sigma = 0$, $\lim_{n\to\infty} \|\varphi_n - \varphi\|_{\mathcal{L}^{q''}(\mathcal{F}D)} = 0$. Esiste $u'_n \in C^1(D)$ tale che: $\operatorname{div} u'_n = 0$ in D, $u'_n \times \nu = \varphi_n$ su $\mathcal{F}D$. Esiste $u' = \mathcal{L}^{q''}(D)$ che ha per divergenza debole su D la misura $\mu(B) = \int_{B\cap\mathcal{F}D} \varphi\, d\sigma$ (teor. XXXV se $q'' < +\infty$ e teor. XLI se $q'' = +\infty$) e sarà (cfr. $(53)_{r,r',r''}$ se $q'' < +\infty$ e (57) se $q'' = +\infty$) .

$$\inf_{u_0 \in \mathcal{U}_{q''}} \|u' - u'_n + u_0\|_{\mathcal{L}^{q''}(D)} \le K \|\varphi - \varphi_n\|_{\mathcal{L}^{q''}(\mathcal{F}D)},$$

ove $\mathcal{U}_{q''}$ è l'insieme dei vettori di $\mathcal{L}^{q''}(D)$ che hanno divergenza debole su D identicamente nulla. Dal teor. LI segue che si può modificare u'_n in guisa tale che riesca $u'_n \in C^1(D)$,

$$\operatorname{div} u'_n = 0, \quad u'_n \times \nu = \varphi_n \text{ su } \mathcal{F}D, \quad \lim_{n\to\infty} \|u' - u'_n\|_{\mathcal{L}^{q''}(D)} = 0 .$$

Sia $u'' = u - u'$. Riesce: $u'' \in \mathcal{L}^{\min(q,q'')}(D)$. Esiste $u''_n \in \overset{\circ}{C}{}^1(D)$ tale che (teor. L) :

$$\lim_{n\to\infty} \|u''_n - u''\|_{\mathcal{L}^{\min(q,q'')}(D)} = 0, \quad \lim_{n\to\infty} \|\operatorname{div} u''_n - f\|_{\mathcal{L}^{q'}(D)} = 0.$$

La successione $\{u'_n + u''_n\}$ è quella di cui alla tesi del teorema.

LV . Sia D di classe 2. Condizione necessaria e sufficiente perchè $u \in \mathcal{L}^q(D)$, $1 \leq q \leq +\infty$, abbia divergenza debole su D rappresentata dalla misura $\mu(B) = \int_{B \cap \mathcal{F}D} \varphi \, d\sigma$ con $\varphi \in \mathcal{L}^{q''}(\mathcal{F}D)$, $1 \leq q'' \leq +\infty$ è che esista una successione $\{u_n\}$, con $u_n \in C^1(D)$, tale che:

$$\lim_{n \to \infty} \int_D (u_n - u) \times \operatorname{grad} v \, dx = 0, \quad v \in \mathcal{H}^{\infty, p}(D)$$

$$\operatorname{div} u_n = 0, \quad \lim_{n \to \infty} \| u_n \times \nu - \varphi \|_{\mathcal{L}^{q''}(\mathcal{F}D)} = 0.$$

convenendo di assumere come $\mathcal{L}^{\infty}$ lo spazio C^0

La sufficienza è ovvia. Proviamo la necessità.

Siano $\varphi_n \in C^2(\mathcal{F}D)$ e $u_n \in C^1(D)$ tali che:

$$\operatorname{div} u_n = 0, \quad u_n \times \nu = \varphi_n \text{ su } \mathcal{F}D, \quad \lim_{n \to \infty} \| \varphi_n - \varphi \|_{\mathcal{L}^{q''}(\mathcal{F}D)} = 0$$

Poichè

$$\int_D u_n \times \operatorname{grad} v \, dx = - \int_{\mathcal{F}D} \varphi_n v \, d\sigma, \quad v \in \mathcal{H}^{\infty, p}(D)$$

e

$$\int_D u \times \operatorname{grad} v \, dx = - \int_{\mathcal{F}D} \varphi v \, d\sigma, \quad v \in \mathcal{H}^{\infty, p}(D),$$

si avrà

$$\lim_{n \to \infty} \int_D (u_n - u) \times \operatorname{grad} v \, dx = 0, \quad v \in \mathcal{H}^{\infty, p}(D),$$

donde la tesi.

17. Alcuni teoremi di caratterizzazione per le funzioni dotate di gradiente debole.

LVI. Condizione necessaria e sufficiente perchè $u \in \mathcal{L}^{p'}(D)$ abbia per gradiente debole su $D - \mathcal{F}D$ (su D) il vettore $f \in \mathcal{L}^p(D)$, $1 \leq p' \leq +\infty$, $1 \leq p \leq +\infty$ è che sia soddisfatta la seguente condizione:

a) per ogni $v \in \mathcal{L}^q(D)$ ed avente divergenza debole su D (su $D - \mathcal{F}D$) contenuta in $\mathcal{L}^{q'}(D)$, riesca: $\int_D u \operatorname{div} v \, dx = - \int_D v f \, dx$.

La condizione a) è ovviamente sufficiente. Proviamone la necessità. Nel caso $1 \leq p < +\infty$, $1 \leq p' < +\infty$, la tesi è immediata conseguenza del teor. XLIII (teor. XLIV).
Nei casi $1 < p \leq +\infty, p' = +\infty$; $p = +\infty, 1 < p' \leq +\infty$ la necessità segue dal teor. L (teor. XLVII). Sia ora $p = +\infty$, $p' = 1$; ci si riduce subito al caso precedente $p = +\infty$, $p' > 1$ ricordando che deve essere $u \in C^0(D)$.
Rimane da considerare solo il caso: $p = 1$, $p' = +\infty$. E' facile constatare che la successione $\{u_n\}$ di cui al teor. XLIII (teor. XLIV) e costruita con il procedimento indicato da Gagliardo, nel caso attuale $p = 1$, $p' = +\infty$, gode delle seguenti proprietà:

$$\lim_{n \to \infty} \| \operatorname{grad} u_n - f \|_{\mathcal{L}^1(D)} = 0, \quad \lim_{n \to \infty} \| u_n - u \|_{\mathcal{L}^1(D)} = 0$$

ed inoltre le funzioni u_n risultano equipseudolimitate cioè esiste una costante L tale che, q.o. risulti $|u_n| \leq L$ [30]. Si ha allora, in corrispondenza ad ogni $v \in \mathcal{L}^\infty(D)$ avente divergenza debole su D (su $D - \mathcal{F}D$) contenuta $\mathcal{L}^1(D)$, in virtù del teorema di Lebesgue sul passaggio al limite sotto segno di integrale:

$$\lim_{n \to \infty} \int_D u_n \operatorname{div} v \, dx = \int_D u \operatorname{div} v \, dx,$$

donde si trae facilmente l'asserto.

LVII. <u>Condizione necessaria e sufficiente perchè</u> $u \in \mathcal{L}^{p'}(D)$ $1 \leq p' \leq +\infty$, <u>abbia per gradiente debole su</u> $D - \mathcal{F}D$ (<u>su</u> D) <u>una misura</u> μ <u>è che esista</u> $F \in [\mathcal{L}^\infty(D)]^*$ <u>tale che</u>

---

(30) Crr. E. Gagliardo, loc cit. in (11); per controllare questa proposizione, basta tener presente che le successioni costruite con gli operatori regolarizzatori di Friedrichs godono delle suddette proprietà.

(75) $$\int_D u \operatorname{div} v \, dx = - F(v)$$

per ogni $v \in \mathcal{L}^{\infty}(D)$ avente divergenza debole su D (su $D-\mathcal{F}D$) contenuta in $\mathcal{L}^{q'}(D)$.

La condizione è ovviamente sufficiente. Proviamone la necessità. Consideriamo il caso in cui $u \in \mathcal{L}^{p'}(D)$ sia dotato di gradiente debole su $D-\mathcal{F}D$, rappresentato da una misura $\mu \in V(D-\mathcal{F}D)$. La dimostrazione è analoga a quella del teor. VIII. Si costruisce, con lo stesso procedimento là indicato, una successione $\{u_n\}$ di funzioni di $C^1(D)$ la quale gode delle seguenti proprietà: comunque si consideri un dominio $D' \subset D-\mathcal{F}D$, riesce

$$\lim_{n\to\infty} \| u_n - u \|_{\mathcal{L}^{p'}(D')} = 0 \quad \text{se } p' < +\infty ;$$

se $p' = +\infty$ si ha $\lim_{n\to\infty} \| u_n - u \|_{\mathcal{L}^1(D')} = 0$, e le funzioni di $\{u_n\}$ sono equipseudolimitate [31]. Ne viene, per ogni $v \in \mathcal{L}^{\infty}(D')$ avente divergenza debole su D' appartenente a $\mathcal{L}^{q'}(D')$:

(76) $$\lim_{n\to\infty} \int_{D'} u_n \operatorname{div} v \, dx = \int_{D'} u \operatorname{div} v \, dx .$$

Inoltre riesce:

(77) $$\int_{D'} | \operatorname{grad} u_n | \, dx \leq V_{\mu}(D-\mathcal{F}D) , \quad n > n_{D'} .$$

Per ogni n e per ogni $v \in \mathcal{L}^{\infty}(D')$ avente divergenza debole su D' appartenente a $\mathcal{L}^{q'}(D')$, risulta:

(78) $$\int_{D'} v \times \operatorname{grad} u_n \, dx = - \int_{D'} u_n \operatorname{div} v \, dx .$$

Da qui, tenento conto di (76), viene l'esistenza del limite (finito):

---

[31] Nel caso $p' = +\infty$, le proprietà di $\{u_n\}$ seguono subito dalla definizione di operatore regolarizzatore secondo Friedrichs.

$$\lim_{n\to\infty} \int_{D'} v \times \text{grad}\, u_n\, dx = -F_{D'}(v).$$

Per la (77) si ha che $F_{D'}$ risulta essere una funzionale lineare e continuo su $\mathcal{L}^{\infty}(D')$ e quindi, passando al limite per $n\to\infty$ in (78) (e ricordando (76)), si ottiene l'asserto, relativamente ad un qualsiasi dominio $D' \subset D - \mathcal{F}D$. La condizione $D' \subset D - \mathcal{F}D$ non è, però, restrittiva, ai fini del conseguimento della nostra tesi, dal momento che u può sempre essere prolungato in un dominio $\tilde{D}$ contenente D nel suo interno, in modo che la u così prolungata appartenga a $\mathcal{L}^{p'}(\tilde{D})$ ed abbia gradiente debole su $\tilde{D} - \mathcal{F}\tilde{D}$ rappresentato da una misura $\in V(\tilde{D} - \mathcal{F}\tilde{D})$.

Esaminiamo ora il caso in cui u sia dotata di gradiente debole su D rappresentato da una $\mu \in V(D)$. Sia D_ρ, $0 < \rho < \rho_0$ il dominio di cui alla definizione di dominio propriamente regolare. La u è, ovviamente, dotata di gradiente debole su D_ρ, per quasi tutti i $\rho \in (0, \rho_0)$, rappresentato dalla misura:

$$\mu_\rho(B) = \mu[B \cap (D_\rho - \mathcal{F}D_\rho)] + \int_{B \cap \mathcal{F}D_\rho} u \times \nu_{\mathcal{F}D_\rho}\, d\sigma$$

Per quasi tutti i $\rho \in (0, \rho_0)$ si ha dunque:

$$V_{\mu_\rho}(D_\rho) \leq V_\mu(D_\rho - \mathcal{F}D_\rho) + \int_{\mathcal{F}D_\rho} |u|\, d\sigma.$$

Poniamo:

$$u_\rho(x) \begin{cases} = u(x) & x \in D_\rho \\ = 0 & x \notin D_\rho . \end{cases}$$

Conveniamo poi di prolungare la misura μ_ρ su tutti i boreliani contenuti in D definendola eguale a zero sui boreliani privi di punti in comune con D_ρ. Si ha dunque, per quasi tutti i $\rho \in (0, \rho_0)$ abbastanza piccoli :

$$V_{\mu_\rho}(D) \leq V_\mu(D_\rho - \mathcal{F}D_\rho) + \int_{\mathcal{F}D_\rho} |u|\, d\sigma$$

$$V_{\mu_\rho}(D) \leq V(D - \mathcal{F}D) + \int_{\mathcal{F}D} |u^*|\, d\sigma + 1$$

avendo indicato con u^* la traccia di u su $\mathcal{F}D$ e ricordando che:

$$\lim_{\rho \to 0^+} \int_{\mathcal{F}D_\rho} |u|\, d\sigma = \int_{\mathcal{F}D} |u^*|\, d\sigma$$

(il limite è fatto prescindendo da un conveniente insieme di misura nulla per ρ). La funzione $u_\rho(x)$, per quasi tutti i $\rho \in (0, \rho_0)$, ha per gradiente debole su D (e quindi anche su $D - \mathcal{F}D$) la misura $\mu_\rho \in V(D)$. Se ora poniamo :

$$u_\rho^{(n)}(x) = \int_{D_\rho} \mathcal{K}_{1/n}(x-\xi)\, u(\xi)\, d\xi \equiv \int_D \mathcal{K}_{1/n}(x-\xi)\, u_\rho(\xi)\, d\xi$$

(sarà $u_\rho^{(n)} \in \overset{\circ}{C}{}^1(D)$ per $n >$ di un certo n_ρ), ripetendo, in relazione al dominio D ed alla funzione u_ρ , un ragionamento fatto nella dimostrazione del teor. VIII, si vede che esiste finito il limite

$$\lim_{n \to \infty} \int_D v \times \operatorname{grad} u_\rho^{(n)}\, dx \equiv F_\rho(v)$$

per ogni $v \in \mathcal{L}^{q'}(D)$ ed avente divergenza debole su D contenuta in $\mathcal{L}^\infty(D)$; inoltre : $F_\rho \in [\mathcal{L}^\infty(D)]^*$ e risulta:

$$\|F_\rho\|_{[\mathcal{L}^\infty(D)]^*} \leq V_{\mu_\rho}(D) \leq V_\mu(D - \mathcal{F}D) + \int_{\mathcal{F}D} |u^*|\, d\sigma + 1 = M$$

per tutti i $\rho \in (0, \rho_0)$ abbastanza piccoli. D'altra parte si ha $(\rho < \rho' < \rho_0)$:

$$\lim_{n \to \infty} \|u_\rho^{(n)} - u_\rho\|_{\mathcal{L}^{p'}(D_{\rho'})} = 0 \qquad \text{se } p' < +\infty$$

e $\lim_{n\to\infty} \| u_\varsigma^{(n)} - u_\varsigma \|_{\mathcal{L}^1(D_{\varsigma'})} = 0$ e $u_\varsigma^{(n)}$ equipseudolimitate (al variare di ς e di n), se $p' = +\infty$.

Ne viene: $\lim_{n\to\infty} \int_D u_\varsigma^{(n)} \operatorname{div} v \, dx = \int_D u_\varsigma \operatorname{div} v \, dx \equiv \int_{D_\varsigma} u \operatorname{div} dx$ per ogni v del tipo detto, e quindi : $\int_{D_\varsigma} u \operatorname{div} v \, dx = -F_\varsigma(v)$.

Poichè, per ogni v del tipo sudetto, riesce: $\lim_{\varsigma\to 0^+} \int_{D_\varsigma} u \operatorname{div} v \, dx = \int_D u \operatorname{div} v \, dx$ si ha allora che esiste finito il limite $\lim_{\varsigma\to 0^+} F_\varsigma(v) \equiv F(v)$; risulta $\int_D u \operatorname{div} v \, dx = -F(v)$, con v appartenente alla predetta classe. D'altra parte si ha $|F_\varsigma(v)| \leq M \| v \|_{\mathcal{L}^\infty(D)}$ donde : $|F(v)| \leq M \| v \|_{\mathcal{L}^\infty(D)}$. Ne viene $F \in [\mathcal{L}^\infty(D)]^*$. L'asserto è così completamente provato.

LVIII. Sia D di classe 2 . Condizione necessaria e sufficiente perchè $u \in \mathcal{L}^{p'}(D)$ abbia per gradiente debole su D la misura $\mu(B) = \int_B f \, dx + \int_{B \cap \mathcal{F}D} \varphi \, d\sigma$ con $f \in \mathcal{L}^p(D)$, $\varphi \in \mathcal{L}^{p''}(\mathcal{F}D)$ è che, per ogni $v \in \mathcal{L}^q(D)$, avente per divergenza debole su D la misura

$$\alpha(B) = \int_B f' \, dx + \int_{B \cap \mathcal{F}D} \varphi' \, d\sigma ,$$

con $f' \in \mathcal{L}^{q'}(D)$, $\varphi' \in \mathcal{L}^{q''}(\mathcal{F}D)$, riesca : $\int_D u f' \, dx = -\int_D v f \, dx - \int_{\mathcal{F}D} \varphi \varphi' \, d\sigma$, $1 \leq p < +\infty$, $1 \leq p' < +\infty$, $1 \leq p'' < +\infty$.

La tesi segue subito dal teor. XLV.

Il teor. LVIII è vero anche nei seguenti casi: $3 < p \leq +\infty$, $1 < p' \leq +\infty$, $1 < p'' \leq +\infty$; $p = 3$, $p' = +\infty$, $p'' = +\infty$; $p = 3$, $1 < p' \leq +\infty$, $1 < p'' < +\infty$; $1 < p < 3$, $1 < p' \leq +\infty$, $1 < p'' \leq \frac{2p}{3-p}$, come segue subito dal teor. LIII, oppure $1 < p \leq +\infty$, $1 < p' \leq +\infty$, $1 < p'' \leq +\infty$, $p'' \leq p$, come segue dal teor. LIV ; ed è vero anche senza alcuna limitazione per p, p', p'', se $\mathcal{L}^\infty$ si identifica con C^0.

18. Alcuni teoremi di caratterizzazione per le funzioni dotate di divergenza debole

LIX. Condizione necessaria e sufficiente perchè $u \in \mathcal{L}^{q}(D)$ abbia per divergenza debole su $D-\mathcal{F}D$ (su D) la funzione $f \in \mathcal{L}^{q'}(D)$ è che sia soddisfatta la seguente condizione, per $1 \leq q \leq +\infty, 1 \leq q' \leq +\infty$:

b) per ogni $v \in \mathcal{L}^{p'}(D)$ ed avente gradiente debole su D (su $D-\mathcal{F}D$) contenuto in $\mathcal{L}^{p}(D)$, riesce: $\int_D u \times \operatorname{grad} v \, dx = -\int_D v f \, dx$.

La condizione b) è ovviamente sufficiente. Proviamone la necessità. Nel caso $1 \leq q < +\infty, 1 \leq q' < +\infty$ la tesi è immediata conseguenza del teor. XLVII (teor. L). Nei casi $1 < q \leq +\infty, q' = +\infty$; $q = +\infty, 1 < q' \leq +\infty$, la necessità segue dal teor. XLIV (teor. XLIII). Sia ora $q = 1$, $q' = +\infty$. Esiste una successione $\{u_n\}$ costituita di funzioni di $C^1(D)$ (di $\mathring{C}^1(D)$) tale che $\lim_{n\to\infty} \| u_n - u \|_{\mathcal{L}^1(D)} = 0$, $\lim_{n\to\infty} \| \operatorname{div} u_n - f \|_{\mathcal{L}^1(D)} = 0$, per il teor. XLII (teor. L). Passando al limite, per $n \to \infty$, nella identità: $\int_D u_n \times \operatorname{grad} v \, dx = -\int_D v \operatorname{div} u_n \, dx$, ove $v \in \mathcal{L}^1(D)$ ha gradiente debole su D (su $D-\mathcal{F}D$) contenuto in $\mathcal{L}^{\infty}(D)$ – e quindi è $v \in C^0(D)$ – si ha l'asserto.

Sia, da ultimo, $q = +\infty, q' = 1$. Esiste una successione $\{u_n\}$ di funzioni di $C^1(D)$ (di $\mathring{C}^1(D)$) tale che: $\lim_{n\to\infty} \| u_n - u \|_{\mathcal{L}_1(D)} = 0$, $\lim_{n\to\infty} \| \operatorname{div} u_n - f \|_{\mathcal{L}^1(D)} = 0$ ed inoltre le funzioni u_n sono equipseudolimitate in D; per ottenere $\{u_n\}$ basta prendere in considerazione la successione introdotta nella dimostrazione del teorema XLVIII (teor. L); è facile vedere che essa gode delle proprietà ora dette nella ipotesi $u \in \mathcal{L}^{\infty}(D)$. Allora, passando al limite, per $n \to \infty$, nella identità $\int_D u_n \times \operatorname{grad} v \, dx = -\int_D v \operatorname{div} u_n \, dx$, che sussiste per ogni $v \in \mathcal{L}^{\infty}(D)$ avente gradiente debole su D (su $D-\mathcal{F}D$) contenuto in $\mathcal{L}^1(D)$, si ottiene la tesi.

LX. Condizione necessaria e sufficiente perchè $u \in \mathcal{L}^q(D), 1 \leq q \leq +\infty$ abbia per divergenza debole su $D - \mathcal{F}D$ (su D) una misura μ è che esista $F \in [\mathcal{L}^\infty(D)]^*$ tale che:

(79) $$\int_D u \times \operatorname{grad} v \, dx = -F(v)$$

per ogni $v \in \mathcal{L}^\infty(D)$ avente gradiente debole su D (su $D - \mathcal{F}D$) contenuto in $\mathcal{L}^p(D)$.

La condizione è ovviamente sufficiente. Mostriamone la necessità. Consideriamo dapprima il caso in cui u abbia divergenza debole su $D - \mathcal{F}D$. Si costruisca una successione $\{u_n\}$ di vettori di $C^1(D)$ che goda di proprietà analoghe a quelle della successione $\{u_n\}$ indicata nella prima parte della dimostrazione del teor. LVII; naturalmente la (76) deve essere ora sostituita con la seguente:

(76') $$\lim_{n \to \infty} \int_{D'} u_n \times \operatorname{grad} v \, dx = \int_{D'} u \times \operatorname{grad} v \, dx$$

che sussisterà per ogni funzione $v \in \mathcal{L}^\infty(D')$ avente gradiente debole su D' appartenente a $\mathcal{L}^p(D')$; la (77) andrà sostituita con la

(77') $$\int_{D'} |\operatorname{div} u_n| \, dx \leq V_\mu(D - \mathcal{F}D), \quad n > n_{D'} .$$

Si perviene così all'esistenza di un $F_{D'} \in [\mathcal{L}^\infty(D')]^*$ tale che, in corrispondenza ad ogni v del tipo detto, risulta :

(80) $$\int_{D'} u \times \operatorname{grad} v \, dx = -F_{D'}(v)$$

e tale che:

(81) $$\| F_{D'} \|_{[\mathcal{L}^\infty(D')]^*} \leq V_u(D - \mathcal{F}D).$$

Supponiamo ora $q > 1$ e quindi $p < +\infty$. Sia $v \in \mathcal{L}^{\infty}(D)$ avente gradiente su D appartenenete a $\mathcal{L}^{p}(D)$. Poniamo:

$$v_{\rho}(x) \begin{cases} = v(x) & x \in D_{\rho} \\ = 0 & x \in D - D_{\rho/2}\,; \end{cases}$$

per $x \in D_{\rho/2} - D_{\rho}$, porremo:

$$v_{\rho}[\xi + t\lambda(\xi)] = v[\xi + (2t-\rho)\lambda(\xi)]\quad, \xi \in \mathcal{F}D,\ \rho/2 \leq t < \rho.$$

Dato che v ha traccia nulla su $\mathcal{F}D$, è subito visto che $v_{\rho}(x)$ è assolutamente continua secondo Tonelli in D. Inoltre $v_{\rho} \in \mathcal{L}^{\infty}(D)$ ed anzi:

(82) $$\| v_{\rho} \|_{\mathcal{L}^{\infty}(D)} = \| v \|_{\mathcal{L}^{\infty}(D)}\,.$$

Si ha poi $\operatorname{grad} v_{\rho} \in \mathcal{L}^{p}(D)$ ed anzi esiste una costante M tale che, per ogni $\rho \in (0, \rho_0)$, risulta:

$$\| \operatorname{grad} v_{\rho} \|_{\mathcal{L}^{p}(D-D_{\rho})} \leq M \| \operatorname{grad} v \|_{\mathcal{L}^{p}(D-D_{\rho})}\,.$$

Se ne deduce:

(83) $$\lim_{\rho \to 0} \| \operatorname{grad} v_{\rho} - \operatorname{grad} v \|_{\mathcal{L}^{p}(D)} = 0\,.$$

Da (80) si trae:

(84) $$\int_{D_{\rho/2}} u \times \operatorname{grad} v_{\rho}\, dx = - F_{D_{\rho/2}}(v_{\rho})\,.$$

Da (83) viene:

(85) $$\lim_{\rho \to 0} \int_{D_{\rho/2}} u \times \operatorname{grad} v_{\rho}\, dx = \int_{D} u \times \operatorname{grad} v\, dx\,.$$

Esiste quindi finito il limite: $\lim_{\rho \to 0} F_{D_{\rho/2}}(v_{\rho}) \equiv F(v)$. Da (81) (82) si trae:

$$|F_{D_{\rho/2}}(v)| \leq \| F_{D_{\rho/2}} \|_{[\mathcal{L}^{\infty}(D_{\rho/2})]^{*}} \| v_{\rho} \|_{\mathcal{L}^{\infty}(D_{\rho/2})} \leq \| \mu \|_{V(D-\mathcal{F}D)} \| v \|_{\mathcal{L}^{\infty}(D)}\,.$$

Se ne deduce $F \in [\mathcal{L}^{\infty}(D)]^*$. Passando al limite, per $\rho \to 0$, in (84), e tenendo presente (85), si ha l'asserto, per $Q > 1$. Se poi è $Q = 1$, allora le "funzioni di prova" risultano appartenere a $\mathcal{H}^{\infty}(D) \cap C^{0}(D)$; in tal caso, servendosi del metodo indicato da Gagliardo [32] è possibile costruire una successione $\{v_n\}$ di funzioni di $\mathring{C}^{1}(D)$ tale che : $\lim_{n\to\infty} \| v_n - v \|_{C^0(D)} = 0$, $\lim_{n\to\infty} \| \mathrm{grad}(v_n - v) \|_{\mathcal{L}^1(D)} = 0$, $|\mathrm{grad}\, v_n| \leq L$, ove L è una costante opportuna, indipendente da n. Se allora si passa al limite, per $n \to \infty$, nella relazione $\int_D u\, \mathrm{grad}\, v_n\, dx = \int_{D-\partial D} v_n\, d\mu$ si ottiene l'asserto per $Q = 1$.

Consideriamo ora il caso in cui u abbia divergenza debole su D. Sia $\tilde{D}$ un dominio propriamente regolare tale che $D \subset \tilde{D} - \mathcal{F}\tilde{D}$. Prolungando u e μ fuori di D a tutto $\tilde{D}$, con il valore zero, si vede che la u, così prolungata, ha per divergenza debole su $\tilde{D}$ la μ, prolungata nel modo detto. Con il procedimento indicato sopra si viene a costruire una successione $\{u_n\}$ di funzioni di $C^{1}(D)$ le quali, tra l'altro, godono della seguente proprietà : se D' è tale che $D \subset D' - \mathcal{F}D'$, $D' \subset \tilde{D} - \mathcal{F}\tilde{D}$ definitivamente si ha $\mathrm{spt}\, u_n \subset D' - \mathcal{F}D'$ e quindi ;

(86) $$\int_{D'} u_n \times \mathrm{grad}\, v\, dx = - \int_{D'} v\, \mathrm{div}\, u_n\, dx$$

in corrispondenza ad ogni $v \in \mathcal{L}^{\infty}(D)$ avente gradiente debole su $D - \mathcal{F}D$ contenuto in $\mathcal{L}^{p}(D)$, preventivamente prolungato a tutto $\tilde{D}$ in modo che risulti: $v \in \mathcal{L}^{\infty}(\tilde{D})$, $\mathrm{grad}\, v \in \mathcal{L}^{p}(\tilde{D})$. Si ha inoltre, per le proprietà della successione $\{u_n\}$:

(87) $$\lim_{n\to\infty} \int_{D'} u_n \times \mathrm{grad}\, v\, dx = \int_{D'} u \times \mathrm{grad}\, v\, dx \equiv \int_{D} u \times \mathrm{grad}\, v\, dx .$$

(32) Cfr. loc. cit. in (11) .

Da (86) viene allora che esiste il limite $\lim_{n\to\infty} \int_{D'} v \operatorname{div} u_n \, dx = F(v)$

Sia D_n il supporto di u_n. Si ha, per ogni v del tipo ora detto :

$$|F(v)| = \lim_{n\to\infty} \left| \int_{D'} v \operatorname{div} u_n \, dx \right| = \lim_{n\to\infty} \left| \int_{D_n} v \operatorname{div} u_n \, dx \right| \leq$$

$$\leq \lim_{n\to\infty} \|v\|_{\mathcal{L}^\infty(D_n)} \int_{D_n} |\operatorname{div} u_n| \, dx \leq \|\mu\|_{V(D)} \lim_{n\to\infty} \|v\|_{\mathcal{L}^\infty(D_n)} .$$

Poichè il prolungamento di v da D a $\tilde{D}$ può essere fatto in modo che

$$\|v\|_{\mathcal{L}^\infty(D)} = \|v\|_{\mathcal{L}^\infty(\tilde{D})} \geq \|v\|_{\mathcal{L}^\infty(D_n)} ,$$

si avrà : $\quad |F(v)| \leq \|\mu\|_{V(D)} \|v\|_{\mathcal{L}^\infty(D)}$.

Ne viene $F \in [\mathcal{L}^\infty(D)]^*$ e quindi, passando al limite, nella (86), per $n \to \infty$ si ha l'asserto .

LXI . Sia D di classe 2. Condizione necessaria e sufficiente perchè $u \in \mathcal{L}^q(D)$ abbia per divergenza debole su D la misura $\mu(B) = \int_B f \, dx + \int_{B \cap \partial D} \varphi \, d\sigma$ con $f \in \mathcal{L}^{q'}(D)$ e $\varphi \in \mathcal{L}^{q''}(\partial D)$, è che, in corrispondenza ad ogni $v \in \mathcal{L}^{p'}(D)$ avente per gradiente debole su D la misura $\alpha(B) = \int_B f' \, dx + \int_{B \cap \partial D} \varphi' \, d\sigma$ con $f' \in \mathcal{L}^p(D)$, $\varphi' \in \mathcal{L}^{p''}(\partial D)$, riesca : $\int_D u f' \, dx = -\int_D v f \, dx - \int_{\partial D} \varphi \varphi' \, d\sigma$.

La tesi segue subito dal teor. XLV. Il teor LXI è vero anche nei seguenti casi $1 \leq q < \frac{3}{2}$, $1 \leq q' < +\infty$, $1 \leq q'' < +\infty$; $q = \frac{3}{2}$, $1 \leq q' < +\infty$, $1 \leq q'' < +\infty$; $q = \frac{3}{2}$, $q' = 1$, $q'' = 1$; $\frac{3}{2} < q < +\infty$, $1 \leq q' < +\infty$, $\frac{2}{3} q \leq q'' < +\infty$ come segue subito dal teor. LIII, e nel caso $1 \leq q < +\infty$, $1 \leq q' < +\infty$, $1 \leq q'' < +\infty$, $q \leq q''$, come segue dal teor. LIV ed è vero anche senza alcuna limitazione per q, q', q'' se si identifica $\mathcal{L}^\infty$ con C^0.

19. Alcune definizioni relative ai poliedri ed alle famiglie quasi totali di poliedri.

Per poliedro di E^3 qui si intenderà sempre un dominio limitato di E^3, che indicheremo con P, avente per frontiera una unica superficie regolare semplice e chiusa contenuta nell'unione di un numero finito di piani; l'intersezione di $\mathcal{F}P$ con ciascuno di tali piani è costituito da un numero finito di dominii bidimensionali connessi, a due a due disgiunti; ognuno di tali dominii (bidimensionali connessi) è una faccia di P e sarà indicata con $\mathcal{F}_k P$ $(k=1,\dots,n$ se n sono le facce di P nel senso ora detto$)$; il piano contenente $\mathcal{F}_k P$ sarà denotato con $\Sigma_k(P)$ (naturalmente non è escluso che riesca $\Sigma_k(P) \equiv \Sigma_h(P), k \neq h$); ogni faccia $\mathcal{F}_k P$ di P è un poligono del piano $\Sigma_k(P)$ cioè è un dominio limitato del piano $\Sigma_k(P)$, avente per frontiera un'unica curva regolare semplice chiusa contenuta nell'unione di un numero finito di rette; la frontiera del dominio piano $\mathcal{F}_k P$ sarà chiamata bordo della faccia $\mathcal{F}_k P$ e indicata con $\mathcal{B}_k P$; i punti di $\mathcal{B}_k P$, nei quali la curva $\mathcal{B}_k P$ è sprovvista di tangente, si chiamano i vertici di $\mathcal{B}_k P$ o vertici della faccia $\mathcal{F}_k P$; diconsi vertici di P i punti che sono vertici di almeno una faccia $\mathcal{F}_k P$ di P; la curva $\mathcal{B}_k P$ è costituita da un numero finito di segmenti; ogni segmento appartenente a $\mathcal{B}_k P$ e congiungente due vertici distinti di $\mathcal{B}_k P$ dicesi lato del bordo $\mathcal{B}_k P$ od anche lato della faccia $\mathcal{F}_k P$: l'intersezione di due facce distinte di P, se non è vuota, è costituita da un insieme finito (eventualmente vuoto) di punti isolati (che sono vertici per una almeno delle due facce) e da un insieme finito (eventualmente vuoto) di segmenti; ognuno di questi segmenti che sia contenuto nell'intersezione di due facce distinte $\mathcal{F}_k P$ e $\mathcal{F}_h P$ e che non sia contenuto propriamente in alcun segmento appartenente a $(\mathcal{F}_k P) \cap (\mathcal{F}_h P)$

dicesi spigolo del poliedro $\mathcal{P}$; i due estremi di uno spigolo di $\mathcal{P}$ sono due vertici di $\mathcal{P}$ e ciascuno di essi è vertice di almeno una delle due facce di $\mathcal{P}$ che si intersecano nel detto spigolo (ma non necessariamente di entrambe le facce) . Ogni poliedro è ovviamente un dominio propriamente regolare nel senso di Fichera (cfr.loc.cit. in [3]) . Con $n_K(\mathcal{P})$ indicheremo sempre la retta passante per lo zero di E^3 e ortogonale al piano $\Sigma_K(\mathcal{P})$. Parlando di porzione di spigolo del poliedro $\mathcal{P}$ intenderemo sempre , nel seguito, un qualunque segmento (di lunghezza positiva) contenuto in uno spigolo di $\mathcal{P}$.

Sia $\mathcal{P}$ un fissato poliedro di E^3 . Sia μ una misura (vettoriale, a tre componenti) definita su tutti i boreliani di E^3 , identicamente nulla su ogni boreliano che non abbia punti in comune con $\mathcal{P}$. Sia $\overline{\pi}_K$ una misura (vettoriale a tre componenti) definita sui boreliani del piano $\Sigma_K(\mathcal{P})$ e identicamente nulla su ogni boreliano che non abbia punti in comune con la faccia $\mathcal{F}_K\mathcal{P}$. Con $\{\overline{\pi}\}$ indicheremo la famiglia delle misure $\overline{\pi}_K$ testè introdotte (si tratta di una famiglia costituita da un numero finito di elementi). Sia inoltre $\{\pi_\Sigma\} \equiv \{\pi\}$ una famiglia di misure (vettoriali a tre componenti) tale che:

1°) indicato con Σ il generico piano orientato di E^3 , π_Σ è ina misura definita sui boreliani di Σ , identicamente nulla su tutti quei boreliani che non hanno alcun punto in comune con $\mathcal{P}$;

2°) indicato con $-\Sigma$ il piano che ha orientazione opposta a quella di Σ si ha $\pi_{-\Sigma} = -\pi_\Sigma$;

3°) se Σ_t denota il piano ottenuto traslando Σ del vettore νt (ove $|\nu|=1$, $\nu \perp \Sigma$, t è un numero reale) e se B_t è il boreliano di Σ_t ottenuto traslando il boreliano $\mathcal{B}$ di Σ del vettore νt, la funzione di t: $V_{\pi_{\Sigma_t}}(\mathcal{B}_t)$ è localmente sommabile .

Diremo che un poliedro $P \subset \mathcal{P}$ è <u>"asportabile interno" in relazione a</u> $\mathcal{P}, \mu, \{\pi\}$ se :

I) nessun vertice, nessuno spigolo e nessuna porzione di spigolo di P giace su $\mathcal{F}\mathcal{P}$ e nessun piano $\Sigma_K(P)$ può coincidere con qualcuno dei piani $\Sigma_h(\mathcal{P})$:

II) per nessun K il piano $\Sigma_K(P)$ è di discontinuità per μ ;

III) nessuna delle rette di $\Sigma_K(P)$ che contenga almeno un lato di $\mathcal{B}_K P$ è di discontinuità per $\pi_{\Sigma_K(P)}$

Diremo che il poliedro $P \subset \mathcal{P}$ è <u>"asportabile di frontiera" in</u> relazione a $\mathcal{P}, \mu, \{\pi\}, \{\overline{\pi}\}$ se:

1) la frontiera $\mathcal{F}P$ di P è composta di n facce delle quali le prime m (con $n \geq m \geq 1$) sono contenute in $\mathcal{F}\mathcal{P}$ mentre le rimanenti $n-m$ non hanno alcuno dei propri punti interni in comune con $\mathcal{F}\mathcal{P}$;

2) esiste un poliedro $\widetilde{P}$ di E^3 che gode delle seguenti proprietà :

3) $\widetilde{P} \cap \mathcal{P} = P$;

4) nessuna delle facce $\mathcal{F}_K\widetilde{P}$ di $\widetilde{P}$ appartiene a piani del tipo $\Sigma_k(\mathcal{P})$, nessuno dei vertici nessuno degli spigoli e nessuna delle porzioni di spigolo di $\widetilde{P}$ giace su $\mathcal{F}\mathcal{P}$;

5) nessuno degli spigoli e nessuno dei vertici di $\mathcal{P}$ appartiene a piani del tipo $\Sigma_K(\widetilde{P})$;

6) nessun pian o $\Sigma_K(\widetilde{P})$ è di discontinuità per μ ;

7) nessuna retta di $\Sigma_K(\widetilde{P})$ che contenga qualche lato di $\mathcal{B}_K(\widetilde{P})$

é di discontinuità per la misura $\pi_{\Sigma_k(\tilde{P})}$;

8) nessuno dei piani $\Sigma_k(\tilde{P})$ contiene "concentrazioni" di qualcuna delle misure $\overline{\pi}_h$ (33)

E' evidente che $\mathcal{P}$ é un poliedro "asportabile di frontiera" in relazione a $\mathcal{P}, \mu, \{\pi\}, \{\overline{\pi}\}$.

Una famiglia $\{P\}_{\mathcal{P}}$ di poliedri di E^3 contenuti in $\mathcal{P}$ dicesi quasi totale in relazione a $\mathcal{P}, \mu, \{\pi\}, \{\overline{\pi}\}$ se:

1° su ogni retta r passante per lo zero di E^3 esiste un insieme $\tilde{N}_r$, di misura lineare lebesguiana nulla, siffatto che $\{P\}_{\mathcal{P}}$ contenga tutti e soli i poliedri di E^3 per i quali si verifica che:

2° P é un poliedro "asportabile interno" relativamente a $\mathcal{P}, \mu, \{\pi\}$ oppure "asportabile di frontiera" in relazione a $\mathcal{P}, \mu, \{\pi\}, \{\overline{\pi}\}$;

3° se P é "asportabile interno", $\Sigma_k(P)$ interseca la retta $n_k(P)$ in punti che non appartengono a $\tilde{N}_{n_k(P)}$;

4° se P é "asportabile di frontiera" esiste un poliedro $\tilde{P}$ di E^3 verificante, rispetto a $P, \mathcal{P}, \mu, \{\pi\}, \{\overline{\pi}\}$ le 3) .. 8), tale che $\Sigma_k(\tilde{P})$ non interseca $n_k(\tilde{P})$ in punti di $\tilde{N}_{n_k(\tilde{P})}$.

20. Equilibrio di un corpo limitato.

Prenderemo in considerazione soltanto "corpi limitati" che siano schematizzabili con poliedri $\mathcal{P}$ (del tipo precisato nel paragrafo precedente). In accordo con le considerazioni svolte e con le definizioni

(33) Con ciò intendiamo che il piano $\Sigma_k(P)$ non contiene alcun boreliano bidimensionale B, di misura lebesguiana bidimensionale nulla, che sia contenuto anche in qualche piano del tipo $\Sigma_h(\mathcal{P})$ e tale che $V_{\overline{\pi}_h}(B) \neq 0$.

date in relazione ai corpi "indefinitamente estesi" nel lavoro citato in (1), si supporrà di schematizzare la forza di massa agente sul corpo $\mathcal{C}$ rappresentato dal poliedro $\mathcal{P}$, con una misura μ (vettoriale a tre componenti) definita su tutti i boreliani $\mathcal{B}$ contenuti in $\mathcal{P}$ e che penseremo, poi, definita su tutti i boreliani di E^3 prolungandola con il valore zero fuori di $\mathcal{P}$; supporremo inoltre che le "forze esterne di superficie" agenti su $\mathcal{F}\mathcal{P}$ possano chematizzarsi con delle misure vettoriali $\bar{\pi}_k$ la k-esima delle quali é definita sui boreliani di $\Sigma_k(\mathcal{P})$ ed é identicamente nulla sui boreliani che non hanno punti in comune con $\mathcal{F}_k\mathcal{P}$; da ultimo ammetteremo che gli "sforzi interni" attraverso porzioni piane contenute in $\mathcal{P}$ si possano schematizzare mediante una famiglia di misure $\{\pi\}$ verificante le 1°) 2°) 3°) del paragrafo prec., nel senso che se $\mathcal{F}_k P$ é la faccia di un poliedro "asportabile interno" relativamente a $\mathcal{P}, \mu, \{\pi\}$ (o "di frontiera" in relazione a $\mathcal{P}, \mu, \{\pi\}, \{\bar{\pi}\}$) lo "sforzo" attraverso le porzioni di $\mathcal{F}_k P$ sia rappresentabile con la misura $\pi_{\Sigma_k(P)}$.

Il corpo $\mathcal{C}$ limitato, schematizzato da $\mathcal{P}$, sarà detto "in equilibrio sotto l'azione della forza di massa μ, delle forze di superficie $\{\bar{\pi}\}$, e degli sforzi $\{\pi\}$", se esiste una famiglia $\{P\}_{\mathcal{P}}$ quasi totale di poliedri relativa a $\mathcal{P}, \mu, \{\pi\}, \{\bar{\pi}\}$ tale che per ogni $P \in \{P\}_{\mathcal{P}}$ risulti:

(88) $$\mu(P) = \Sigma_k \, \pi_{\Sigma_k(P)}(\mathcal{F}_k P)$$

(89) $$\int_P x \wedge d\mu = \Sigma_k \int_{\mathcal{F}_k P} x \wedge d\pi_{\Sigma_k(P)}$$

se P é "asportabile interno" in relazione a $\mathcal{P}, \mu, \{\pi\}$, e

(90) $$\mu(P) = \sum_{1}^{m}{}_k \, \bar{\pi}_{e_k}(\mathcal{F}_k P) + \sum_{m+1}^{n}{}_k \, \pi_{\Sigma_k(P)}(\mathcal{F}_k P)$$

(91) $$\int_{\underline{P}} \chi \wedge d\mu = \sum_{1}^{m}{}_k \int_{\mathcal{F}_k \underline{P}} \chi \wedge d\bar{\pi}_{\ell_k} + \sum_{m+1}^{n}{}_k \int_{\mathcal{F}_k \underline{P}} \chi \wedge d\pi_{\Sigma_k(\underline{P})}$$

se $\underline{P}$ é "asportabile di frontiera" in relazione a $\mathcal{P}, \mu, \{\pi\}, \{\bar{\pi}\}$, ove $\Sigma_{\ell_k}(\mathcal{P}) \equiv \Sigma_k(\underline{P})$ ed ove si è convenuto di scegliere l'orientamento dei piani $\Sigma_k(\underline{P}), \Sigma_{\ell_k}(\mathcal{P})$ verso l'interno rispettivamente di $\underline{P}$ e di $\mathcal{P}$.

Questa definizione di equilibrio per un "corpo limitato "può ricondursi a quella relativa ad un "corpo indefinitamente esteso" (cfr. lavoro citato in[1]) mediante il seguente teorema :

LXII . <u>Condizione necessaria e sufficiente perché esista una famiglia $\{\underline{P}\}_{\mathcal{P}}$ quasi totale di poliedri in relazione a $\mathcal{P}, \mu, \{\pi\}, \{\bar{\pi}\}$, su ogni poliedro della quale siano soddisfatte le (88) (89) oppure (90) (91) (secondo che il poliedro sia "asportabile interno" o "di frontiera") é che, posto</u>

$$\bar{\mu}(B) = \mu(B) - \Sigma_k \, \bar{\pi}_k [B \cap \Sigma_k(\mathcal{P})] ,$$

<u>esista una famiglia quasi totale di poliedri $\{\underline{P}\}$ relativa a $\bar{\mu}$ e $\{\pi\}$</u> [34] <u>tale che , per ogni poliedro $\underline{P} \in \{\underline{P}\}$, il quale non abbia né vertici né spigoli né porzioni di spigolo contenute in $\mathcal{F}\mathcal{P}$, risulti</u> :

(92) $$\bar{\mu}(\underline{P}) = \Sigma_k \; \pi_{\Sigma_k(\underline{P})}(\mathcal{F}_k \underline{P})$$

(93) $$\int_{\underline{P}} \chi \wedge d\bar{\mu} = \Sigma_k \int_{\mathcal{F}_k \underline{P}} \chi \wedge d\pi_{\Sigma_k(\underline{P})} .$$

Mostriamo la necessità. Sia $\{\underline{P}\}_{\mathcal{P}}$ la famiglia quasi totale relativa a $\mathcal{P}, \mu, \{\pi\}, \{\bar{\pi}\}$, di cui all'ipotesi. Come famiglia $\{\underline{P}\}$ quasi

---

[34] Cfr. definizione di famiglia quasi totale rispetto a $\bar{\mu}, \{\pi\}$ in loc. cit. in (1) p. 212 .

totale relativa a $\bar{\mu}, \{\pi\}$ di cui alla tesi assumiamo la famiglia quasi totale determinata da $\bar{\mu}$, da $\{\pi\}$ e dagli insiemi N_r di misura nulla sulla retta passante per l'origine (di cui alla definizione di famiglia quasi totale $\{P\}$ rispetto a $\bar{\mu}, \{\pi\}$) ottenuti aggiungendo a ciascun $\tilde{N}_r$ (relativo alla famiglia $\{P\}_{\mathcal{P}}$) l'insieme N'_r costituito dalle intersezioni con r di ciascuno dei piani ortogonali ad r e contenenti almeno una faccia $\mathcal{F}_k\mathcal{P}$ di $\mathcal{P}$ oppure almeno uno spigolo di $\mathcal{P}$ oppure almeno un vertice di $\mathcal{P}$, nonchè dalle intersezioni con r di ciascuno dei piani ortogonali ad r e di discontinuità per la misura μ, o che contenga "concentrazioni" di qualcuna delle $\bar{\pi}_h$.

Sia $P \in \{P\}$ ed inoltre supponiamo che P non abbia né spigoli né porzioni di spigolo né vertici contenuti in $\mathcal{F}\mathcal{P}$. Se $P \subset \mathcal{P}$, allora, il poliedro P dovrà essere contenuto in $\mathcal{P}-\mathcal{F}\mathcal{P}$ (infatti , nessun piano $\Sigma_k(P)$ può coincidere con qualcuno dei $\Sigma_l(\mathcal{P})$ - per la definizione di N_r -, nessun vertice , nessun spigolo e nessuna porzione di spigolo di $\mathcal{P}$ può giacere su $\mathcal{F}P$ - ancora per la definizione di N_r -, nessun vertice , nessun spigolo e nessuna porzione di spigolo di P può giacere su $\mathcal{F}\mathcal{P}$ - per l'ipotesi fatta su P -) ; inoltre nessun piano $\Sigma_k(P)$ sarà di discontinuità per μ - in forza dell'ipotesi fatta su N_r - ; infine, nessuna retta di $\Sigma_k(P)$ contenente qualche lato di $\mathcal{B}_k P$, sarà di discontinuità per $\pi_{\Sigma_k(P)}$ - come segue dal fatto che $\{P\}$ è una famiglia quasi totale relativa a $\{\pi\}$ oltre che a $\bar{\mu}$ - . Pertanto P sarà "asportabile interno" in relazione a $\mathcal{P}, \mu, \{\pi\}$.

Inoltre , nessun piano $\Sigma_k(P)$ interseca $m_k(P)$ in punti di $\tilde{N}_{m_k(P)}$

dato che $\tilde{N}_r \subset N_r$ per costruzione. Allora sarà $P \in \{P\}_{\mathcal{P}}$, e quindi, in corrispondenza ad esso, saranno verificate le (88) (89) che coincidono , in tal caso, rispettivamente con le (92) (93) dato che $P \subset \mathcal{P} - \mathcal{F}\mathcal{P}$ e $\mu(B) = \bar{\mu}(B)$ per ogni $B \subset \mathcal{P} - \mathcal{F}\mathcal{P}$. Se $P \subset (\mathcal{C}\mathcal{P}) \cup (\mathcal{F}\mathcal{P})$ le (92) (93) sono verificate perché , in corrispondenza a ciascuna di esse, ambo i membri della eguaglianza sono nulli; infatti, in base al ragionamento fatto poco sopra, dovrà essere $P \subset \mathcal{C}\mathcal{P}$. Consideriamo , da ultimo, il caso in cui sia $P \cap (\mathcal{P} - \mathcal{F}\mathcal{P}) \neq \emptyset$, $P \cap \mathcal{C}\mathcal{P} \neq \emptyset$ e supponiamo dapprima che $P \cap \mathcal{P}$ sia connesso . Facciamo vedere che $P \cap \mathcal{P}$ é un poliedro "asportabile di frontiera" in relazione a $\mathcal{P}, \mu, \{\pi\}, \{\bar{\pi}\}$ assumendo come $\tilde{P}$ il poliedro P stesso. Sia ha intanto che $P \cap \mathcal{P}$ é un poliedro verificante la proprietà 1) del paragrafo prec. in conseguenza del fatto che: nessuna faccia $\mathcal{F}_k P$ di P appartiene a piani del tipo $\Sigma_h(\mathcal{P})$ - in forza della definizione di N_r - , nessuno spigolo o porzione di spigolo, e nessun vertice di $\mathcal{P}$ può giacere su $\mathcal{F}P$ - ancora in forza della definizione di N_r - , nessuno spigolo o porzione di spigolo e nessun vertice di P può giacere su $\mathcal{F}\mathcal{P}$ - in forza dell'ipotesi fatta su P. Le 4) 5) 6) 7) del paragrafo prec. sono soddisfatte in modo evidente. La 8) del paragr. prec. segue dal fatto che $\Sigma_k(\tilde{P}) \equiv \Sigma_k(P)$ non contiene "concentrazioni" di $\bar{\pi}_h$ (per la definizione di N_r) per alcun valore di h . Inoltre, nessun piano $\Sigma_k(P) \equiv \Sigma_k(\tilde{P})$ interseca $m_k(P)$ in punti di $\tilde{N}_{m_k(P)}$ dato che $\tilde{N}_{m_k(P)} \subset N_{m_k(P)}$ per costruzione e dato che, essendo $P \in \{P\}$, nessun $\Sigma_k(P)$ interseca $m_k(P)$ in punti di $N_{m_k(P)}$. Si ha quindi $P \cap \mathcal{P} \in \{P\}_{\mathcal{P}}$. Saranno allora soddisfatte in corrispondenza ad esso le (90) (91) e quindi , in particolare,

(90)' $\mu(P\cap\mathcal{P})=\sum_1^m{}_K\,\bar{\pi}_{e_K}(\mathcal{F}_K\,P\cap\mathcal{P})+\sum_{m+1}^n{}_K\,\pi_{\Sigma_K(P\cap\mathcal{P})}(\mathcal{F}_K P\cap\mathcal{P})$

ove $\Sigma_{e_K}(\mathcal{P})\equiv\Sigma_K(P\cap\mathcal{P})$. Si noti ora che (per la propr. 5))

$$\sum_1^m{}_K\,\bar{\pi}_{e_K}(\mathcal{F}_K\,P\cap\mathcal{P})=\Sigma_e\,\bar{\pi}_e\,[P\cap\mathcal{P}\cap\Sigma_e(\mathcal{P})]\ ,$$

e quindi ;

(90)'' $\mu(P\cap\mathcal{P})-\sum_1^m{}_K\,\bar{\pi}_{e_K}(\mathcal{F}_K\,P\cap\mathcal{P})=\mu(P\cap\mathcal{P})-$

$$-\Sigma_e\,\bar{\pi}_e\,[P\cap\mathcal{P}\cap\Sigma_e(\mathcal{P})]=\bar{\mu}(P\cap\mathcal{P}).$$

Poiché ogni $\pi\equiv\pi_\Sigma$ ha il supporto $\subset\mathcal{P}$, risulta ;

$$\sum_{m+1}^n{}_K\,\pi_{\Sigma_K(P\cap\mathcal{P})}(\mathcal{F}_K\,P\cap\mathcal{P})=\Sigma_K\,\pi_{\Sigma_K(P)}(\mathcal{F}_K P)\,.$$

Da qui , da (90)' e da (90)'' segue (92) . In modo analogo si ottiene (93) .

Se $P\cap\mathcal{P}$ non è connesso si ragionerà sulle sue singole componenti connesse come si è ora ragionato su $P\cap\mathcal{P}$. La necessità é così provata.

Mostriamo, adesso, la sufficienza.

Sia $\{P\}$ la famiglia di poliedri quasi totale rispetto a $\bar{\mu}$ ed a $\{\pi\}$ di cui all'ipotesi. Come famiglia $\{P\}_\mathcal{P}$ quasi totale relativa a $\mathcal{P},\mu,\{\pi\},\{\bar{\pi}\}$, di cui alla tesi, assumiamo la famiglia quasi totale determinata da $\mathcal{P},\mu,\{\pi\},\{\bar{\pi}\}$ e dagli insiemi $\tilde{N}_r$ di misura lineare lebesguiana nulla sulle rette r passanti per l'origine (di cui alla definizione di famiglia quasi totale

$\{P\}_{\mathcal{P}}$) ottenuti aggiungendo a ciascun $N_{\mathcal{R}}$ (relativo alla famiglia $\{P\}$) l'insieme costituito dalle intersezioni con $\mathcal{R}$ di ciascuno dei piani ortogonali ad $\mathcal{R}$ che sono di discontinuità per $\bar{\mu}$ nonché dalle intersezioni con $\mathcal{R}$ dei piani ortogonali ad $\mathcal{R}$ e contenenti almeno una faccia di $\mathcal{P}$ o almeno uno spigolo di $\mathcal{P}$ o almeno un vertice di $\mathcal{P}$. Sia ora $\underline{P} \in \{P\}_{\mathcal{P}}$ e supponiamo che $\underline{P}$ sia "asportabile interno"; allora dovrà essere $\underline{P} \subset \mathcal{P} - \mathcal{F}\mathcal{P}$ (come si vede con un ragionamento già fatto nella dimostrazione della necessità). Per la definizione di $\tilde{N}_{\mathcal{R}}$ si ha che nessun piano $\Sigma_K(\underline{P})$ è di discontinuità per $\bar{\mu}$; inoltre, dato che è anche $\underline{P} \in \{P\}_{\mathcal{P}}$, nessuna retta di $\Sigma_K(\underline{P})$ contenente lati di $\mathcal{B}_K\underline{P}$ é di discontinuità per $\pi_{\Sigma_K(\underline{P})}$; infine, dato che $N_{\mathcal{R}} \subset \tilde{N}_{\mathcal{R}}$, nessun piano $\Sigma_K(\underline{P})$ interseca $m_K(\underline{P})$ in punti di $N_{m_K(\underline{P})}$. Allora si ha $\underline{P} \in \{P\}$ e inoltre $\underline{P}$ non ha né vertici né spigoli né porzioni di spigolo contenute in $\mathcal{F}\mathcal{P}$ (essendo $\underline{P}$ "asportabile interno"). Saranno quindi verificate, in corrispondenza al detto $\underline{P}$, le (92) (93) le quali coincidono, rispettivamente, con (88) (89) dato che μ e $\bar{\mu}$ possono differire tra loro solo su boreliani di $\mathcal{F}\mathcal{P}$ mentre ora si ha $\underline{P} \subset \mathcal{P} - \mathcal{F}\mathcal{P}$. Sia adesso $\underline{P} \in \{P\}_{\mathcal{P}}$ e supponiamo che $\underline{P}$ sia "asportabile di frontiera". Esisterà allora $\tilde{\underline{P}}$ verificante 4) ... 8). Facciamo vedere che si può imporre a $\tilde{\underline{P}}$ l'ulteriore condizione di appartenere a $\{\underline{P}\}$. Intanto si ha che il piano $\Sigma_K(\tilde{\underline{P}})$ interseca $m_K(\tilde{\underline{P}})$ fuori di $\tilde{N}_{m_K(\tilde{\underline{P}})}$ e quindi anche fuori di $N_{m_K(\tilde{\underline{P}})}$ dato che $N_{m_K(\tilde{\underline{P}})} \subset \tilde{N}_{m_K(\tilde{\underline{P}})}$. Inoltre $\Sigma_K(\tilde{\underline{P}})$ non é di discontinuità per μ (propr. 6)) e non contiene " concentrazioni" di alcuna delle $\bar{\pi}_h$ (propr. 8)) e quindi non è di discontinuità per $\bar{\mu}$. Inoltre, per la 7), nessuna retta di $\Sigma_K(\tilde{\underline{P}})$ che contenga qualche lato di $\mathcal{B}_K(\tilde{P})$, é di discontinuità per $\pi_{\Sigma_K}(\tilde{P})$. Pertanto si ha $\tilde{P}\in\{P\}$. Inoltre, per la 4), $\tilde{P}$ non ha né vertici né spigoli né porzioni di spigolo giacenti su $\mathcal{F}\mathcal{P}$. Allora, in corrispondenza a $\tilde{P}$ sono verificate le (92) (93). In particolare si ha:

$$\bar{\mu}(\tilde{P}) = \Sigma_h \pi_{\Sigma_h(\tilde{P})}(\mathcal{F}_h \tilde{P}).$$

Ma :

$$\bar{\mu}(\tilde{P}) = \mu(\mathcal{P}\cap\tilde{P}) - \Sigma_K \bar{\pi}_K[\tilde{P}\cap\Sigma_K(\mathcal{P})] = \mu(P) - \sum_1^m {}_K \bar{\pi}_{h_K}(\mathcal{F}_K P)\ .$$

in forza della 5). Si ha poi $\Sigma_h \pi_{\Sigma_h(\tilde{P})}(\mathcal{F}_h \tilde{P}) = \sum_{m+1}^n {}_K \pi_{\Sigma_K(P)}(\mathcal{F}_K P)$ come é ben evidente, e quindi risulta soddisfatta la (90) in relazione a P.

Analogamente si controlla che sussiste la (91) in relazione al detto P.

Il teorema LXII é così dimostrato.

Dal teor. LXII testé dimostrato, e dai teorr. I.6, II.6, I.10; II.10 di loc. cit. in (1) si deduce subito il seguente teorema che fornisce la traduzione in "equazioni deboli" delle equazioni dell'equilibrio (88) (89) (90) (91).

LXIII. <u>Condizione necessaria e sufficiente perché, in corrispondenza a $\mathcal{P}, \mu, \{\bar{\pi}\}$ esista $\{\pi\}$ tale che siano soddisfatte (88) ... (91) su ogni poliedro P di una opportuna famiglia quasi totale di poliedri relativa a $\mathcal{P}, \mu, \{\pi\}, \{\bar{\pi}\}$, è che esistano tre misura vettoriali α_j (di componenti α_{ij}) tali che:</u>

(94) $$\int_{\mathcal{P}} \sum_1^3 {}_i \frac{\partial v}{\partial x^i} d\alpha_{ij} = -\int_{\mathcal{P}} v\, d\mu_j \quad , v\in C^1(\mathcal{P})$$

(95) $$\alpha_{ij} = \alpha_{ji}$$

<u>essendo $\alpha_j(B)=0$ su ogni B che non ha punti in comune con</u>

$\mathcal{P}$, e

$$\bar{\mu}(B) = \mu(B) - \sum_K \bar{\pi}_K [B \cap \Sigma_K(\mathcal{P})].$$

Se μ e $\{\bar{\pi}\}$ sono tali che, in corrispondenza ad essi esiste $\{\alpha_{ij}\}$ verificante (94) (95) , un sistema $\{\pi\}$ soddisfacente le (88) ... (91) (in riferimento ad una opportuna famiglia quasi totale $\{P\}_{\mathcal{P}}$ relativa a $\mathcal{P}, \mu, \{\bar{\pi}\}, \{\pi\}$) si ottiene assumendo come γ- esima componente di π_Σ sui boreliani B_Σ di Σ la misura : $\sum_1^3 {}_i \cos(\nu, x^i) \frac{\partial \alpha_{i\gamma}}{\partial \nu}(B_\Sigma; 1_\Sigma)$ (35)

Se μ e $\bar{\pi}$ sono tali che, in corrispondenza ad essi esiste $\{\pi\}$ verificate (88) ... (91) (per ogni poliedro di una opportuna famiglia quasi totale $\{P\}_{\mathcal{P}}$ relativa a $\mathcal{P}, \mu, \{\bar{\pi}\}, \{\pi\}$), un sistema $\{\alpha_{ij}\}$ verificante (94) (95) si ottiene assumendo :

$$\alpha_{ij} = \int_{a^i}^{b^i} \pi_{ij}(I_{y^i}; x^i)\, dx^i .$$

Da qui e dai teoremi sull'operatore di divergenza debole (dei paragrafi precedenti) si trae, in particolare, che, assegnati ad arbitrio la forza di massa μ e la forza esterna di superficie $\bar{\pi}$, con $\mu(\mathcal{P}) = \sum_K \bar{\pi}_K [\Sigma_K(\mathcal{P})]$, esiste un sistema di pressioni interne $\{\pi\}$ che verifica, in relazione a μ ed a $\bar{\pi}$ la prima equazione di equilibrio (88) o (90) , e tale inoltre che, in corrispondenza ad ogni piano Σ la misura π_Σ sia assolutamente continua con densità $\in \mathcal{L}^q$, $1 \leq q < 3/2$. E' ora evidente che tutti i teoremi di regolarizzazione e di traccia per le funzioni dotate di divergenza debole, dati nei paragrafi precedenti , si traducono in altrettanti teoremi di regolarizzazione all'interno e sul contorno per le so-

---

(35) Per le notazioni qui usate cfr. loc. cit. in [1]

luzioni $\{\pi\}$ della prima equazione dell'equilibrio (88) (90).

Inoltre, dai risultati del paragrafo 14 del loc. cit. in (1) e dal teor. LXII si deduce che :

LXIV . Condizione necessaria e sufficiente perché in corrispondenza a $\mathcal{P}, \mu, \{\bar{\pi}\}$ esista $\{\pi\}$ verificante (88) .. (91) su ogni poliedro P di una opportuna famiglia quasi totale di poliedri relativa a $\mathcal{P}, \mu, \{\bar{\pi}\}, \{\pi\}$, e tale inoltre che riesca

$$\pi_\Sigma(B_\Sigma) = \nu_\Sigma \, |\pi_\Sigma(B_\Sigma)| \tag{96}$$

ove ν_Σ é il versore normale a Σ, è che esista una funzione $p(x) \in \mathcal{L}^1(\mathcal{P})$ che abbia per gradiente debole su $\mathcal{P}$ la misura $\bar{\mu}$ con $\bar{\mu}(B) = \mu(B) - \Sigma_K \bar{\pi}_K[B \cap \Sigma_K(\mathcal{P})]$

Se μ e $\{\bar{\pi}\}$ sono tali che, in corrispondenza ad essi, esiste $p(x) \in \mathcal{L}_0^1(\mathcal{P})$ avente per gradiente debole su $\mathcal{P}$ la misura μ, esiste uno ed un solo sistema soddisfacente le (88) ... (91) (in riferimento ad una opportuna famiglia quasi totale $\{P\}_\mathcal{P}$ relativa a $\mathcal{P}, \mu, \{\bar{\pi}\}, \{\pi\}$) e la (96), ed esso é determinato dalla relazione $\pi_\Sigma(B_\Sigma) = \int_{B_\Sigma} p(x)\, d\Sigma$ (ove si pensi di prolungare $p(x)$ fuori di $\mathcal{P}$ con il valore zero).

Se μ e $\bar{\pi}$ sono tali che, in corrispondenza ad essi, esiste $\{\pi\}$ verificante le (88) ... (91) (in riferimento ad una opportuna famiglia quasi totale $\{P\}_\mathcal{P}$ relativa a $\mathcal{P}, \mu, \{\bar{\pi}\}, \{\pi\}$) e la (96) , esiste una ed una sola funzione $p(x) \in \mathcal{L}^1(\mathcal{P})$ che ha per gradiente debole su $\mathcal{P}$ la misura $\bar{\mu}$ ed essa é determinata dalla relazione $p(x) = \frac{d\pi_\Sigma}{d\Sigma}$.

LXV. Condizione necessaria e sufficiente perché, in corrispondenza a $\mathcal{P}, \mu$ e $\{\bar{\pi}\}$ il sistema (88) .. (91) (96) ammetta soluzione (nell'incognita $\{\pi\}$) é che $\int_{\mathcal{P}} v\, d\bar{\mu} = 0$ per ogni $v \in C^1(\mathcal{P})$ avente divergenza nulla.

I due teoremi ora dati possono interpretarsi come teoremi relativi all'equilibrio di un corpo "fluido" ideale (schematizzato dal poliedro $\mathcal{P}$) ove, in accordo con le definizioni date nel n. 14 del loc. cit. in [(1)], si intenda per corpo "fluido" (ideale) un corpo per il quale le "pressioni interne" sono schematizzabili con misure π_Σ soddisfacenti (96), cioè con misure vettoriali che sono sempre normali alla superficie attraverso la quale le dette "pressioni" agiscono. Allora, i teoremi di regolarizzazione e di traccia per le funzioni dotate di gradiente debole (dati nei paragrafi precedenti) non sono altro che teoremi di regolarizzazione all'interno e sulla frontiera per la soluzione del sistema (88) ... (91) (96).

Gaetano Fichera

Problemi elastostatici con ambigue condizioni al contorno.

Gli argomenti esposti in questa conferenza sono stati pubblicati nei seguenti lavori di G. Fichera:

[1] Problemi elastostatici con vincoli unilaterali: il problema di Signorini con ambigue condizioni al contorno - atti Acc. Naz. dei Lincei (Memorie), serie VIII, v. VII, 1964.

[2] Elastostatics problem with unilateral constraints: the Signorini problem with ambiguous boundary conditions - Seminari dell'Istituto Nazionale di Alta Matematica 1962-63, Ediz. Cremonese, Roma 1965.

[3] Semicontinuity of Multiple Integrals in Ordinary Form - Archive for Rational Mechanics and Analysis - vol. 17, n. 5, 1964.

[4] Un teorema generale di semicontinuità per gli integrali multipli e sue applicazioni alla Fisica Matematica - Atti del Simposio Lagrangiano Acc. Scienze Torino, 1963.

[5] The Signorini elastostatics problem with ambiguous boundary conditions - Proceedings of the Intern. Symposium "Applications of the theory of functions in Continuum Mechanics" Tbilisi (URSS), 1963.

La ricerca cui la conferenza si riferisce trovasi compiutamente esposta in [1]. Il lavoro [2] é una traduzione in lingua inglese di [1]. Il lavoro [3] contiene un teorema generale di semi-continuità, sul quale si fondano i teoremi di esistenza contenuti in [1]. Le note [4] e [5] sono esposizioni riassuntive, tenute in Congressi, del lavoro [1]. La conferenza tenuta in Bressanone, al Corso C.I.M.E., è stata conforme a quanto esposto in [4] e [5].

CENTRO INTERNAZIONALE MATEMATICO ESTIVO
(C. I. M. E.)

G. GRIOLI

SISTEMI A TRASFORMAZIONI REVERSIBILI

Corso tenuto a Bressanone dal 31 maggio al 9 giugno
1965

SISTEMI A TRASFORMAZIONI REVERSIBILI

di

Giuseppe Grioli (Università di Padova)

I

Qualche premessa di Cinematica delle deformazioni finite

1) Richiami indispensabili di teoria delle matrici. Quando occorra, userò il linguaggio della teoria delle matrici. Per chiarezza, richiamerò gli elementi fondamentali di cui farò uso qualche volta.

Siano

(1) $$a \equiv \left| a_{ih} \right| ; \qquad b \equiv \left| b_{ih} \right|$$

due matrici quadrate di ordine 3. Per prodotto s'intenderà la matrice quadrata

(2) $$ab \equiv \left| a_{ih} b_{hl} \right|$$

In generale, risulta $ab \neq ba$

Indicherò con $\bar{a}$ la matrice coniugata di a :

(3) $$\bar{a} \equiv \left| a_{hi} \right|$$

Si ha sempre $\overline{ab} = \bar{b}\bar{a}$.

Indicherò con Ca la matrice complementare di a , cioè quella matrice quadrata che ha per elementi i complementi A_{ih} degli elementi a_{ih} di a . Si ha $C(ab) = CbCa$.

Con a^{-1} indicherò la matrice inversa di a , cioè la matrice quadrata soddisfacente all'uguaglianza

$$(4) \qquad aa^{-1} \equiv 1 \equiv \begin{vmatrix} 1 & 0 & 0 \\ 0 & 1 & 0 \\ 0 & 0 & 1 \end{vmatrix} \equiv a^{-1}a\ .$$

Da $a_{ih}A_{lh} = {}_{il}A$, con δ_{il} simbolo di Kronecker e A determinante di a , segue

$$(5) \qquad a^{-1} = \frac{\overline{Ca}}{A}$$

Si ha inoltre,

$$(6) \qquad (ab)^{-1} = b^{-1}a^{-1}, \qquad Ca^{-1} = (Ca)^{-1}, \qquad \text{Det}\ .\ a^{-1} = \frac{1}{\text{Det}\ a}\ .$$

Se un vettore $\underline{v}$ si rappresenta mediante la matrice

$$(7) \qquad \underline{v} \equiv \begin{vmatrix} v_1 \\ v_2 \\ v_3 \end{vmatrix} \ ,$$

il prodotto $a\underline{v}$ è il vettore $\underline{w}$ espresso dalla matrice

$$(8) \qquad \underline{w} \equiv \left| a_{il}v_l \right| \equiv \begin{vmatrix} a_{1l}\ v_l \\ a_{2l}\ v_l \\ a_{3l}\ v_l \end{vmatrix}\ .$$

Se a è simmetrica esistono, come è ben noto, almeno tre direzioni mutuamente ortogonali, caratterizzate dai versori $\underline{u}^{(i)}$, per le quali il vettore $a\underline{u}^{(i)}$ è parallelo a $\underline{u}^{(i)}$. Ognuno dei vettori $\underline{u}^{(i)}$ è soluzione del sistema $a_{rl}u_l = \lambda u_r$. I vettori $\underline{u}^{(i)}$ caratterizzano le direzioni unite della matrice a .

La matrice ρ è un rotore quando il vettore $\rho\,\underline{v}$, qualunque sia $\underline{v}$, si ottiene da $\underline{v}$ mediante uno spostamento rigido con un punto unito. Sussiste il teorema: condizione necessaria e sufficiente affinchè ρ sia un rotore è che risulti

(9) $$\|\rho\| = 1 \, , \qquad \rho^{-1} = \bar{\rho} \longrightarrow \rho_{rl}\,\rho_{sl} = \delta_{rs}$$

Per la matrice a sono invarianti le quantità

(10) $$I_1 a = \sum_i a_{ii} \, , \qquad I_2 a = \sum_i A_{ii} \, , \qquad I_3 a = \| a \| = A \, ,$$

e si ha, evidentemente,

(11) $$I_3(ab) = I_3 a \, . \, I_3 b, \qquad I_3 \rho = 1$$

2) Sulla corrispondenza tra due distinte configurazioni di un medesimo sistema continuo. Di un medesimo sistema continuo S si considerino due configurazioni C, C'. Siano P e P' due punti corrispondenti e y_i, x_i le loro coordinate rispetto alla medesima terna trirettangola levogira di riferimento di origine O.

La corrispondenza tra C e C' è caratterizzata dalle relazioni

(12) $$x_i = x_i\,(y_1, y_2, y_3),$$

eventualmente dipendenti dal tempo, oltre che da altri parametri.

Le relazioni invertibili (12) soddisfano a tutte le ben note condizioni di regolarità imposte dal loro significato meccanico. In particolare risulta

(13) $$D = \| x_{i,h} \| > 0 \, ,$$

ove la virgola denota derivazione rispetto alle y_h.

Posto

(14) $$dP \equiv (dy_i)\, d_i \, , \qquad dP' \equiv (dx_i)\, d_i$$

risulta

(15) $$dP' = adP$$

ove a è la matrice

(16) $$a \equiv \left| x_{i,h} \right|$$

Si consideri la matrice

(17) $$b = \bar{a}\, a \equiv \left| x_{r,l}\, x_{r,h} \right| .$$

Si ha, evidentemente,

(18) $$\left\| b_{rs} \right\| = D^2 > 0 , \qquad b_{rs} = b_{sr} , \qquad b_{rr} > 0 .$$

Ne segue l'esistenza di almeno tre direzioni unite mutuamente trirettangole per b. Sia T' una (o la) terna trirettangola unita di b. Rispetto ad essa risulta $b_{rs} = 0 (r \neq s)$, $b_{rr} = b'_r > 0$. Si può dunque porre

(19) $$d_r = \sqrt{b'_r} > 0 ,$$

se con b'_r si denotano i coefficienti principali di b, cioè gli elementi diagonali della matrice b rispetto al riferimento T'.

In generale la determinazione dei b'_r (e dei d_r) presuppone la risoluzione di un'equazione di terzo grado.

Tutto ciò indica l'esistenza di una matrice d determinata dalla uguaglianza

(20) $$d^2 = dd = b ,$$

per la quale risulta

(21) $$\left\| d_{rs} \right\| = \left\| x_{r,s} \right\| = D > 0, \qquad \bar{d} = d .$$

Si consideri la matrice

(22) $$\rho = ad^{-1}$$

Per essa risulta $\| \rho \| = 1$ e inoltre

(23) $$\bar{\rho}\rho = d^{-1}\,\bar{a}ad = d^{-1}bd^{-1} = d^{-1}ddd^{-1} = 1.$$

Si ha dunque anche $\bar{\rho} = \rho^{-1}$ e ρ è un rotore per il quale è

(24) $$a = \rho d$$

Nella corrispondenza caratterizzata da (12) d caratterizza la deformazione (allungamenti, variazione di volume, estensioni superficiali, ecc.), ρ la rotazione locale. La determinazione di d e ρ non è facile. Signorini lo ha fatto in qualche caso.

Accanto all'uso delle matrici b e d è fondamentale quello della matrice ε definita dall'uguaglianza

(25) $$1 + 2\varepsilon = b$$

Le componenti di ε, se con $\underline{u}$ si denota lo spostamento PP', sono

(26) $$\varepsilon_{rs} = \frac{1}{2}\left[u_{r,s} + u_{s,r} + u_{i,r}u_{i,s} \right],$$

che per semplicità chiamerò caratteristiche di deformazione (nonostante tale denominazione spetti alle ε_{rr} e alle $2\varepsilon_{rs}$ ($r \neq s$)).

3) Una proprietà di minimo.(1) Sia c un intorno sferico del generico punto P di C e Q un qualunque punto di c. Sia Q' il corrispondente di Q nella corrispondenza caratterizzata da (12). Si supponga che il raggio r di c sia tanto piccolo da poter ritenere che in c la matrice a sia costante ed espressa dalla sua determinazione in P. Lo stesso accadrà per la rotazione locale ρ e per la deformazione pura d. Sia R un qualunque spostamento rigido a partire da C e Q'' il cor-

---

(1). G. GRIOLI "Una proprietà di minimo nella Cinematica delle deformazioni finite" Boll. U. M. I. S. II, A. II, 5 p. 452 (1940).

rispondente di Q in R . Chiamasi divario relativo ai due spostamenti considerati (quello effettivo e R) e alla porzione c la quantità

(27) $$D = \int_c |Q'Q''|^2 \, dc$$

Sussiste il teorema : Tra tutti gli spostamenti rigidi operanti su c quello che rende minimo il divario D è lo spostamento rigido locale in P contenuto nello spostamento effettivo CC'.

In altri termini lo spostamento rigido minimante D è il prodotto della traslazione $\underline{t}$ = PP' per la rotazione ρ valutata in P .

Per rendersene conto basta osservare che essendo ρ e d costanti entro c , si ha

(28) $$QQ' = -PQ + \underline{t} + \rho\, dPQ \,, \qquad QQ'' = -PQ + t' + \rho' PQ,$$

se $\underline{t}'$ e ρ' sono la traslazione e il rotore che compongono R .
Ne segue , data la costanza in c di ρ e d ,

(29) $$D = (\underline{t}' - \underline{t})^2 c + \int_c |(\rho' - \rho d)\, PQ|^2 dc$$

Si riconosce facilmente che il minimo di D si consegue per $\underline{t}' = \underline{t}$ e $\rho' = \rho$.

G. Grioli

II

Sul potenziale termodinamico

1) Concetto di reversibilità. Parlando di sistemi a trasformazioni reversibili intenderò riferirmi a corpi nello schema del continuo per i quali in corrispondenza a ogni loro trasformazione infinitesima è soddisfatta (con riferimento all'unità di volume di uno stato di riferimento) l'uguaglianza

(1) $$\frac{dQ}{T} = dE$$

ove dQ denota la quantità di calore assorbita, T la temperatura assoluta alla quale avviene la trasformazione e dE il differenziale esatto dell'entropia E. In altri termini, $\frac{1}{T}$ è il fattore integrante di dQ. E' ben noto come, introdotto il potenziale termodinamico

(2) $$J = U - ET \quad ,$$

detto energia libera (ove U è l'energia interna), da (1), (2) e dal primo principio della termodinamica segua

(3) $$dl^{(i)} + EdT = -dJ \, ,$$

ove $dl^{(i)}$ rappresenta il lavoro delle forze interne corrispondente alla trasformazione considerata.

Se l'espressione di $dl^{(i)}$ è del tipo

(4) $$dl^{(i)} = y_r d\xi_r$$

ne segue

$$(5) \qquad y_r = -\frac{\partial J}{\partial \xi_r}, \qquad E = -\frac{\partial J}{\partial T}.$$

oltre alla dipendenza di J dalle ξ_r e da T.

2) <u>Corpi elastici.</u> Nell'ambito della teoria delle deformazioni finite si danno varie definizioni di corpo elastico. In senso molto lato, si suole chiamare corpo elastico un sistema continuo per il quale lo stress dipende ed è determinato dallo stato attuale, precisamente, dalla conoscenza delle caratteristiche di deformazione e della temperatura. Più restrittivamente, può dirsi corpo elastico un corpo per il quale non solo si presenta la circostanza precedente, ma inoltre lo stress deriva da un potenziale. Da alcuni Autori tali corpi sono detti iperelastici e non è detto che le due categorie non si identifichino.

Io userò un concetto di corpo elastico consono a quello abituale nella scuola italiana (Signorini) che in realtà, pur essendo formalmente più restrittivo di quelli sopra enunciati, sembra più vicino al corrispondente concetto fisico. Si dirà, cioè; corpo elastico un corpo a trasformazioni reversibili che ammette una configurazione di equilibrio spontaneo - cioè, in assenza di forze esterne attive (ma non necessariamente di vincoli) - a temperatura T^o tale che per ogni spostamento a partire da quella configurazione e per ogni T^o contenuta in un certo intervallo - il lavoro risulti nullo se lo spostamento è rigido, negativo in ogni altro caso.

Una tale definizione implica l'esistenza di un potenziale termodinamico da cui lo stress deriva, cioè l'iperelasticità e contiene in effetti una condizione di stabilità che può non essere verificata. Ciò accade, ad es., nel caso di un cilindro costretto da vincoli agenti sulle basi ad avere una lunghezza diversa da quella naturale - esente, questa, da stress - . Se la variazione di lunghezza rispetto allo stato naturale ha raggiunto un certo valore si presenta un caso di instabilità che non permette di soddisfare la condizione richiesta dalla definizione sopra data. Naturalmente , si può parlare, più semplicemente di materiale (non di corpo) elastico, intendendo con ciò un materiale con cui si può costruire un corpo elastico.

Si può osservare che un materiale elastico non può avere l'energia libera dipendente dalle caratteristiche di deformazioni solo per tramite del determinante funzionale D. Infatti, se così fosse il lavoro delle forze interne riuscirebbe nullo per ogni trasformazione isoterma senza variazione di volume. In particolare, sotto la definizione data, i fluidi non viscosi, in particolare i gas, non sono corpi elastici.

3) Una diversa definizione di sistema a trasformazioni reversibili.

E' interessante osservare che nella definizione prima data di corpo elastico si può sopprimere la condizione che il sistema sia a trasformazioni reversibili . Intendo dire che è possibile dimostrare che l'ipotesi dell'esistenza di una configurazione a partire dalla quale il lavoro delle forze interne è negativo per ogni spostamento isotermo non rigido implica come come conseguenza che il sistema sia a trasformazioni reversibili , cioè la (1) .

Sia A una configurazione soddisfacente a tale condizione e B una configurazione non raggiungibile da A mediante uno spostamento rigido.

Risulterà

$$(6) \qquad \int_{\widehat{AB}} dl^{(i)} < 0 , \qquad \int_{\overline{AB}} dl^{(i)} < 0 ,$$

per due distinti cammini $\widehat{AB}$, $\overline{AB}$. Sarà certamente

$$(7) \qquad \int_{\widehat{AB}} dl^{(i)} = \int_{\overline{AB}} dl^{(i)}$$

Infatti , sia , per assurda ipotesi,

$$(8) \qquad \int_{\widehat{AB}} dl^{(i)} > \int_{\overline{AB}} dl^{(i)} .$$

Ne segue

$$(9) \qquad \int_{\widehat{AB}} dl^{(i)} + \int_{\overline{BA}} dl^{(i)} > 0 ,$$

che per motivi di continuità implica

$$(10) \qquad \int_{ABA'} dl^{(i)} > 0$$

per un A' sufficientemente vicino ad A, contrariamente all'ipotesi. Dunque il lavoro delle forze interne nel passagio da A a B non solo è negativo ma neppure dipende dal cammino percorso. Di conseguenza il lavoro per ogni ciclo chiuso passante per A è nullo . Si riconosce allora facilmente che lo stesso capita per ogni ciclo chiuso, anche se non passante per A . Ne segue l'esistenza di una funzione potenziale per la quale risulta

$$(11) \qquad dl^{(i)} = - d F .$$

Quanto è stato mostrato costituisce, in effetti, un teorema già dimostrato da Caprioli [2] ma si può andare un pò avanti e dimostrare che in realtà che F non può differire dall' energia libera altro che per una funzione della sola temperatura assoluta e che il sistema è a trasformazioni reversibili. A tal fine si richiamino le relazioni dovute ai primi due principi della termodinamica :

(12) $$dQ + dF = dU, \qquad \frac{dQ}{T} \leqslant dE.$$

Si può sempre porre

(13) $$dQ = T(dE - d^{(i)}E),$$

se con $d^{(i)}E \geqslant 0$ si rappresenta l'irreversibile accrescimento dell'entropia. Da (12), (13) segue

(14) $$T(dE - d^{(i)}E) + dF - dU = 0.$$

Supposto E, F, U dipendenti da un certo numero di parametri di stato ξ_i e da T da (14) segue, per ogni trasformazione isoterma

(15) $$\left(T\frac{\partial E}{\partial \xi_i} + \frac{\partial F}{\partial \xi_i} - \frac{\partial U}{\partial \xi_i}\right)\partial \xi_i = T\, d^{(i)}E \geqslant 0.$$

Insieme alla scelta $\partial \xi_i$ può considerarsi la scelta $-\partial \xi_i$. Da (15) segue allora $d^{(i)}E = 0$.

Il sistema è, pertanto, a trasformazioni reversibili. Da (15) se-

(2) L. CAPRIOLI "Su un criterio per l'esistenza dell'energia di deformazione" Boll. Un. Mat. Ital. S. III, 10, p. 481 (1955)

gue inoltre ,

$$(16) \qquad \frac{\partial F}{\partial \xi_i} = \frac{\partial J}{\partial \xi_i} , \qquad F = J + \varphi(T) .$$

Il significato della funzione della sola temperatura $\varphi(T)$ è espresso dalla uguaglianza

$$(17) \qquad \frac{d\varphi}{dT} = E + \frac{\partial F}{\partial T} .$$

4) Sulla rappresentazione euleriana e lagrangiana dello stress.

Lo schema più semplice di sistema continuo tridimensionale è quello in cui la configurazione è a ogni istante individuata dalla conoscenza di un vettore rappresentante lo spostamento del generico punto di una configurazione di riferimento . Possono concepirsi tipi più generali si sistemi continui, ad es. , quelli il cui stato attuale è individuato oltre che da un vettore spostamento anche dalla conoscenza dell'orientamento di un triedro associato al generico elemento materiale, senza neppure escludere che nel passaggio da una configurazione a un'altra tale triedro possa deformarsi.

La necessità della conoscenza di tali elementi si presenta necessaria per la individuazione dello stress attuale, inteso in senso molto lato, cioè senza una necessaria simmetria delle caratteristiche di tensione e, anzi , senza che neppure queste da sole siano sufficienti a individuare l'effettivo stato tensionale nel corpo.

La Meccanica di tali sistemi - in particolare, delle microstrutture - è molto complessa e, sostanzialmente, ancora da farsi. Io nel seguito mi riferirò , generalmente,- con la locuzione materiale semplice - al caso più semplice che la conoscenza dello spostamento e della temperatura sia sufficiente per la determinazione dello stress, supponendo, anzi, che questo sia individuato dalla conoscenza delle sole caratteristiche di tensione supposto simmetriche, caso che, del resto, non è affatto semplice nell'ambito delle deformazioni finite. Ho detto generalmente perchè in qualche punto considererò sistemi più complessi il cui studio oggi è di attualità. Per essi lo stato tensionale è individuato da due matrici non simmetriche, quella delle caratteristiche di tensione e quella delle coppie di contatto.

Per ogni materiale semplice a trasformazioni reversibili l'energia libera dipende dalla temperatura e dalle caratteristiche di deformazione . Lo stesso accade per la matrice simmetrica lagrangiana dello stress:

(18) $$\beta \equiv \left| Y_{rs} \right| ,$$

ove le Y_{rs} sono le caratteristiche di tensione di Piola-Cosserat . Invece la matrice simmetrica che rappresenta lo stress nello stato attuale :

(19) $$\beta' \equiv \left| X_{rs} \right| ,$$

ove le X_{rs} sono le caratteristiche euleriane di tensione, dipende in generale, dalla rotazione locale, tenuto conto del legame tra β e β' :

(20) $$\beta' = \frac{1}{D} a \beta \bar{a} .$$

Si ponga

(21) $$q = \frac{1}{D} d \beta d .$$

Tenuto conto di (I, 24), da (20), (21) segue

(22) $$\beta' = \rho q \rho^{-1}$$

Comincio con l'osservare che se una direzione $\underline{v}$ è unita per la matrice q, allora la direzione $\rho \underline{v}$ è unita per la matrice β'. Infatti, supposto

(23) $$q \underline{v} = \lambda \underline{v}$$

ove λ è un coefficiente numerico e posto $\underline{w} = \rho \underline{v}$, si ha

(24) $$\beta' \underline{w} = \rho q \rho^{-1} \rho \underline{v} = \rho q \underline{v} = \lambda \rho \underline{v} = \lambda \underline{w} .$$

Di conseguenza : Condizione necessaria e sufficiente affinchè una terna trirettangola T_ρ sia unita per β' è che la terna ottenuta da T_ρ mediante la rotazione ρ^{-1} sia unita per q.

In altri termini, la terna unita di β' si ottiene da quella di q - che esiste certamente, dato che q è simmetrica - mediante la rotazione ρ. Ne segue che β' e q hanno gli stessi coefficienti principali :

(25) $$X_i = Q_i .$$

Per convincersene, basta ricordare che se $\underline{i}_s$ sono i versori degli assi della terna di riferimento, gli elementi di una qualunque matrice $\nu \equiv |\nu_{rs}|$ sono espressi da $\nu_{rs} = \underline{i}_r \times \nu \underline{i}_s$. Di conseguenza, se $\underline{c}_i$ sono i versori degli assi della terna unita di q e $\underline{\bar{c}}_i = \rho\, \underline{c}_i$ quelli della corrispondente di β', si ha

$$(26)\quad X_i = \underline{\bar{c}}_i \times \beta' \underline{\bar{c}}_i = \rho\, \underline{c}_i \times \beta' \rho\, \underline{c}_i = \underline{c}_i \times \rho^{-1} \beta' \rho\, \underline{c}_i = \underline{c}_i \times q\, \underline{c}_i = Q_i$$

Se le Y_{rs} dipendono da certi parametri ξ_i, ne segue che mentre le X_{rs} dipendono dagli ξ_i dalle ε_{rs} e, inoltre dalla rotazione locale, invece le tensioni principali X_i dipendono solo dai parametri ξ_i e dalle ε_{rs}. In particolare, nei sistemi semplici le tensioni principali X_i dipendono - oltre che dalla temperatura - dalle sole caratteristiche di deformazione, come le Y_{rs}.

Si presenta (per le sole X_i) una proprietà analoga a quella dei fluidi non viscosi per i quali la pressione attuale dipende solo dalle caratteristiche di deformazione (per tramite di D).

Ciò appare naturale, quando si pensi che per un fluido non viscoso ogni direzione è direzione unita per lo stress euleriano, riducendosi la β' a un'omotetia.

5) <u>Sul concetto di isotropia</u>. E' ben noto che il concetto di isotropia, come è abitualmente presentato, esprime, sostanzialmente, l'indipendenza dalla direzione delle relazioni costitutive che legano stress e strain. Ciò si traduce nell'invarianza di tali relazioni rispetto alle rotazioni e, nel caso di sistemi che ammettono un

potenziale da cui deriva lo stress, alla dipendenza di esso dai parametri fondamentali solo attraverso un certo numero di invarianti.

Tuttavia è possibile dare una condizione di isotropia alla quale, se assunta come definizione, può indubbiamente attribuirsi una maggiore ampiezza poichè non può escludersi che essa abbracci una più vasta categoria di sistemi materiali, almeno per esprimere una condizione di parziale isotropia, e, comunque, può rendere più agevole la conoscenza della determinazione della struttura analitica del potenziale termodinamico.

Si supponga che d e β abbiano le medesime direzioni unite. Segue che anche q ha le medesime direzioni unite di d. Poichè le direzioni unite di β' si ottengono, in base a (21), (22) da quelle di q mediante la rotazione ρ, si deduce che l'immagine in C' di una qualunque direzione unita di d è direzione unita di β' e viceversa.

Questa osservazione permette di dare la seguente definizione di sistema isotropo [3]: Con riguardo a una determinata configurazione di riferimento, C, un sistema è isotropo se ogni direzione unita di deformazione x ha per immagine in C' una direzione unita di tensione.

Tale definizione è completamente equivalente a quella esprimente la dipendenza dell'energia libera dalle caratteristiche di

---

[3] Su tale modo di caratterizzare l'isotropia dei materiali semplici vedi A. SIGNORINI "Trasformazioni termoelastiche finite" Memoria I, Ann. Mat. Pura e Appl. S. IV., 22, p. 136 (1943) e I. GASPARINI "Sopra una proprietà caratteristica dei sistemi isotropi" Boll. Un. Mat. Ital. 2, 13-18 (1943).

deformazione attraverso i tre loro invarianti principali nel caso di sistemi semplici a trasformazioni reversibili ma anche in questo caso è concettualmente più vasta in quanto formalmente indipendente dalle equazioni costitutive , mentre sembrerebbe spontaneo nel dare la definizione di isotropia riferirsi ad esse, affermandone la loro invarianza rispetto a un certo gruppo di trasformazioni (come, del resto, è abituale fare [4]) .

Si consideri un sistema semplice la cui energia libera dipenda dalle ε_{rs} solo attraverso i tre invarianti principali. Il sistema è , pertanto, isotropo nel senso abituale. Denotando con I, II, l'invariante lineare della matrice ε e quello quadratico (definiti concordemente a (I, 10)) e tenuto conto di

$$\frac{\partial D}{\partial \varepsilon_{rs}} = D \left[(1 + 2\varepsilon)^{-1} \right]_{rs} \tag{27}$$

da (5) si trae facilmente

$$\beta = - \frac{\partial J}{\partial I} - \frac{\partial J}{\partial II} (I - \varepsilon) - \frac{\partial J}{\partial D} D (1 + 2\varepsilon)^{-1} \tag{28}$$

Si riconosce, così subito che β e ε hanno le medesime direzioni unite (e così pure d) . Di conseguenza anche q (vedi (21)) ha le medesime direzioni unite di β e d e le direzioni unite di β' sono le immagini in C' di quelle di ε , d e β .

---

(4) Vedi , ad es. , Handbuch Der Physik , Band III/1 , pp. 700-701 (C. TRUESDELL and TOUPIN

Si supponga, invece, che ogni direzione unita di deformazione abbia per immagine in C' una direzione unita per la matrice β'. Se per l'elemento considerato ci si riferisce alla terna unita T_u di ε e si denotano con E_i i coefficienti principali della ε, l'energia libera si può esprimere mediante gli E_i e tre parametri φ_i che individuano l'orientamento della T_u rispetto a una presupposta terna di riferimento (dato che i tre invarianti principali sono in corrispondenza biunivoca con i tre E_i).

Di conseguenza per il lavoro delle forze interne -per ogni trasformazione isoterma- si ha l'espressione.

(29) $$dl^{(i)} = -\frac{\partial J}{\partial E_r} dE_r - \frac{\partial J}{\partial \varphi_r} d\varphi_r = Y_{rs}\, d\varepsilon_{rs}.$$

Da $E_r = \underline{c}_r \times \varepsilon\, \underline{c}_r$ segue, nel passaggio dalla configurazione attuale a una vicinissima,

(30) $$d E_r = d\, \varepsilon_{rr}.$$

La matrice β ha le medesime direzioni unite di q e d (vedi (21)). Di conseguenza, con riferimento alla T , detti Y_r i coefficienti principali di β, si ha

(31) $$dl^{(i)} = Y_r\, d E_r .$$

Dal confronto di (29) con (31) si deduce

(32) $$\frac{\partial J}{\partial \varphi_r} = 0 .$$

L'energia libera dipende, cioè, solo dalle E_r. Quindi

solo dagli invarianti principali, c.d.d. .

6) <u>Qualche osservazione sull'isotropia di sistemi con coppie di contatto.</u> Nel caso di un sistema continuo con possibilità di uno stato tensionale determinato dalla matrice non simmetrica Y_{rs} e da una seconda, pure non simmetrica che rappresenta la distribuzione delle coppie di contatto, si riconosce che l'energia libera - nell'ipotesi che il sistema sia, dal punto di vista cinematico, completamente descritto dalla conoscenza del vettore spostamento - non può dipendere, oltre che dalla temperatura, altro che dalle caratteristiche di deformazione e dalle variabili $_{rs}$ definite dalle uguaglianze [5].

(33) $$\eta_{rs} = x_{i\,r}\mu_{is} ,$$

con

(34) $$\eta_{is} = \frac{1}{2\,D} \sum_m \left[u_{i+2,\,sm} A_{i+1m} - u_{i+1,\,sm} A_{i+2m} \right]$$

essendo A_{lm} il complemento algebrico di x_{lm} nel determinante D . Se $T \equiv \| T_{rs} \|$ è la matrice simmetrica associata alla β , $T = \frac{1}{2} (\beta + \tilde{\beta})$, il lavoro delle forze interne ha un'espressione del tipo

(35) $$dl^{(i)} = T_{rs}\, d\varepsilon_{rs} + Q_{rs}\, d\eta_{rs} .$$

---

(5) A. BRESSAN "Sui sistemi continui nel caso asimmetrico" Ann. Mat. Pura e Appl. S. IV , V . LXII (1963) .

La parte simmetrica della matrice euleriana dello stress, $T' = \frac{1}{2}(\beta' + \bar{\beta}')$, verifica l'uguaglianza

$$(36) \qquad T' = \frac{1}{D} \, a \, T \, \bar{a} \, ,$$

analoga alla (20).

Si supponga che ogni direzione unita della d abbia come immagine in C' una direzione unita di T' (il che implica che d e T abbiano le medesime direzioni unite). Con riferimento a una terna unita di d si ha allora

$$(37) \qquad d\,l^{(i)} = T_r \, dE_r + Q_{rs} \, d\eta_{rs} \, ,$$

mentre l'energia libera si può pensare dipendente, oltre che dalla temperatura e dalle η_{rs}, dalle E_r e da tre parametri φ_i aventi il medesimo significato che nel numero precedente.

Si pensi a uno spostamento a partire dalla configurazione attuale soddisfacente alla condizione

$$(38) \qquad d\eta_{rs} = dx_{i,r} \, \mu_{is} + x_{i,r} \, d\mu_{is} = 0$$

ma non irrotazionale $d(u_{r,s} - u_{s,r}) \neq 0$), in modo che i $d\varphi_r$ non siano tutti e tre nulli. Un ragionamento completamente analogo a quello svolto nel numero precedente, tenuto conto che in (35) manca ora il termine dipendente dalle Q_{rs}, porta alla conclusione che l'energia libera dipende dalle ε_{rs} solo per tramite dei tre invarianti principali.

Tenuto conto di

$$Q_{rs} = - \frac{\partial J}{\partial \eta_{rs}} \tag{39}$$

vale il viceversa .

Il comportamento rispetto alla proprietà di isotropia di un tale tipo di continuo è , pertanto, analogo a quello che si ha nel caso di un sistema semplice, per quanto concerne la parte simmetrica della matrice degli sforzi . Anzi, se si trattasse di un sistema senza coppie di contatto la questione dell'isotropia sarebbe con ciò esaurita, in quanto la parte non simmetrica di quella matrice non interviene nell'espressione del lavoro delle forze interne ed è determinata dall'equazione dei momenti .

Qualche osservazione conviene fare in riguardo alla matrice delle coppia di contatto. Detta

$$\Psi \equiv | \Psi_{rs} | \tag{40}$$

la matrice euleriana generalmente non simmetrica di tali coppie , si ha [6]

$$\Psi_{rs} = \frac{1}{D} \lambda_{rl} \, x_{s,l} \, , \tag{41}$$

con

$$\lambda_{rl} = - \tau A_r - \frac{\partial J}{\partial \mu_{rl}} \tag{42}$$

ove τ è un parametro costitutivo . Poichè l'energia libera J dipende dalle μ_{rl} solo per tramite delle η_{pq} , si ha

---

[6] G. GRIOLI "Elasticità asimmetrica"Ann. Mat. Pura e Appl. IV, V. 4, 389--418 (1960)

$$(43) \qquad \frac{\partial J}{\partial u_{rl}} = \frac{\partial J}{\partial \eta_{pl}} x_{r,p}$$

e da (41) , (42) , (43) segue

$$(44) \qquad \Psi_{rs} = - \frac{\tau}{D} x_{s,l} A_{rl} - \frac{1}{D} x_{r,p} x_{s,l} \frac{\partial J}{\partial \eta_{pl}}$$

che si semplificano in

$$(45) \qquad \Psi_{rs} = - \tau \delta_{rs} - x_{r,p} x_{r,l} \frac{\partial J}{\partial \eta_{pl}} \frac{1}{D} .$$

Le (45) , tenuto conto di (39) , danno

$$(46) \qquad \Psi = - \tau + \frac{1}{D} a Q \bar{a} .$$

Un ragionamento analogo a quelli precedenti mostra che condizione necessaria e sufficiente affinchè una direzione sia unita per la Ψ è che essa sia l'immagine in C' di una direzione unita di d Q d . Soltanto che adesso non si può essere sicuri che la matrice Q ammetta tre direzioni unite (mutuamente ortogonali), dato che essa non è simmetrica. Certo che se esistono tre direzioni unite per la Q, cioè se la Q è simmetrica, con riferimento alle coppie di contatto si può assumere una definizione di isotropia del tutto analoga a quella valida per i sistemi semplici . E' lecita una analoga definizione in caso contrario , con riferimento all'unica direzione unita (certamente esistente) di Q ? Forse è da presumersi che nel caso delle coppie di contatto non sempre sia lecito parlare di isotropia in senso lato, in senso lato, come per i sistemi semplici ma solo in modo parziale con riferimento all'unica direzione unita di dQd .

7) Su qualche possibile tipo di struttura analitica dell'energia libera. La determinazione sperimentale dell'energia libera è praticamente impossibile se non si hanno indicazioni preliminari che ne definiscano in qualche modo la struttura analitica, anche nel caso meno complesso dei sistemi semplici isotropi. E' noto che le esperienze di trazione e pressione - che sono le più facilmente realizzabili - possono determinare [7] il potenziale termodinamico isotermo soltanto nell'interno di una linea dello spazio tridimensionale (I, II, D).

Molti lavori per la determinazione del potenziale termodinamico isotermo a temperatura T definito, nel caso dei sistemi semplici isotropi di cui solo mi occuperò in questo numero, dalla uguaglianza

$$(47) \qquad W_T = \mathcal{J}(\varepsilon_{r,s}, T, T) - \mathcal{J}(o, T, T),$$

sono legati ai nomi di molti Autori [8] quali ad es., De Saint-Venant, Voigt, Kotter, Southwell, Hencky, Seth, Murnaghan, Zvolinski, Riz, Novozilov, Baker, Eircksen, Hanin and Reiner, Csonka, Chu Boa-Teh Smith, Rivlin, ecc. In Italia, in questo campo, i lavori più

---

(7) A. SIGNORINI "Trasformazioni termoelastiche finite", Memoria I, Ann. Mat. Pura e Appl., S. IV, 30 p. 27 (1949).

(8) Per un'esauriente bibliografia in proposito vedi C. TRUESDELL "The mechanical foundations of elasticity and fluid dynamics" J. Rat. Mech. 2X 1 125-300 (1952 e "Corrections and additions to the Mechanical foundations of elasticity and fluid dynamics" J. Rat. Mech. 2, 593 616 (1953)

interessanti sono dovuti a Signorini. Nel Suo indirizzo sono altresì da ricordare i lavori di Tolotti[9], Bordoni[10] e Manacorda[11].

Una naturale tendenza porta a costruire per l'energia libera (o per il potenziale termodinamico isotermo) uno sviluppo in serie di potenze, arrestandolo a un certo termine. In tal modo si costruiscono espressioni polinomiali approssimate di W_T, $\mathcal{J}$ e dello stress.

Tuttavia, si riconosce la possibilità di costruire qualche tipo di espressione esatta dell'energia libera che dia luogo a effettive espressioni polinomiali per le caratteristiche euleriane dello stress.

Innanzitutto, ricordo che, sulla base delle (20), si riconosce che mentre le caratteristiche di tensione lagrangiane Y_{rs} dipendono unicamente dalle ε_{rs} (e dalla temperatura), lo stesso non accade per la matrice euleriana $|X_{rs}|$. Come è stato osservato, ciò capita solo per le tensioni principali, X_i. Per l'intera matrice $|X_{rs}|$ la dipendenza dalle sole ε_{rs} (e da T) si ha solo nel caso di spostamenti irrotazionali : $\rho = 1$, $u_{r,s} - u_{s,r} = 0$.

---

(9) C. TOLOTTI "Deformazioni elastiche finite: onde ordinarie di discontinuità e casi tipici di solidi elastici isotropi "Ren. Mat. Appl. S. 7, 4 34-59 (1943).

(10) P. G. BORDONI "Deduzione di un'equazione di stato dei solidi dalla teoria delle trasformazioni termoelastiche finite "Rend. Acc. Naz. Lincei, S. VIII, V. XIV, 6, 1-7 (1953).

(11) T. MANACORDA X "Sul legame sforzi deformazione nelle trasformazioni finite di un mezzo continuo isotropo"Riv. Mat. Univ. di Parma, 4 31-42 (1953). Non trattasi, in questo caso, di mezzi elastici; tuttavia, sussiste un legame sforzi deformazione del tipo (48).

In generale, da (20), (28) si deducono, nel caso di isotropia, le relazioni equivalenti a quelle di Finger :

$$(48) \qquad X_{rs} = -\frac{1}{D}\left[l\delta_{rs} + 2m\,\varepsilon^{(\rho)}_{rs} + n\,\varepsilon^{(\rho)}_{rt}\,\varepsilon^{(\rho)}_{st}\right]$$

ove si è posto

$$(49) \qquad \varepsilon^{(\rho)}_{rs} = \frac{1}{2}(u_{r,s} + u_{s,r} + u_{r,l}\,u_{s,l}),$$

$$l = D^{(\rho)}\frac{\partial J}{\partial D^{(\rho)}} + \frac{\partial J}{\partial I^{(\rho)}} + I^{(\rho)}\frac{\partial J}{I I^{(\rho)}}$$

$$(50) \qquad m = \frac{\partial J}{\partial I^{(\rho)}} + I^{(\rho)}\frac{\partial J}{\partial I I^{(\rho)}} - \frac{1}{2}\frac{\partial J}{\partial I I^{(\rho)}}$$

$$n = -2\frac{\partial J}{\partial I I^{(\rho)}},$$

essendo $I^{(\rho)}$, $II^{(\rho)}$ e $D^{(\rho)}$ gli invarianti principali dell matrice $\varepsilon^{(\rho)}$.

E' possibile esprimere le caratteristiche euleriane di tensione come funzioni delle caratteristiche di deformazione dello spostamento inverso :

$$(51) \qquad \varepsilon^{(i)}_{rs} = -\frac{1}{2}(u_{r,s} + u_{s,r} - u_{l,r}\,u_{l,s}).$$

Nel caso isotropo, si ottengono le espressioni

$$(52) \qquad X_{rs} = l^{(i)}\,\delta_{rs} + 2\,m^{(i)}\,\varepsilon^{(i)}_{rs} + n^{(i)}\,\varepsilon^{(i)}_{rt}\,\varepsilon^{(i)}_{ts}$$

ove è

$$
\begin{aligned}
& l^{(i)} = D^{(i)} \left[D^{(i)} \frac{\partial \mathcal{J}}{\partial D^{(i)}} + \frac{\partial \mathcal{J}}{\partial I^{(i)}} + I^{(i)} \frac{\partial \mathcal{J}}{\partial I I^{(i)}} \right] , \\
(53) \qquad & m^{(i)} = D^{(i)} \left[\frac{\partial \mathcal{J}}{\partial I^{(i)}} + I^{(i)} \frac{\partial \mathcal{J}}{\partial I I^{(i)}} - \frac{1}{2} \frac{\partial \mathcal{J}}{\partial I I^{(i)}} \right] , \\
& n^{(i)} = - 2 D^{(i)} \frac{\partial \mathcal{J}}{\partial I I^{(i)}} ,
\end{aligned}
$$

e, naturalmente , si pensa l'energia libera espressa mediante gli invarianti principali $I^{(i)}$, $I I^{(i)}$, $D^{(i)}$ della $\varepsilon^{(i)}$.

Signorini ha mostrato [12] come siano possibili per le X_{rs} delle espressioni esatte di secondo grado nelle $^{(i)}_{rs}$. Invece le X_{rs} non possono essere lineari nelle $\varepsilon^{(i)}_{rs}$, in quanto in tal caso il potenziale termodinamico isotermo avrebbe l'espressione

$$
(54) \qquad W^{(i)}_{T} = \alpha \left[\frac{1 + I^{(i)}}{D^{(i)}} - 1 \right] ,
$$

con α costante, la quale non soddisfa alla condizione necessaria di essere definita positiva, com'é facile riconoscere [13] .

In generale , la condizione che le X_{rs} siano dei polinomi di grado n nelle $\varepsilon^{(i)}_{rs}$, si traduce in quella che $l^{(i)}$, $m^{(i)}$ e $n^{(i)}$ siano funzioni tali degli invarianti principali da risultare rispettivamente di grado n,n-1 e n-2 nelle $\varepsilon^{(i)}_{rs}$. Tali espres-

---

(12) A. SIGNORINI . Loco cit . nota (7) .

(13) A. SIGNORINI . Loco cit. nota (7) .

sioni dovranno soddisfare alle condizioni di integrabilità del sistema che immediatamente si deduce da (53) :

$$\frac{\partial J}{\partial I^{(i)}} = \frac{1}{D^{(i)}} \left[m^{(i)} + \frac{n^{(i)}}{2} \quad (I^{(i)} - \frac{1}{2}) \right] ,$$

(55) $$\frac{\partial J}{\partial I I^{(i)}} = - \frac{n^{(i)}}{2 D^{(i)}} ,$$

$$\frac{\partial J}{\partial D^{(i)}} = \frac{1}{D^{(i)2}} \left[l^{(i)} - m^{(i)} + \frac{n^{(i)}}{4} \right] .$$

Una questione analoga si può porre sulla base delle (48) , (49), (50)ed è interessante constatare che al contrario di quanto accade quando ci si riferisca alle $\varepsilon_{rs}^{(i)}$, per le X_{rs} sono invece possibili espressioni lineari nelle $\varepsilon_{rs}^{(\rho)}$, almeno per quanto riguarda la condizione che il potenziale termodinamico riesca definito positivo.

In tal caso nelle (48) si deve assumere

(56) $$= (\alpha_1 \quad I^{(\rho)} + \alpha_2) D^{(\rho)} , \quad m = \alpha_3 D^{(\rho)} , \qquad n = 0 ,$$

con α_1, α_2 , α_3 costanti (supposto il sistema omogeneo) . Le condizioni di integrabilità dedotte da (50)' , portano - nell'ipotesi che lo stato di riferimento sia naturale (esente da stress) - subto alle espressioni

(57) $$l = \gamma D^{(\rho)} \quad I_1^{(\rho)} , \qquad m = \gamma D^{(\rho)} ,$$

alle quali corrisponde il potenziale termodinamico

$$W_T^{(\rho)} = \gamma D^{(\rho)} \left[I^{(\rho)} - 1 + \frac{1}{D^{(\rho)}} \right] . \tag{58}$$

Al potenziale termodinamico (58) corrisponde lo stress euleriano

$$\beta = - \gamma \left[I^{(\rho)} + 2 \varepsilon^{(\rho)} \right] . \tag{59}$$

Si riconosce subito che il coefficiente γ deve essere positivo[14] Infatti, tenuto conto che gli invarianti principali e i coefficienti principali della $\varepsilon^{(\rho)}$ coincidono rispettivamente con quelli della ε, si pensi a una deformazione del corpo per la quale si abbia $I^{(\rho)} = 0$; in corrispondenza risulta $D^{(\rho)} < 1$. La condizione che la $W_T^{(\rho)}$ risulti definita positiva implica allora $\gamma > 0$. Tale condizione è anche sufficiente . Infatti , posto

$$e_i = 1 + 2 E_i^{(\rho)} , \tag{60}$$

se è $\gamma > 0$, la condizione che $W_T^{(\rho)}$ sia definita positiva si traduce in quella che per ogni scelta delle e_i positive e non tutte uguali a uno la funzione

$$\psi = e_1 + e_2 + e_3 + \frac{2}{\sqrt{e_1 e_2 e_3}} - 5 > 0 \tag{61}$$

(14) Vale la pena di osservare che se si sviluppa l'espressione (58) di $W_T^{(\rho)}$ sino alle potenze di secondo grado nelle $u_{r,s}$ e quella (59) di β sino a quelle di primo grado si ottengono le note formule della teoria classica lineare, pur di identificare γ con quello dei due coefficienti di Hamé che è sempre positivo e di supporre uguale a 0, 25 il valore del coefficiente di Poisson . E' quasi superfluo rimarcare che proprio questo è il valore di quel coefficiente (o presso a poco) per un gran numero di solidi elastici .

sia positiva.

Osservo subito che da

$$\frac{\partial \psi}{\partial e_i} = 1 - \frac{1}{e_i \sqrt{e_1 e_2 e_3}} \tag{62}$$

si deduce che le tre derivate parziali primo della ψ si annullano allora e solo allora che sia $e_1 = e_2 = e_3 = 1$. Inoltre, per tali valori delle e_i si ha

$$\left(\frac{\partial^2 \psi}{\partial e_p \partial e_q}\right) \xi_p \xi_q = \frac{1}{2}\left[(\xi_1 + \xi_2)^2 + (\xi_2 + \xi_3)^2 + (\xi_3 + \xi_1)^2 + \right.$$
$$\left. + \xi_1^2 + \xi_2^2 + \xi_3^2\right]. \tag{63}$$

Si conclude, pertanto che in $e_1 = e_2 = e_3 = 1$ la ψ che assume valore nullo, ha l'unico punto di stazionarietà che è un minimo, il che dimostra la sufficienza della condizione $\gamma > 0$.

Può avere interesse vedere quali siano le espressioni corrispondenti alla (58) del potenziale termodinamico e dello stress euleriano mediante le caratteristiche dello spostamento inverso.

A tal fine comincio con l'osservare che per la matrice $\varepsilon^{(i)} = \left|\varepsilon^{(i)}_{rs}\right|$ si ha

$$1 + 2\varepsilon^{(i)} = \bar{a}^{-1} a^{-1} = (a\bar{a})^{-1} = (\rho d\, d\rho^{-1})^{-1} \tag{64}$$

ed è

$$\rho d\, d\rho^{-1} = \rho(1 + 2\varepsilon)\rho^{-1} = 1 + 2\varepsilon^{(\rho)} \tag{65}$$

Ne segue

$$1 + 2\varepsilon^{(\rho)} = (1 + 2\varepsilon^{(i)})^{-1} \tag{66}$$

Si ha , inoltre ,

$$
(67)\quad \begin{cases} I\,(1+2\varepsilon^{(i)})^{-1} = \dfrac{I\,I\,(1+2\varepsilon^{(i)})}{I\,I\,I(1+2\varepsilon^{(i)})} = \dfrac{3+4\,I^{(i)} + 4\,I\,I^{(i)}}{D^{(i)\,2}} , \\ \\ I\,(i+2\varepsilon^{(\rho)}) = 3 + 2\,I^{(\rho)} . \end{cases}
$$

Dall'uguaglianza di matrici espressa da (66) segue l'uguaglianza dei rispettivi invarianti . Di conseguenza, da (67) si deduce

$$
(68)\qquad I^{(\rho)} = \frac{1}{2}\left[\frac{3 + 4\,I^{(i)} + 4\,II^{(i)}}{D^{(i)\,2}} - 3\right] .
$$

Basta allora sostituire in (58) la trovata espressione di $I^{(\rho)}$ e tenere presente che è $D^{(\rho)} = \dfrac{1}{D^{(i)}}$, per dedurre la richiesta espressione di $W_T^{(i)}$:

$$
(69)\qquad W_T^{(i)} = \gamma\left[1 - \frac{5}{2D^{(i)}} + \frac{3+4\,I^{(i)} + 4\,II^{(i)}}{2D^{(i)\,3}} \right] .
$$

Corrispondentemente, sulla base delle (52) , (53) si ottiene l'espressione di β' :

$$
(70)\qquad \beta' = \gamma\left\{ \frac{5}{2}\left(1 - \frac{1}{D^{(i)2}}\right) - \frac{2}{D^{(i)2}}\left(2\,I^{(i)} + 3\,I\,I^{(i)}\right) + \right.
$$

$$
\left. + \frac{2}{D^{(i)2}}\left(1+2\,I^{(i)}\right)\varepsilon^{(i)} - \frac{4}{D^{(i)2}}\varepsilon^{(i)}\varepsilon^{(i)} \right\} .
$$

8) Sull'energia libera come funzione di stato caratteristica delle capacità meccaniche di un sistema materiale. Di solito, la conoscenza delle capacità meccaniche di reazione di un sistema continuo determina le variabili da cui l'energia libera può e deve dipendere ma è interessante osservare che vale il viceversa. Intendo dire che se in qualche modo si è potuto stabilire da quali variabili - oltre la temperatura - l'energia libera dipende, ciò determina, dal punto di vista meccanico, il tipo di sistema continuo. Rimane precisamente determinato quali tipi di reazioni esso è capace di esplicare attraverso gli elementi superficiali.

Si consideri, ad es., un sistema continuo la cui energia libera dipenda - oltre che dalla temperatura - esclusivamente dal determinante $D(\varepsilon_{rs})$. Posto

$$(71) \qquad p = \frac{\partial J(D)}{\partial D}$$

segue, per ogni trasformazione isoterma

$$(72) \qquad dl^{(i)} = -dJ = p\, dD = p \frac{\partial D}{\partial \varepsilon_{rs}} = p \frac{\partial D}{\partial \varepsilon_{rs}} x_{i,r}\, du_{i,s}\,.$$

Per l'interno corpo si ha, quindi,

$$(73) \qquad d\mathcal{L}^{(i)} = \int_C dl^{(i)}\, dC = -\int_C \left[p \frac{\partial D}{\partial \varepsilon_{rs}} x_{i,r} \right]_{,s} du_i\, dC - \\ - \int_\sigma p \frac{\partial D}{\partial \varepsilon_{rs}} x_{i,r}\, du_i\, n_s\, d\sigma\,.$$

Da (27) segue

(74) $$\frac{\partial D}{\partial \varepsilon_{rs}} x_{i,r}\, x_{l,s} = D\left[a(1+2\varepsilon)^{-1}\bar{a}\right]_{i,l} = D\left[\rho d\, d^{-1} d^{-1} d \rho^{-1}\right]_{i,l} = D\,\delta_{il}\,.$$

Quindi

(75) $$p\frac{\partial D}{\partial \varepsilon_{rs}} x_{i,r} = p\, A_{is}\,.$$

Da (73) , (75) segue

(76) $$d\mathcal{L}^{(i)} = -\int_C (pA_{is})_{,s}\, d u_i\, d C - \int_\sigma pA_{is}\, n_s\, d u_i\, d\sigma\,,$$

che tenuto presente che è $A_{is,s} = 0$, dà , in definitiva,

(77) $$d\mathcal{L}^{(i)} = -\int_C \frac{\partial p}{\partial x_l} A_{is}\, x_{l,s}\, d u_i\, dC - \int_\sigma pA_{is}\, n_s d u_i\, d\sigma =$$

$$= -\int_{C'} \frac{\partial p}{\partial x_i}\, d u_i\, d C' - \int_{\sigma'} p\, n'_i\, d u_i\, d\sigma'\,.$$

Il principio dei lavori virtuali porta, allora, subito alle equazioni

(78) $$\frac{\partial p}{\partial x_i} = F'_i \;(\text{ in } C') \qquad p\, n'_i = f'_i\,,\;(\text{ su } \sigma')\,,$$

da cui si riconosce che il sistema ha solo la capacità di reagire con delle pressioni e la impossibilità di uno stato di equilibrio con forze superficiali esterne che non abbiano questo carattere.

Ho voluto considerare il caso semplice dei fluidi non viscosi a titolo di esempio, per chiarire quanto primo affermato. In modo altrettanto semplice è possibile riconoscere che se l'energia libera dipende solo dalle caratteristiche di deformazione trattasi di un sistema semplice.

Vorrei, ora invece, considerare in modo un pò più esteso il caso che l'energia libera dipenda dallo spostamento solo per tramite delle sue derivate prime e seconde rispetto alle coordinate dei punti dello stato di riferimento. In modo completamente equivalente si può supporre che l'energia libera dipenda dalle $x_{r,s}$ e dalle variabili $\bar{x}_{r,lm}$ definite dall'eguaglianze

$$(79) \qquad \bar{x}_{r,lm} = \frac{x_{r,lm}}{D} .$$

Supporrò, cioè,

$$(80) \qquad \mathcal{J} = \mathcal{J}(x_{r,s}, \bar{x}_{r,lm}; T', T) .$$

Dall'essere il sistema a trasformazioni reversibili si deduce il fatto essenziale che esistono delle funzioni ν_{rs}, η_{rlm}, per le quali in corrispondenza a ogni elemento di volume dC dello stato di riferimento il lavoro delle forze interne, per ogni trasformazione isoterma assume l'espressione

$$(81) \qquad dC\, dl^{(i)} = \nu_{rs}\, du_{r,s} + \eta_{rlm}\, d\bar{x}_{r,lm} ;$$

$$\left[\nu_{rs} = -\frac{\partial \mathcal{J}}{\partial u_{r,s}} , \eta_{rlm} = -\frac{\partial \mathcal{J}}{\partial \bar{x}_{r,lm}} \right] .$$

Si ponga

$$(82) \qquad \begin{aligned} \xi_{rs} &= \nu_{rl}\, x_{s,l} \\ \psi_{rlm} &= \frac{1}{D}\, x_{l,p}\, x_{m,q}\, \eta_{rqp} . \end{aligned}$$

Tenuto conto delle uguaglianze

$$
(83) \qquad \begin{cases} (dx_r)_{/s} = \dfrac{A_{sl}\, du_{rl}}{D} , \\ (dx_r)_{/st} = \dfrac{A_{sp} A_{tq}}{D^2} \left[du_{r,pq} - \dfrac{A_{st}}{D} x_{s,pq}\, du_{rt} \right] = \dfrac{A_{sp} A_{tq}}{D} d(\bar{x}_{r,pq}), \end{cases}
$$

ove la sbarretta indica derivazione rispetto alle coordinate x_s dei punti dello stato attuale, con qualche trasformazione si riconosce che alla espressione lagrangiana (81) del lavoro delle forze interne corrisponde; per l'elemento dC' di volume dello stato attuale, l'espressione euleriana

$$
(84) \qquad dC' d l^{(i)} = \left[\xi_{rs} (du_r)_{/s} + \Psi_{rlm} (du_r)_{/lm} \right] dC' .
$$

Il lavoro delle forze interne per l'intero corpo, con riferimento alla configurazione attuale e a uno spostamento infinitesimo a partire da essa è, pertanto,

$$
(85) \qquad d\mathcal{L}^{(i)} = \int_{C'} \left[\xi_{rs} (du_r)_{/s} + \Psi_{rlm} (du_r)_{/lm} \right] dC' =
$$

$$
= \int_{C'} (\Psi_{rlm,lm} - \xi_{rs,s})\, du_r\, dC' + \int_{\sigma'} (\Psi_{rlm,m} - \xi_{rl})\, du_r\, n'_l\, d\sigma' - \int_{\sigma'} \Psi_{rlm} (du_r)_{/l}\, n'_m\, d\sigma'
$$

Supporrò adesso, quantunque ciò non sia necessario, che sul corpo agiscano nello stato attuale delle forze di massa $\underline{F}'dC'$, superficiali $\underline{f}'d\sigma'$ e dei momenti di massa e superficiali espressi, rispettivamente da $\underline{M}'dC'$, $\underline{m}'d\sigma'$. Supporrò, inoltre, che la rotazione $d'\underline{\omega}$ locale sia determinata dalla conoscenza del vettore $d\underline{u}$ che caratterizza il passaggio dalla configurazione C' a una vicinissima:

$$
(86) \qquad (d'\omega)_r = \frac{1}{2} \varepsilon_{rpq} (du_q)_{/p} ,
$$

ove ε_{rpq} è il tensore di Ricci dello spazio euclideo tridimensionale. Conseguentemente, il lavoro delle forze esterne è espresso da

$$
(87) \qquad d\mathcal{L}^{(e)} = \int_{C'} \left[F'_r du_r + \frac{1}{2} M'_r \varepsilon_{rpq} (du_q)_{/p} \right] dC' + \int_{\sigma'} \left[f'_r du_r + \frac{1}{2} m'_r \varepsilon_{rpq} (du_q)_{/p} \right] d\sigma' .
$$

In base a (85), (87) e al principio dei lavori virtuali, la condizione di equilibrio nella configurazione C' si traduce nella condizione analitica

$$\int_{C'}(\Psi_{rlm,ml} - \xi_{rl,l} + F'_r)du_r dC' + \frac{1}{2}\int_{C'} M'_r \varepsilon_{rpq}(du_q)/_p \, dC' +$$

(88)

$$+\int_{\sigma'}[(\Psi_{rlm,m} - \xi_{rl})n'_l + f'_r]\, d\sigma' + \int_{\sigma'}\left[\frac{1}{2} m_r \varepsilon_{rpq} - \Psi_{qpm} n'_m\right](du_q)/_p \, d\sigma' = 0$$

da ritenersi valida per ogni spostamento virtuale $d\underline{u}$.

Da (88) si trae il sistema equivalentc di equivalente di equazioni:

(89) $$\Psi_{rlm,lm} - \xi_{rs,s} - \frac{1}{2}\varepsilon_{lpr} M^q_{l,p} + F'_r = 0 \quad , \quad (\text{in } C') ,$$

(90) $$\begin{cases} (\Psi_{rlm,m} - \xi_{rl})n'_l - \frac{1}{2}\varepsilon_{tlr} M'_t + f'_r = 0 \\ \Psi_{rlm} n'_m - \frac{1}{2}\varepsilon_{tlr} m'_t = 0 \end{cases}$$

Comincio con l'osservare che la seconda delle (90) mostra che su σ' Ψ_{rlm} è emisimmetrico rispetto agli indici r, l, per ogni m:

(91) $$\left[\Psi_{rlm}\right]_{r=l} = 0 \quad , \quad \Psi_{rlm} = -\Psi_{lrm} \quad .$$

Poichè le (89), (90) sono valide anche per ogni porzione del corpo, ove, naturalmente, $\underline{f}'$, $\underline{m}'$, si interpretino quali sollecitazioni esterne per quella porzione, pur essendo dovute ad altri elementi del corpo, si deduce che le (91) valgono in ogni punto di C'.

Esisterà allora una matrice Ψ_{rs} soddisfacente alle uguaglianze

(92) $$\Psi_{rlm} = \frac{1}{2}\varepsilon_{rpl}\Psi_{pm} \quad .$$

Si consideri la matrice emisimmetrica

(93) $$\bar{\xi}^{(e)} = \frac{1}{2}\left[\Psi_{pl,l} - M_p\right]\varepsilon_{rps} .$$

Posto

(94) $$X_{rs} = \xi_{rs} + \bar{\xi}^{(e)}_{rs}$$

si riconosce facilmente che le (89), (90) si scrivono

(95) $$\begin{cases} X_{rs/s} - F'_r = o & \text{(in } C'\text{)} \\ \dot{X}_{rs}n'_s - f'_r = o \qquad \Psi_{pl}n'_l - m'_p = o & \text{(su } \sigma'\text{)} \end{cases}$$

dalle quali si deduce facilmente che le azioni meccaniche che si trasmettono attraverso un qualunque elemento superficiale $d\sigma'$ di normale $\underline{n}$ sono traducibili nella forza $X_{rs}n'_s d\sigma'$ più la coppia di momento $\Psi_{rs}n'_s d\sigma'$. In altri termini, lo stato tensionale in ogni punto della configurazione attuale è caratterizzato dalle matrici non simmetriche $|X_{rs}|$, $|\Psi_{rs}|$.

Non può certamente pensarsi che il sistema possa reagire a rotazioni locali che siano indipendenti dallo spostamento $\underline{u}$, cioè indipendenti dalla matrice ρ. Infatti, se così fosse nella espressione (87) delle forze esterne si dovrebbero aggiungere i termini

(96) $$\int_{C'} M'_r \overline{d\omega}_r dc' \qquad \int_{\sigma'} m'_r \overline{d\omega}_r d\sigma' ,$$

ove $\overline{d\omega}_r$ rappresenta una rotazione locale aggiuntiva e indipendente da $\underline{u}$. L'applicazione del principio dei lavori virtuali, in tale caso, implicherebbe per l'equilibrio

(97) $$M'_r = 0, \qquad m'_r = 0$$

E' altresì da osservarsi che la emisimmetria del sistema ψ_{rlm} è dovuta all'ipotesi che la rotazione locale sia unicamente quella espressa dalla matrice ρ. La possibilità di equilibrio in presenza di sollecitazioni esterne con coppie superficiali è subordinata dunque alla condizione che l'energia libera sia una tale funzione delle $x_{r,s}$, $\bar{x}_{r,lm}$ da dar luogo, in base a (87) a espressioni delle ψ_{rlm} soddisfacenti alle (91). Se così non fosse sembrerebbe che il sistema continuo possa stare in equilibrio solo con sollecitazioni esterne prive di coppie superficiali ($m'_r = 0$), pur non essendo escluso che in tal caso vi siano delle coppie interne di contatto. La questione andrebbe, però, ulteriormente chiarita, almeno dal punto di vista analitico.

III

Questioni analitiche connesse con una visione integrale del problema fondamentale dell'elastostatica lineare. Vincoli in superficie

1) Premesse. La formulazione del problema analitico connesso a quello meccanico dell'equilibrio elastico discende naturalmente da una impostazione di tipo integrale derivante dalla diretta applicazione dei principi fondamentali della Meccanica. Tuttavia, la sua esplicitazione viene fatta in forma differenziale e sembrerebbe sostanzialmente impossibile fare diversamente, per lo meno in quei problemi nei quali i dati al contorno riguardano gli spostamenti, per la presenza di vincoli in superficie.

Naturalmente, tale esplicitazione presuppone la derivabilità delle caratteristiche di tensione, fatto questo che non discende affatto dai principi fondamentali della Meccanica. Una trattazione che possa, pertanto, prescindere da tale proprietà di derivabilità è, quindi, da preferirsi a quella in forma differenziale.

Tale possibilità, mi pare si presenti senz'altro almeno nel caso linearizzato classico, pur di stabilire la possibilità di inversione di un classico teorema - dopo averlo opportunamente precisato in presenza di vincoli in superficie - : il teorema di Menabrea. Si viene così a stabilire, tra l'altro, una particolare forma del teorema di esistenza.

Si consideri in forma euleriana il sistema fondamentale di equilibrio :

(1) $$X_{rs,s} = F'_r \qquad (\text{in } C') ,$$

$$X_{rs} n'_s = f'_r \qquad (\text{su } \sigma') .$$

Posto

$$(2) \qquad b^{(r)}_{\tau_1\tau_2\tau_3} = \frac{1}{C'}\left\{\int_{C'} F'_r x_1^{\tau_1} x_2^{\tau_2} x_3^{\tau_3} dC' + \int_{\sigma'} f'_r x_1^{\tau_1} x_2^{\tau_2} x_3^{\tau_3} d\sigma'\right\} ,$$

si può dimostrare che ogni soluzione delle (1) soddisfa le infinite equazioni integrali

$$(3) \qquad C' b^{(r)}_{\tau_1\tau_2\tau_3} = \tau_1 \int_{C'} X_{r1} x_1^{\tau_1 - 1} x_2^{\tau_2} x_3^{\tau_3} dC' + \tau_2 \int_{C'} X_{r2} x_1^{\tau_1} x_2^{\tau_2 - 1} x_3^{\tau_3} dC' + \tau_3 \int_{C'} X_{r3} x_1^{\tau_1} x_2^{\tau_2} x_3^{\tau_3 - 1} dC' ,$$

ove τ_1, τ_2, τ_3 assumono tutti i possibili valori positivi interi o il valore zero.

Se le X_{rs} soddisfacenti alle (3) sono derivabili, vale il viceversa , cioè è possibile dimostrare che ogni soluzione derivabile di (3) verifica le equazione (1) .

In altri termini c'é la piena equivalenza tra il sistema delle equazioni di Cauchy e quello delle equazioni integrali (3) nella classe degli stress derivabili ma la classe delle soluzioni del sistema (3) è certamente più vasta di quella delle soluzioni di quadrato sommabile del sistema di Cauchy.

Poichè, come è stato osservato, la derivabilità dello stress non è prevista, in generale, dai principi fondamentali, sembra naturale sostituire al sistema (1), il sistema (3) che deve considerarsi come la traduzione analitica delle conseguenze delle leggi fondamentali della Meccanica.

Naturalmente, lo studio del sistema (3), come, del resto, anche quello delle (1), presuppone la conoscenza del campo d'integrazione, fatto che non si verifica nel caso delle deformazioni finite. Se esso fosse assegnato sarebbe possibile dalle (3) dedurre delle limitazioni per lo stress ma ciò non capita in generale. Invece, nel caso della teoria lineare classica l'inconveniente non si presenta in quanto, notoriamente, il campo d'integrazione viene identificato - com'é lecito - con la nota configurazione di riferimento , C.

D'ora in poi mi riferirò appunto a tale caso. Nella trattazione identificherò, quindi le $X_{r,s}$ con le componenti lagrangiane Y_{rs} e le x_i con le y_i (ai fini della derivazione o dell'intrgrazione). Supporrò, pertanto, che lo stress reale appartenga alla classe Γ delle soluzioni dela sistema

$$(4) \quad \tau_1 \int_C Y_{r1}\, y_1^{\tau_1-1} y_2^{\tau_2} y_3^{\tau_3} dC + \tau_2 \int_C Y_{r2}\, y_1^{\tau_1} y_2^{\tau_2-1} y_3^{\tau_3} dC + \tau_3 \int_C Y_{r3}\, y_1^{\tau_1} y_2^{\tau_2} y_3^{\tau_3-1} dC = C b^{(r)}_{\tau_1\tau_2\tau_3}$$

con le $b^{(r)}_{\tau_1\tau_2\tau_3}$ espresse da

$$(5) \quad \dot{b}^{(r)}_{\tau_1\tau_2\tau_3} = -\frac{1}{C}\left\{\int_C F_r\, y_1^{\tau_1} y_2^{\tau_2} y_3^{\tau_3} dC + \int_\sigma f_r\, y_1^{\tau_1} y_2^{\tau_2} y_3^{\tau_3} d\sigma\right\} .$$

Lo stress reale è quello congruente e nel caso di forze ovunque assegnate al contorno può individuarsi, nella classe Γ , quale quello che rende minimo un certo funzionale. Trattasi in realtà dell'applicazione del classico teorema di Menabrea.

Se, però , esistono vinooli in superficie o, in particolare, sono ivi su una parte assegnati gli spostamenti, come se ne potrà tener conto? Si tratterà di assicurarsi che è ancora possibile costruire un funzionale operante unicamente sull'incognito stress il quale - se ammette minimo in una certa classe Γ_V contenuta in Γ - dà luogo con lo stress minimante alla soluzione del problema di equilibrio elastico.

Intendo dire che lo stress così determinato non solo è congruente ma dà anche luogo -attraverso note formule - a spostamenti soddisfacenti le condizioni imposte dai vincoli. E' da osservarsi, tra l'altro, che ciò costituisce una precisazione del teorema di esistenza : condizione necessaria e sufficiente affinchè il problema analitico connesso a quello meccanico ammetta soluzione è che un certo funzionale operante sullo stress - da precisarsi di volta in volta in dipendenza dei vincoli in superficie - ammetta minimo nella classe Γ_v.

Trattasi, in un certo senso, di un postulazione integrale del problema fondamentale, in quanto non c'é - come si è osservato - equivalenza, in generale, tra il sistema di equazioni integrali (3) associato al funzionale da minimizzare e il problema al contorno sull'incognito spostamento che si deduce da (1) esprimendo lo stress mediante le sue derivate e precisando le condizioni di vincolo superficiale.

Per concludere questa premessa, osservo che se $\{P_t\}$ è la successione completa dei polinomi in $y_1^{\tau_1}\ y_2^{\tau_2}\ y_3^{\tau_3}$ ortonormali in C, ogni stress della classe Γ_v - cioè , soluzione di (4) - è esprimibile mediante la serie

$$Y_{rs} = \sum_{i=1}^{\infty} \alpha_i^{(rs)} P_i \tag{6}$$

ove le

$$\alpha_i^{(rs)} = \int_C Y_{rs} P_i \, dC \tag{7}$$

sono opportune combinazioni lineari di soluzioni di (4) . In altri termini, al sistema (4) è sostituibile il sistema di sestuple (6) convergenti verso funzioni di quadrato sommabile .

2) Sull'inversione del teorema di Menabrea. Conseguense analitiche.
Ammetterò che lo stress minimante il funzionale che di volta in volta sarà considerato, nella classe Γ_v , sia congruente. La dimostrazione di tale proprietà è, sostanzialmente, analoga a quella ben nota valida nel caso che sulla frontiera siano ovunque assegnate le forze superficiali, caso che neppure considererò .

Si supponga, invece che su una parte $\bar{\sigma}$ della frontiera del corpo sia imposto un determinato spostamento, $\underline{\bar{u}}$, mentre sulla rimanente parte siano note le forze . Se $\underline{\bar{u}} = 0$ si è nel caso del vincolo di incastro. Occorre caratterizzare, in ogni caso che sarà considerato, la classe Γ'_v delle reazioni vincolari di cui il vincolo è capace. Come avviene nel caso dell'incastro, supporrò che adesso il vincolo possa esplicare qualunque sistema di reazioni vincolari ; la classe Γ'_v cioè , contiene ogni possibile sistema di vettori applicati in punti di $\bar{\sigma}$.

Il funzionale da associare alle (4) è

$$A(Y) = \int_C W(Y)dC - \int_{\bar{\sigma}} \Phi_r \bar{u}_r \, d\bar{\sigma} \quad . \tag{8}$$

$W(Y)$ è la forma quadratica definita positiva nelle Y_{rs} che esprime l'energia potenziale elastica e Φ_r una reazione vincolare. E' da tenersi presente che adesso le $b^{r(r)}_{\tau_1 \tau_2 \tau_3}$ vanno costruite in base a (5) ove le f_r si identifichino su $\bar{\sigma}$ con le reazioni Φ_r. Il minimo di $A(Y)$ va cercato nella classe Γ_V di tutti gli stress Y_{rs} soddisfacenti alle (4), al variare comunque di Φ_r subordinatamente alla sola condizione di essere in equilibrio con le forze esterne assegnate (tale condizione è, del resto, espressa dalle stesse (4) per $\tau_1=\tau_2=\tau_3=0$ e da talune di esse per $\tau_1+\tau_2+\tau_3=1$.

Non è difficile riconoscere che la condizione di minimo di si traduce in quella che risulti

$$(9) \qquad \int_C W(\xi_{rs})\,dC + \int_{\bar{\sigma}} q_r v_r \,d\bar{\sigma} \geq 0 \quad ,$$

ove è $v_r = u'_r - \bar{u}_r$, essendo u'_r lo spostamento indotto dalle Y_{rs} minimante e la (9) vale per ogni scelta delle ξ_{rs} simmetriche e delle q_r verificanti il sistema

$$(10) \qquad \tau_1\int_C \xi_{r1} y_1^{\tau_1-1} y_2^{\tau_2} y_3^{\tau_3} dC + \tau_2\int_C \xi_{r2} y_1^{\tau_1} y_2^{\tau_2-1} y_3^{\tau_3} dC + \tau_3\int_C \xi_{r3} y_1^{\tau_1} y_2^{\tau_2} y_3^{\tau_3-1} dC = C\, b^{(r)}_{\tau_1\tau_2\tau_3}$$

$$b^{(r)}_{\tau_1\tau_2\tau_3} = -\frac{1}{C}\int_{\bar{\sigma}} q_r\, y_1^{\tau_1} y_2^{\tau_2} y_3^{\tau_3}\, d\bar{\sigma} \;,$$

E' facile riconoscere che l'imposta condizione di minimo impone l'annullarsi dell'integrale

$$(11) \qquad \varphi = \int_{\bar{\sigma}} q_r v_r \,d\bar{\sigma} \quad .$$

Ciò significa che il vettore $\underline{v}$ è su $\bar{\sigma}$ ortogonale a ogni vettore $\underline{q}$ equilibrato su $\bar{\sigma}$. In una visione integrale della questione ciò è sufficiente, significando che lo spostamento $\underline{u}'$ subordinato dallo stress minimante, in media su $\bar{\bar{\sigma}}$ differisce da quello prescritto al più per uno spostamento rigido ma è interessante osservare che se, restringendo la classe ove si cerca la soluzione si ammette la derivabilità delle Y_{rs} e di $\underline{u}'$, le condizioni di vincolo sono verificate localtmente. Ciò, in particolare, si riconosce subito nel caso che l'assegnato vettore $\bar{\underline{u}}$ sia la traccia su $\bar{\sigma}$ di un vettore definito in C e su $\bar{\bar{\sigma}}$, derivabile in C.

Infatti, si supponga che le ξ_{rs}, q_r verifichino il sistema differenziale

$$
(12) \quad \begin{cases} \xi_{rs,s} = 0 \\ \xi_{rs} n_s = \begin{cases} q_r & (\text{su } \bar{\sigma}) \\ 0 & (\text{su } \sigma - \bar{\sigma}) . \end{cases} \end{cases}
$$

e $\underline{v}$ può considerarsi come la traccia su $\bar{\bar{\sigma}}$ di un vettore definito in C e su $\bar{\bar{\sigma}}$ e derivabile, al pari di $\underline{u}$ e $\underline{u}'$ Da (11), (12) segue allora

$$
(13) \quad \varphi = \int_\sigma \xi_{rs} n_s v_r d\sigma = -\frac{1}{2}\int_C \xi_{rs}(v_{r,s} + v_{s,r}) dC ,
$$

che data la larga arbitrarietà di ξ_{rs} implica l'annullarsi di $v_{r,s} + v_{s,r}$ e quindi la coincidenza di $\underline{v}$ con uno spostamento rigido.

E' interessante considerare il caso che su una parte, $\bar{\bar{\sigma}}$ di σ vi sia un appoggio liscio unilaterale che per semplicità supporrò su supporto rigido. In tal caso è da minimizzare il funzionale

$$
(14) \quad B(Y) = \int_C W(Y) dC ,
$$

nella classe delle soluzioni delle (4), (5) ove le f_r su $\bar{\sigma}$ si identifichino con la reazione $\Phi_r = \Phi n_r$. Il modulo Φ di tale reazione deve soddisfare unicamente alla condizione

$$\Phi \geq 0 \tag{15}$$

e a quella che esprime la condizione di equilibrio di tutte le forze esterne.

La condizione di minimo di $B(Y)$ si traduce nella relazione

$$\int_C W(\xi_{rs})\,dC + \int_{\bar{\sigma}_1} q_r u'_r\,d\bar{\sigma}_1 + \int_{\bar{\sigma}_2} q_r u'_r\,d\bar{\sigma}_2 \geq 0\,, \tag{16}$$

ove $\bar{\sigma}_1$ donota la parte di $\bar{\sigma}$ ove la reazione minimante è non nulla e $\bar{\sigma}_2 = \bar{\sigma} - \bar{\sigma}_1$ e $\underline{u}'$ denota come prima lo spostamento subordinato dallo stress minimante. La (16) è da ritenersi valida per ogni scelta delle ξ_{rs} simmetriche e di q_r soddisfacenti alle (10) , ove, però, il vettore $\underline{q}$ è parallelo a $\underline{n}$ su tutta $\bar{\sigma}$ e concorde ad essa sulla parte $\bar{\sigma}_2$ ove la reazione minimante risulta nulla. Si riconosce facilmente che la (16) implica che siano soddisfatte le relazioni

$$\int_{\bar{\sigma}_1} q\,n_r u'_r\,d\bar{\sigma}_1 = 0\,, \qquad \int_{\bar{\sigma}_2} q\,n_r u'_r\,d\bar{\sigma}_2 \geq 0\,, \tag{17}$$

ove q ha segno arbitrario su $\bar{\sigma}_1$ e soddisfa alla condizione $q \geq 0$ su $\bar{\sigma}_2$. Le (17) mostrano che lo spostamento subordinato dallo stress minimante è in media ortogonale a ogni vettore ortogonale alla normale alla superficie di appoggio (in equilibrio con le forze esterne) su $\bar{\sigma}_1$ ove non c'é distacco e, invece, forma, in media, angolo non ottuso con ogni vettore parallelo e concorde a quella normale che si è supposta orientata verso la parte consentita dal vincolo ove c'è distacco. Le condizioni di appoggio sono, cioè , soddisfatte in media.

Quando sopra si fonda sull'ipotesi che nello stato di equilibrio esista un'effettiva zona di appoggio, cioè che $\bar{\sigma}_1$ non svanisca. Inoltre, che comunque si divida la $\bar{\sigma}$ in due parti accada che un arbitrario sistema di reazioni parallele e concordi a $\underline{n}$ su una delle due parti possa essere sempre equilibrato da un sistema di reazioni agenti sull'altra parte, parallele a $\underline{n}$. Tali condizioni sono generalmente soddisfatte nei casi concreti, in particolare nel caso di un appoggio piano

Anche ora è interessante osservare che se ci si pone in una classe di funzioni dotate di proprietà di regolarità analoghe a quelle del caso dell'incastro si riesce a dimostrare che lo spostamento subordinato dallo stress minimante verifica in ogni punto di $\bar{\sigma}$ una delle due coppie di condizioni

$$(18)\qquad \begin{cases} u'_r n_r = 0 \\ \Phi > 0 \end{cases} \qquad\qquad \begin{cases} u'_r n_r \geq 0 \\ \Phi = 0 \end{cases},$$

com'è richiesto dal vincolo di appoggio unilaterale .

CENTRO INTERNAZIONALE MATEMATICO ESTIVO

(C. I. M. E.)

Walter NOLL

THE FOUNDATIONS OF MECHANICS

Corso tenuto a Bressanone dal 31 maggio al 9 giugno

1965

THE FOUNDATIONS OF MECHANICS

by

Walter Noll

1. Introduction. "Classical mechanics" is based upon the concepts of absolute space and absolute time. Even the natural philosophers of of the Renaissance realized, however, that space has physical relevance only with respect to a frame of reference. Only relative, not absolute, velocities can be physically detected. Absolute accelerations cannot be detected by purely kinematrical means, but they have a dynamical effect, namely inertia. It is to account for inertia that Newtonian and Eulerian mechanics employ absolute space. It was clearly recognized only recently that this is the only role that absolute space may play, and that the effect of the dependence of space on the frame of reference in all considerations unrelated to inertia must be taken into account by means of principles of frame-indifference.

Use of an absolute space is not the only way of accounting for inertia. It is possible, as Mach has proposed, to regard inertia as an interaction between the bodies of our immediate environment and the remainder of the universe. This is not unnatural, because it is a fact of experience that the inertial frames are just those in which the fixed stars and presumably the other masses of the universe appear more or less at rest. If inertia is thus treated on an equal footing with other physical interactions, it is no longer necessary to introduce an absolute space at all.

In these lectures I shall attempt to show how classical mechanics can be based on a space-time structure in which absolute space does not enter as an ingredient . If this "neo-classical" space-time structure is used, it is no longer necessary to impose principles of frame-indifference, because frames of reference are now introduced only a posteriori; they are not implicit in the basic concepts.

The use of "neo-classical" space-time is especially appropriate in those branches of mechanics in which inertia plays only a minor role, as is the case, for example, in various disciplines of visco-elasticity.

The natural languages reflect the fact that we humans live on this solid earth , which is always available as the frame of reference for our everyday experiences. It was perphaps natural, therefore, that science should have been led to regard space as an absolute. It was Copernicus, by proposing that the earth should be regarded as moving around the sun, who eventually succeeded in convincing the scientific world that the earth should be dethroned from its place as the absolute frame of reference. However, heavy reliance on natural language by the natural philosophers of the Renaissance prevented them from abolishing absolute space altogether. Use of the methods of modern axiomatic mathematics makes it easy to introduce conceptual systems that are free from the prejudices implied by the natural languages. In these lectures, I shall employ the axiomatic language that has become standard in most branches of pure mathematics during the past ten years.

2. Mathematical preliminaries.

A vector space $\mathcal{V}$ is a set of elements u, v, w, ..., which is endowed with a mathematical structure defined by a sum operation and a scalar multiple operation as follows.

(A1) With any two elements u, v $\in \mathcal{V}$ we can associate a sum u + v $\in \mathcal{V}$.

(A2) For any u, v $\in \mathcal{V}$ we have u+v = v+u.

(A3) For any u, v, w $\in \mathcal{V}$ we have (u + v) + w = u + (v + w).

(A4) There exist a zero-element 0 $\in \mathcal{V}$ with the property that u + 0 = u all u $\in \mathcal{V}$.

(A5) With every element u $\in \mathcal{V}$ we can associate an opposite -u $\in \mathcal{V}$ such that u + (-u) = 0.

(S1) With any u $\in \mathcal{V}$ and any $\xi \in \mathcal{R}$ ($\mathcal{R}$ = set of all real numbers) we can associate a scalar multiple ξ u $\in \mathcal{V}$.

(S2) For any u $\in \mathcal{V}$ and ξ, $\eta \in \mathcal{R}$ we have $\xi(\eta$ u) = (ξ η)u.

(S3) For any u $\in \mathcal{V}$ and ξ, $\eta \in \mathcal{R}$ we have (ξ + η)u= ξ u + ηu.

(S4) For any u, v $\in \mathcal{V}$ and $\xi \in \mathcal{R}$ we have ξ (u + v) = ξ u+ ξv .

(S5) For any u $\in \mathcal{V}$ we have

lu = u .

An inner product space $\mathcal{V}$ is a vector space endowed with additional structure defined by an inner product operation as follows :

(I1) With any two elements u, v $\in \mathcal{V}$ we can associate an an inner product u . v $\in \mathcal{V}$.

(I2) For any u, v $\in \mathcal{V}$ we have u . v = v . u.

(I3) For any $\mathring{u}$, v $\in \mathcal{V}$ and $\xi \in \mathcal{R}$ we have ξ(u . v) = (ξu) . v .

(I4) For any u, v, w $\in \mathcal{V}$ we have u . (v + w) = u . v + u . w.

(I5) For any u $\in \mathcal{V}$, u $\neq$ 0 we have u . u > 0 .

The following abbreviations are customary :

$$u - v = u + (-v), \quad u^2 = u \, . \, u, \quad |u| = \sqrt{u^2} \, .$$

The rules of elementary linear algebra are all consequences of the axioms (A1)-(I5) listed above, as is shown in the standard texbooks on the subject . A set of elements $u_1, \ldots, u_n$ is said to be <u>linearly independent</u> if no non-trivial linear combination of the elements is the zero-vector. The number of elements in a maximal linearly independent set can be shown to depend only on the space $\mathcal{V}$ and not on the set. This number dim $\mathcal{V}$ is called the <u>dimension</u> of $\mathcal{V}$.

We note that the axioms (A1)-(A5) , by themselves, give $\mathcal{V}$ the structure of a <u>commutative group</u>.

A <u>Euclidean point-space</u> is a set $\mathcal{E}$, whose elements x, y, ... will be called <u>points,</u> endowed with a structure defined by a <u>distance function</u> d which associates with each pair (x, y) of points a number d(x, y) $\in \mathcal{R}$. The function d will be assumed to satisfy the Euclidean axiom stated below.

The group $\mathcal{A}$ of <u>isometries</u> of d consists of all one-to-one mappings a of $\mathcal{E}$ onto itself which preserve distances, i.e., have the property that

$$d(a(x), a(y)) = d(x, y)$$

holds for every pair (x, y) of points. Of course, the group-product $a \circ b$ of elements in $\mathcal{A}$ is the composition of the mappings

$\mathcal{E}$ and $\mathcal{A}$.

It may happen that $\mathcal{A}$ contains a subgroup $\mathcal{V}$ having the properties (E1)-(E4):

(E1) $\mathcal{V}$ is commutative, i.e., $v \circ u = u \circ v$ when $u, v \in \mathcal{V}$.

(E2) $\mathcal{V}$ is transitive; i.e. for any two points $x, y \in \mathcal{E}$ there is a mapping $v \in \mathcal{V}$ such that $v(x) = y$.

(E3) If $v \in \mathcal{V}$ leaves one point $x_o \in \mathcal{E}$ unchanged, $v(x_o) = x_o$, then it leaves all points $x \in \mathcal{E}$ unchanged, i.e., it is the identity mapping.

(E4) $\mathcal{V}$ can be endowed with the structure of a inner product space such that (i) the vector addition coincides with the composition: $u \circ v = u + v$, and (ii)

$$d(x, y) = |u| \quad \text{if} \quad u(x) = y \tag{2.2}$$

We shall denotate the identity mapping of $\mathcal{E}$ onto $\mathcal{E}$ by o and use the notations

$$u + v = u \circ v, \quad u + x = x + u = u(x) \tag{2.3}$$

It follows from (E1)-(E3) that if $x, y \in \mathcal{E}$ is give, there is exactly one $u \in \mathcal{V}$ such that $u + x = y$. We denote this u by $u = y - x$, so that

$$u = y - x \quad \text{when} \quad y = u + x. \tag{2.4}$$

The rules suggested by the notations (2.3) and (2.4) are valid.

Uniqueness theorem. There is at most one inner product space $\mathcal{V}$ that satisfies the axioms (E1)-(E4).

Proof: Assume that $\mathcal{V}$ and $\hat{\mathcal{V}}$ are two inner product spaces that satisfy (E1)-(E4). We Write

$$
(2.5) \qquad \begin{aligned} u &= y - x \quad \text{when} \quad y = u + x, \quad u \in \mathcal{V} \\ \hat{u} &= y \,\hat{-}\, x \quad \text{when} \quad y = \hat{u} + x, \quad \hat{u} \in \hat{\mathcal{V}}, \end{aligned}
$$

noting that a careful distinction between the point-differences in $\mathcal{V}$ and $\hat{\mathcal{V}}$ is necessary. Choosing $q \in \mathcal{E}$ arbitrarily, we define a mapping f of $\mathcal{V}$ onto $\hat{\mathcal{V}}$ by the condition that

$$
(2.6) \qquad q + u = q \,\hat{+}\, f(u)
$$

It is clear that f is one-to-one and onto, and that

$$
(2.7) \qquad f(0) = 0.
$$

Let $u, v \in \mathcal{V}$ be arbitrary and put

$$
(2.8) \qquad x = q + u, \quad y = q + v, \quad y = x + (v - u)
$$

Putting $\hat{u} = f(u)$, $\hat{v} = f(v)$, it follows from (2.6) that

$$
(2.9) \qquad x = q + \hat{u}, \quad y = q + \hat{v}, \quad y = x + (\hat{v} - \hat{u}).
$$

Applying (2.2) to the cases when u is replaced by $v - u$ and $v - u$ we obtain

$$
(2.10) \qquad (v - u)^2 = (\hat{v} - \hat{u})^2 .
$$

In particular, when $u = \hat{u} = 0$, (2.10) shows that

$$
(2.11) \qquad v^2 = \hat{v}^2 = f(v)^2
$$

holds for all $v \in \mathcal{V}$. Hence, if we expand (2.10)

$$
(2.12) \qquad v^2 - 2v \cdot u + u^2 = v^2 - 2v \cdot u + u^2 ,
$$

the squares concel and we have

$$
(2.13) \qquad v \cdot u = \hat{v} \cdot \hat{u} = f(v) \cdot \hat{u}\, f(v) \cdot f(u) .
$$

Repeated use of (2.13) shows that

(2.14) $$[f(\xi_1 v_1 + \xi_2 v_2) - \xi_1 f(v_1) - \xi_2 f(v_2)] : \hat{u} = 0$$

holds for all $\hat{u} \in \hat{\mathcal{V}}$. Putting $\hat{u}$ equal to the term in brackets in (2.14) we infer from (15) that

(2.15) $$f(\xi_1 v_1 + \xi_2 v_2) = \xi_1 f(v_1) + \xi_2 f(v_2),$$

i.e., that f is a linear transformation of $\mathcal{V}$ onto $\hat{\mathcal{V}}$. Actually, (2.13) shows that f also preserves the inner product, and hence is an inner-product-space isomorphism between $\mathcal{V}$ and $\hat{\mathcal{V}}$. If we can show that f is the identity mapping it will follow that $\mathcal{V}$ and $\hat{\mathcal{V}}$ coincide as inner product spaces.

Let $x \in \mathcal{E}$ and v $\mathcal{V}$ be arbitrary. Put $u = x - q$ so that $x = q + u$. We then derive from (2.6) and (2.15)

(2.16) $$\begin{aligned} x + v &= (q+u) + v = q + (u+v) = q + f(u+v) \\ &= q + (f(u) + f(v)) = (q + f(u)) + f(v) \\ &= (q+u) + f(v) = x + f(v). \end{aligned}$$

Since (2.16) is valid for all $x \in \mathcal{E}$ it follows that $f(v) = v$. Since this is valid for all $v \in \mathcal{V}$ we conclude that f is indeed the identity mapping. Q.E.D.

We now lay dwn the <u>Euclidean axiom</u> : The distance function d is such that its group $\mathcal{A}$ of isometrics admits <u>at least one</u> inner product space $\mathcal{V}$ with the properties (E1)-(E4). The uniqueness theorem above shows that Euclidean point space admits <u>exactly one</u> inner product space $\mathcal{V}$ with the properties (E1)-(E4). This space $\mathcal{V}$ is called the <u>translation space</u> of ; its elements are called <u>spatial vectors.</u>

The isometrics of an Euclidean space $\mathcal{E}$ are characterized by the

Representation theorem : Let $q \in \mathcal{E}$ be chosen arbitrarily. Then every isometry $\mathcal{a}$ of $\mathcal{E}$ has a representation

$$(2.17) \qquad \mathcal{a}(x) = \mathcal{a}(q) + Q(x - q),$$

where Q is an orthogonal transformation of the translation space $\mathcal{V}$.

Proof : The mapping Q defined by

$$(2.18) \qquad Q(v) = \mathcal{a} \circ v \circ \mathcal{a}^{-1}$$

is an isomorphism of the subgroup $\mathcal{V}$ of $\mathcal{A}$ onto its conjugate $\mathcal{V}^{*} = \mathcal{a} \circ \mathcal{V} \circ \mathcal{a}^{-1}$. It is easily shown that $\mathcal{V}^{*}$ has the properties (E1)-(E4). The inner product space structure of $\mathcal{V}^{*}$ required for (E4) can be obtained from that of $\mathcal{V}$ by transport via Q; i.e., one defines $u^{*} \cdot v^{*} = u \cdot v$ when $Q(u) = u^{*}$, $Q(v) = v^{*}$, and $\xi u^{*} = (\xi u)^{*}$ when $Q(u) = u^{*}$, $Q(\xi u) = (\xi u)^{*}$.

The uniqueness theorem implies that $\mathcal{V}$ and $\mathcal{V}^{*}$ must coincide. Hence Q is an inner product space automorphism of $\mathcal{V}$ onto itself ; i.e., Q is an orthogonal transformation of $\mathcal{V}$.

Applying the mapping $\mathcal{a} \circ v = Q(v) \circ \mathcal{a}$ to the point $q \in \mathcal{E}$, we obtain

$$(2.19) \qquad \mathcal{a}(q + v) = \mathcal{a}(q) + Q(v) \ .$$

Substitution of $x - q = v$ into (2.19) yields the desired result (2.17). Q.E.D.

3. Neo-classical space-time.

The event-world of neo-classical space-time is a set $\mathcal{W}$, whose elements e, f, ... are called events, endowed with a mathematical structure defined by a time-lapse function τ and a distance function δ , subject to the axioms (T1)-(T4) , (D1)-(D3) , below.

(T1) With any pair (e, f) of events is associated a time-lapse τ (e, f) $\in \mathcal{R}$.

(T2) For any e, f $\in \mathcal{W}$

(3.1) $$\tau(e, f) = -\tau(f, e)$$

(T3) For any e, f, g $\in \mathcal{W}$

(3.2) $$\tau(e, f) + \tau(f, g) = \tau(e, g)$$

(T4) For any e $\in \mathcal{W}$ and any t $\in \mathcal{R}$ there is a f $\in \mathcal{W}$ such that $\tau(e, f) = t$.

We say that e is earlier or later than if $\tau(e, f) > 0$ or $\tau(e, f) < 0$, respectively. Two events e, f are said to be simultaneous if $\tau(e, f) = 0$. The set of all pairs of simultaneous events

(3.3) $$\mathcal{S} = \{ (e, f) \mid \tau(e, f) = 0 \}$$

is called the simultaneity-relation. It follows from (T2) that (e, e) $\in \mathcal{S}$ for any e $\in \mathcal{W}$ and that (e, f) $\in \mathcal{S}$ if (f, e) $\in \mathcal{S}$. From (T3) it follows that if (e, f) $\in \mathcal{S}$ and (f, g) $\in \mathcal{S}$, then also (e, g) $\in \mathcal{S}$. Therefore, the relation $\mathcal{S}$ is reflexive, symmetric, and transitive : it is an equivalence relation. Thus the simultaneity relation $\mathcal{S}$ determines a partition Γ of the world of events into classes $\mathcal{J}$ of simultaneous events such that

(3.4) $$\mathcal{S} = \bigcup_{\mathcal{J} \in \Gamma} \mathcal{J} \times \mathcal{J} .$$

The classes $\mathcal{J} \in \Gamma$ will be called instantaneous spaces or simply instants. If e $\in \mathcal{J}$, we say that the event e happens at the instant $\mathcal{J}$.

The value $\bar{\tau}$ (e, f) depends only on the instants $\mathcal{J}$ and $\mathcal{J}'$ at which e and f happen. It is possible, therefore , to define a time-lapse function τ between instants by

(3.5) $$\bar{\tau}\,(\mathcal{J}, \mathcal{J}') = \tau\,(e, t) \quad \text{if} \quad e \in \mathcal{J}, \quad f \in \mathcal{J}' .$$

Physically, results of time measurements with clocks are interpreted to be values τ (e, f) of the time-lapse function. The axioms (T1)-(T4) reflect familiar experiences with such measurements.

(D1) The distance function δ associates with each pair (e, f) $\in \mathcal{S}$ of simultaneous events a number δ (e, f) $\in \mathbb{R}$, called the <u>instantaneous distance</u> between e and f.

(D2) For each instant $\mathcal{J}$, the restriction $\delta_{\mathcal{J}}$ of δ to $\mathcal{J} \times \mathcal{J}$ gives $\mathcal{J}$ the structure of an Euclidean point space. The corresponding translation space will be denoted by $\mathcal{V}_{\mathcal{J}}$.

(D3) For all instants $\mathcal{J} \in \Gamma$, $\dim \mathcal{V}_{\mathcal{J}} = 3$.

Physically, the result of distance measurements with measuring sticks are interpreted to be values δ (e, f) of the distance function. The axiom (D1) reflects the fact that each distance measurement is made at a particular instant. The axioms (D2) and (D3) are the abstract of thousands of years of experience with distance measurements.

An <u>automorphism</u> α <u>of the event-world</u> $\mathcal{W}$ is a one-to-one mapping of $\mathcal{W}$ onto itself which preserves time-lapses and distances ; i.e ., an automorphism α satisfies

(3.6) $$\tau\,(\alpha\,(e),\ \alpha(f)\,) = \tau\,(e, f)$$

for all, $e, f \in \mathcal{W}$ and

(3.7) $$\delta(\alpha(e), \alpha(f)) = \delta(e, f)$$

for all $e, f \in \mathcal{W}$.

We write

(3.8) $$\mathcal{U}^{\alpha} = \{\alpha(e) \mid e \in \mathcal{U} \subset \mathcal{W}\}$$

for the set of all images under α of a subset $\mathcal{U}$ of $\mathcal{W}$.
Consider, in particular, instants $\mathcal{J}, \mathcal{J} \in \Gamma$. By (3.5) and (3.6) we have

(3.9) $$\bar{\tau}(\mathcal{J}^{\alpha}, \mathcal{J}^{\alpha}) = \bar{\tau}(\mathcal{J}, \mathcal{J}).$$

Adding $\bar{\tau}(\mathcal{J}, \mathcal{J}^{\alpha})$ to (3.9) and using (3.2) we obtain

(3.10) $$\bar{\tau}(\mathcal{J}, \mathcal{J}^{\alpha}) = \bar{\tau}(\mathcal{J}, \mathcal{J}^{\alpha}) = s_{\alpha},$$

which depends only on α and may be called the <u>time-shift</u> associated with the automorphism α.
It is easily seen that

(3.11) $$\mathcal{J}^{\alpha} = \{f \mid \tau(e, f) = s_{\alpha} \text{ for some } e \in \mathcal{J}\}.$$

The restriction of α to $\mathcal{J}$ is a Euclidean-space isomorphism of $\mathcal{J}$ onto $\mathcal{J}^{\alpha}$.

In general, we define for $\mathcal{J} \in \Gamma$

(3.12) $$\mathcal{J} + s = \{f \mid \tau(e, f) = s \text{ for some } e \in \mathcal{J}\}$$

It is easily seen that $\mathcal{J} + s \in \Gamma$ and that the mapping $\mathcal{J} \to \mathcal{J} + s$ is a mapping of Γ onto itself such that

(3.13) $$\bar{\tau}(\mathcal{J}, \mathcal{J}) = \bar{\tau}(\mathcal{J} + s, \mathcal{J} + s).$$

4. World lines, world motions, frames of reference.

A worldiline ℓ is a subset of $\mathcal{W}$ with the property that for each instant $\mathcal{J} \in \Gamma$,

$$\ell \cap \mathcal{J} = \{\ell(\mathcal{J})\} \tag{4.1}$$

is a singleton, i.e.e a set with only one element.

Let ℓ_1 and ℓ_2 be two world lines, so that

$$\ell_a \cap \mathcal{J} = \{\ell_a(\mathcal{J})\}, \quad a = 1, 2 \tag{4.2}$$

The distance between ℓ_1 and ℓ_2 at the instant $\mathcal{J}$ is defined by

$$d(\mathcal{J}) = \delta(\ell_1(\mathcal{J}), \ell_2(\mathcal{J})). \tag{4.3}$$

We say that ℓ_1 and ℓ_2 are C^n - related (C^∞-related, C^ω-related) if $d(\mathcal{J} + s)$ for some $\mathcal{J} \in \Gamma$ and hence all $\mathcal{J} \in \Gamma$ is n times continuously differentiable (infinitely many times differentiable, analytic) as a function of s.

A world motion of class $C^n (C^\infty, C^\omega)$ is a collection ψ of world lines any two of which are $C^n (C^\infty, C^\omega)$-related and which has the property that the set of events $\mathcal{D}_\psi(\mathcal{J})$ defined by

$$\mathcal{D}_\psi(\mathcal{J}) = \bigcup_{\ell \in \psi} \ell \cap \mathcal{J} = \{\ell(\mathcal{J}) \mid \ell \in \psi\} \tag{4.4}$$

is a closed subset of the instantaneous space $\mathcal{J}$, for each $\mathcal{J} \in \Gamma$.

Note : Two world-lines in a world motion need not be disjoint. The union of two world motions of class C^n need not be of class C^n.

A world motion is called rigid if for any two member world-lines ℓ_1 and ℓ_2, the distance (4.3) is independent of $\mathcal{J}$.

Of course , a rigid world motion is analytic.

A rigid world motion Φ is called a frame of reference if each event $e \in \mathcal{W}$ is a member of exactly one wordline $x \in \Phi$. On a frame of reference, we can define a distance function d by putting

$$d(x,y) = \delta(x(\mathcal{J}), y(\mathcal{J})) \tag{4.5}$$

It is easily see, be considering (D1)-(D3) , that the distance function d endows Φ with the structure of a three-dimensional Euclidean point-space, which is isomorphic to each of the instantaneous spaces. Indeed, the mapping from Φ into $\mathcal{J}$ given by

$$x \to x(\mathcal{J}), \tag{4.6}$$

where $x(\mathcal{J})$ is defined according to (4.1) , is a natural isomorphism from Φ onto $\mathcal{J}$. A frame of reference Φ also gives rise to a natural isomorphism between the translation spaces $\mathcal{V}_\Phi$ and $\mathcal{V}_\mathcal{J}$ of and $\mathcal{J}$. This isomorphism is given by

$$v \to v(\mathcal{J}) = (x+v)(\mathcal{J}) - x(\mathcal{J}). \tag{4.7}$$

Let Φ and Φ^* be two frames of reference, let $\mathcal{J} \in \Gamma$ and $s \in \mathcal{R}$. Assume now that an event $e \in \mathcal{J}$ is given . There is exactly one worldline $x \in \Phi$ such that $x(\mathcal{J}) = e$. Consider the event $x(\mathcal{J}+s) \in \mathcal{J}+s$. There is exactly one worldline $x^* \in \Phi^*$ such that

$$x^*(\mathcal{J}+s) = x(\mathcal{J}+s). \tag{4.8}$$

Let e^* be defined by $e^* = x^*(\mathcal{J}) \in \mathcal{J}$. The mapping $e \to e^*$ defined in this manner clearly is an isometry of the instantaneous space $\mathcal{J}$. Of course, this isometry depends on the choice of s. By the representation theorem of Sect . 2 we have a relation of the form

(4.9) $$e^* = f(s) + Q(s)(e - g),$$

where $f(s) \in \mathcal{J}$ and where $Q(s)$ is an orthogonal transformation of $\mathcal{V}_{\mathcal{J}}$. We have

(4.10) $$Q(0) = 1, \qquad f(0) = g,$$

because for $s = 0$ Eq. (4.9) must reduce to $e^* = e$.

In general, f(s) and Q(s) need have no continuity properties. However, the following result is easily proved:

<u>Theorem</u>: The union of the frames Φ and Φ^* is of class C^n if and only if, for some (and hence all) $\mathcal{J} \in \Gamma$, the functions f(s) and Q(s) in the representation (4.9) are of class C^n. If this is the case, (4.9) is called a <u>change of frame of class C^n</u>.

Assume that a wordline $\mathcal{L}$ and a frame of reference Φ are given. Define $\bar{x}(\mathcal{J})^{\in\Phi}_{\wedge}$ by

(4.11) $$\bar{x}(\mathcal{J})(\mathcal{J}) = \mathcal{L}(\mathcal{J}).$$

We say that $\mathcal{L}$ is <u>C^n relative to</u> Φ if $\bar{x}(\mathcal{J} + s)$ is of class C^n as a function of s. The <u>velocity of</u> $\mathcal{L}$ <u>at</u> $\mathcal{J}$ relative to Φ is defined by

(4.12) $$\bar{v}(\mathcal{J}) = \frac{d}{ds}\bar{x}(\mathcal{J} + s)_{s=0} \in \mathcal{V}_{\Phi}$$

The <u>acceleration of</u> $\mathcal{L}$ at $\mathcal{J}$ <u>relative to</u> is defined by

(4.13) $$\bar{a}(\mathcal{J}) = \frac{d}{ds}\bar{v}(\mathcal{J} + s)\Big|_{s=0} = \frac{d^2}{ds^2}\bar{x}(\mathcal{J} + s)\Big|_{s=0} \in \mathcal{V}_{\Phi}.$$

We also can define <u>instantaneous velocities and acceleration as</u> members of $\mathcal{V}_{\mathcal{J}}$, obtained from $\bar{v}(\mathcal{J})$ and $\bar{a}(\mathcal{J})$ via the natural isomorphism from $\mathcal{V}_{\Phi}$ onto $\mathcal{V}_{\mathcal{J}}$ given by (4.7).

A world-motion ψ is called <u>C^n-regular</u> if (i) it is of class C^n, (ii) no two worldlines in ψ intersect, and (iii) the set $\mathcal{D}_\psi(\mathcal{J}) \subset \mathcal{J}$ defined by (4.4) is a closed piecewise smooth region in $\mathcal{J}$.

Consider a C^n-regular world motion ψ and a frame of reference ϕ. We say that ϕ is <u>C^n-admissible for</u> ψ if $\phi \cup \psi$ is a world motion of class C^n. Any two C^n-admissible frames are related by a change of frame of class C^n.

Let ψ be a C^n-regular world motion and ϕ a C^n-admissible frame. We define

(4.14) $$\hat{\mathcal{D}}(\mathcal{J}) = \{x \mid x \in \phi,\ x(\mathcal{J}) = \ell(\mathcal{J}) \text{ for some } \ell \in \psi\}$$

The region $\hat{\mathcal{D}}(\mathcal{J})$ consist of all those worldlines of ϕ which intersect the worldilines of ψ at the instant $\mathcal{J}$. Also, $\hat{\mathcal{D}}(\mathcal{Y}) \subset \Phi$ is the inverse image of the region $\mathcal{D}_\psi(\mathcal{Y}) \subset \mathcal{Y}$ defined by (4.4) under the natural isomorphism (4.6). If $x \in \hat{\mathcal{D}}(\mathcal{Y})$ is given, in view of the condition (ii) for C^n-regular motions, there is exactly one $\ell \in \Psi$ such that

(4.15) $$x(\mathcal{J}) = \ell(\mathcal{J}).$$

Given any number $s \in \mathcal{R}$, we can then determine $y \in \phi$ uniquely by the condition

(4.16) $$y(\mathcal{J} + s) = \ell(\mathcal{J} + s).$$

We write the dependence of y on x, s, and $\mathcal{J}$ in the form

(4.17) $$y = \chi_{\mathcal{J}}(x, s).$$

Thus we have defined functions

(4.18) $$\chi_{\mathcal{J}} : \hat{\mathcal{D}}(\mathcal{J}) \times \mathcal{R} \to \phi.$$

It is clear that

(4.19) $$\chi_{\mathcal{Y}}(x, 0) = x .$$

The following theorem has a somewhat tedious proof, which will be omitted here.

Theorem : If ψ is a C^n-regular motion and ϕ a C^n-admissible frame for ψ, then the functions $\chi_{\mathcal{Y}}$ are of class C^n and, for each s, the mapping $x \to \chi_{\mathcal{Y}}(x, s)$ is a diffeomorphism of class C^n of the region $\hat{\mathcal{D}}(\mathcal{Y}) \subset \phi$ onto some other region in ϕ.

We call $\chi_{\mathcal{Y}}$ the relative deformation function at the instant of the world motion ψ with respect to the frame ϕ. With the help of the function $\chi_{\mathcal{Y}}$ it is possible to introduce most of the concepts of conventional kinematics. Examples are :

Velocity :

(4.20) $$v_{\mathcal{Y}}(x) = \frac{d}{ds} \chi_{\mathcal{Y}}(x, s) \Big|_{s=0}$$

Acceleration:

(4.21) $$a_{\mathcal{Y}}(x) = \frac{d^2}{ds^2} \chi_{\mathcal{Y}}(x, s) \Big|_{s=0}$$

Velocity gradient :

(4.22) $$L_{\mathcal{Y}}(x) = \nabla v_{\mathcal{Y}} \qquad (x, s)$$

Stretching ;

(4.23) $$D_{\mathcal{Y}}(x) = \frac{1}{2} [L_{\mathcal{Y}}(x) + L_{\mathcal{Y}}(x)^T]$$

Spin :

(4.24) $$W_{\mathcal{Y}}(x) = \frac{1}{2} [L_{\mathcal{Y}}(x) - L_{\mathcal{Y}}(x)^T]$$

It is possible to transport these vectors in $\mathcal{V}_{\phi}$ and tensors over $\mathcal{V}_{\phi}$ into vectors in $\mathcal{V}_{\mathcal{Y}}$ and tensors over $\mathcal{V}_{\mathcal{Y}}$ via the natural isomorphism of $\mathcal{V}_{\phi}$ onto $\mathcal{V}_{\mathcal{Y}}$ defined by (4.7). The resulting instantaneous velocity, acceleration, velocity gradient, and spin will depend on the choice of the frame ϕ . It turns out, however, that the instantaneous stretching is the same for all frames. The proof of this fact is analogous to the proof of frame-indifference for the stretching in conventional kimatics. This non-dependence on the frame suggest that there should be a direct definition for the instantaneous strecthing, a definition which makes no use of any frame.

Let a C^n-regular motion ψ be given and let $e, f \in \mathcal{D}_\psi(\mathcal{J})$. There exists exactly two world lines ℓ, $m \in \psi$ such that

(4.25) $$\ell(\mathcal{J}) = e, \quad m(\mathcal{J}) = f.$$

We define a function $\delta : \mathcal{D}_\psi(\mathcal{J}) \times \mathcal{D}_\psi(\mathcal{J}) \times \mathcal{R} \to \mathcal{R}$ by

(4.26) $$\delta(e, f, s) = \delta(\ell(\mathcal{J} + s), m(\mathcal{J} + s)).$$

It is easily seen that δ is of class C^n.

Put

(4.27) $$\delta_k(e, f) = \frac{d^k}{ds^k} [\delta(e, f, s)]^2 \Big|_{s=0}, \quad k \leq n - 2$$

and define

(4.28) $$A_k(e) = \nabla_e \nabla_f \delta_k(e, f) \Big|_{f=e}.$$

$A_k(e)$ is a symmetric second-order tensor over the instantaneous translation space $\mathcal{V}_\mathcal{J}$. We call $A_k(e)$ the <u>k-th instantaneous Rivlin-Ericksen tensor</u> associated with the motion. It can be shown that $A_1(e)$ is twice the instantaneous stretching and hence provides for a direct definition of this stretching.

5. Bodies

We now introduce a new primitive concept, that of a material universe Ω, consisting of bodies A, B, C, ... The universe Ω is endowed with a structure defined by a relation $<$, subject to the axioms (B1)-(B6), below.

(B1) | Certain pairs of bodies (A, B) are related by $<$; we write $A < B$ if this is the case and say that A is a part of B.

(B2) | $A = B$ if and only if both $A < B$ and $B < A$.

(B3) | If $A < B$ and $B < C$, then $A < C$.

The axioms (B1)-(B3) state that $<$ is a partial ordering of Ω.

If $B < A$ and $C < A$ we call A an envelope of $\{B, C\}$ and write $\{B, C\} < A$. If $\{B, C\} < D$ and if $\{B, C\} < A$ implies $D < A$ we say that D is the least envelope of $\{B, C\}$ and we write $D = B \vee C$, so that

$$\{B, C\} < A \text{ implies } \{B, C\} < B \vee C < A. \tag{5.1}$$

A least envelope may or may not exist, but if it does, it is uniquely determined by B and C. If $A < B$ and $A < C$ we say that A is a common part of $\{B, C\}$ and write $A < \{B, C\}$. If $D < \{B, C\}$ and if $A < \{B, C\}$ implies $A < D$ we say that D is the greatest common part of $\{B, C\}$ and write $D = B \wedge C$, so that

$$A < \{B, C\} \text{ implies } A < B \quad C < \{B, C\}. \tag{5.2}$$

In there is a greatest common part it is unique. Envelopes, the greatest envelope, common parts, and the greatest common part of an arbitrary collection of bodies are defined in a similar manner. Let $\{A_i | i \in I\}$ be such a oollection, its memers A_i marked with indices i taken from an index set I . If the least evelope of the collection exists, we denote it by $\bigvee_{i \in I} A_i$; if the greatest common part exists, we denote it by $\bigwedge_{i \in I} A_i$.

Two bodies A and B are said to be <u>separate</u> if they have no common part . It will be convenient to adjoin to the material universe Ω two <u>improper bodies</u>, the null-body 0 and the universal body ∞ . We extend the relation $<$ to the extended universe

(5.3) $$\Omega' = \Omega \cup \{0, \infty\}$$

by putting

(5.4) $$0 < A < \infty \quad \text{for all} \quad A \in \Omega'.$$

Of course the realtion $<$, when extended by (5.4) , remains a partial ordering . We have

(5.5) $$A \wedge B = 0, \quad A, B \in \Omega$$

if and only if A and B are separate.

The following rules are easily established :

(1) $A < B$ if and only if $A \wedge B = A$ or $A \vee B = B$.

(2) If $A < B$ and if $A \wedge C, B \wedge C$ exist, then

(5.6) $$A \wedge C < B \wedge C .$$

(3) If $A < B$ and if $A \vee C$, $B \vee C$ exist, then

(5.7) $$A \vee C < B \vee C .$$

(4) If $A < B$ and if $B \wedge C = 0$, then $A \wedge C = 0$.

(5) If $A \wedge B$ and $B \wedge C$ exist, then

(5.8) $$(A \wedge B) \wedge C = A \wedge (B \wedge C) = A \wedge B \wedge C ,$$

provided either $(A \wedge B) \wedge C$ or $A \wedge (B \wedge C)$ exist.

It is possible that $A \wedge B \wedge C$ exists but $A \wedge B$ or $B \wedge C$ do not.

We are now ready to state the next axiom :

(B4) For each body $A \in \Omega$ there is exactly one body $A^e \in \Omega$ with the property that $\{A, A^e\}$ have neither a common part nor an envelope ; i.e.,

(5.9) $$A \wedge A^e = 0, \quad A \vee A^e = \infty$$

The body A^e is called <u>the exterior</u> of the body A. If we put $0^e = \infty$, $\infty^e = 0$, then (5.9) is valid for all $A \in \Omega'$. It follows directly from the definition of A^e that

(5.10) $$(A^e)^e = A$$

holds for every $A \in \Omega'$. Also,

(5.11) $$A \prec B \quad \text{implies} \quad A \wedge B^e = 0,$$

which is a consequence of $(5.9)_1$ and rule (4). The next axiom postulates that the c nverse of (5.11) is valid :

(B5) If A and the exterior of B have no common part, then A is a part of B.

It follows from (BC) and rule (5.11) that

(5.12) $$A \prec B \text{ if and only if } A \wedge B^e = 0 .$$

By (5.12) and (5.10) we have $A \prec B$ if and only if $B^e \wedge (A^e)^e = 0$. Hence , using (5.12) again, we find that

(5.13) $$A \prec B \quad \text{if and only if} \quad B^e \prec A^e .$$

The following propositions are corollaries of (5.13) : Let $\{A_i \mid i \in I\}$ be a collection of bodies. If $\bigvee_{i \in I} A_i$ exists, so does $\bigwedge_{i \in I} (A_i^e)$ and

(5.14) $$\bigwedge_{i \in I} (A_i^e) = (\bigvee_{i \in I} A_i)^e .$$

If $\bigwedge_{i \in I} A_i$ exists, so does $\bigvee_{i \in I} (A_i^e)$ and

(5.15) $$\bigvee_{i \in I} (A_i^e) = (\bigwedge_{i \in I} A_i)^e$$

The following result is basic to the theory of bodies :

Lemma: Assume that A_1, A_2, and B are bodies and that $A_1 \wedge B$, $A_2 \wedge B$, and $A_1 \vee A_2$ exist. Then we have

(5.16) $$\{A_1 \wedge B, \ A_2 \wedge B\} < \{B, A_1 \vee A_2\};$$

furthermore, if

(5.17) $$\{A_1 \wedge B, A_2 \wedge B\} < C$$

and

(5.18) $$D < \{B, A_1 \vee A_2\}$$

(5.19) then $$D < C$$

<u>Proof</u>: It follows from the definitions (5.1) and (5.2) that $A_i \wedge C < C$, and $A_i \wedge C < A_i < A_1 \vee A_2$, $i = 1, 2$, which immediately gives (5.16).

Assume now that (5.17) and (5.18) hold. It follows from (5.17) and (5.11) that

(5.20) $$\{A_i \wedge B\} \wedge C^e = 0, i = 1, 2 .$$

Suppose that E is a common part of $\{D, C^e\}$, i.e.

(5.21) $$E < \{D, C^e\}.$$

By (5.18) we have $E < D < B$, and hence

(5.22) $$E < \{B, C^e\} .$$

Now, if G_i is a common part of $\{E, A_i\}$,

(5.23) $$G_i < \{E, A_i\}, . .$$

then $G_i < E < B$ by (5.22) and hence $G_i < \{B, A_i\}$, which is equivalent to

(5.24) $$G_i < A_i \wedge B, i=1,2 .$$

Using rule (4) we infer from (5.24) and (5.20) that $G_i \wedge C^e = 0$. But by (5.23) and (5.22) we also have $G_i < E < C^e$ and hence, by rule (4) , $G_i = G_i \wedge G_i < G_i \wedge C^e = 0$, i.e. $G_i = 0$. We have shown that (5.23) implies $G_i = 0$, which means that

(5.25) $$E \wedge A_i = 0, \quad i = 1, 2 .$$

Applying (5.10) and (5.12) , we infer from (5.25) that $A_i < E^e$, $i = 1, 2$, which is equivalent to $A_1 \vee A_2 < E^e$, or by (5.13) , to

(5.26) $$E < (A_1 \vee A_2)^e .$$

On the other hand , it follows from (5.21) and (5.18) that

(5.27) $$E < D < A_1 \vee A_2 .$$

The relations (5.26) and) (5.27) can both hold only if $E = 0$. We have shown that (5.21) implies $E = 0$, which means that $D \wedge C^e = 0$. But by (5.12) this is equivalent to (5.19) . Q.E.D.

The following four propositions are easy consequences of the Lemma :

(A) If $A_1 \wedge B$, $A_2 \wedge B$, and $A_1 \vee A_2$ exist, then

(5.28) $$(A_1 \vee A_2) \wedge B = (A_1 \wedge B) \vee (A_2 \wedge B) ,$$

provided either the left hand side or the right hand side exists.

(B) If $A_1 \vee B$, $A_2 \vee B$ and $A_1 \wedge A_2$ exist, then

(5.29) $$(A_1 \wedge A_2) \vee B = (A_1 \vee B) \wedge (A_2 \vee B),$$

provided either the left hand side or the right hand side exists.

(C) If $A \wedge B$ and $A^e \wedge B$ exist, then

(5.30) $$B = (A \wedge B) \vee (A^e \wedge B) .$$

(D) We have

(5.31) $$A = B \vee C, \quad B \wedge C = 0$$

if and only if

(5.32) $$B < A, \quad C = A \wedge B^e$$

Proposition (A) is an immediate corollary of the Lemma ; (B) results from applying (A) to A_1^e , A_2^e , and B^e and using (5.14) and (5.15) ; (C) corresponds to the special case $A_1 = A$, $A_2 = A^e$ of (A) . To prove (D) , assume that (5.31) holds. Then $B < A$ follows from $(5.31)_1$. Using (5.12) and (5.10) we infer from $(5.31)_2$ that $C < B^e$, i.e. that $C \wedge B^e = C$. Using $(5.31)_1$ and (A) we obtain

$$A \wedge B^e = (B \vee C) \wedge B^e = (B \wedge B^e) \vee (C \wedge B^e) = 0 \vee C = C,$$

which proves (5.32) . Assume, conversely , that (5.32) holds. We then have $A \wedge B = B$ and hence, by (C) ,

$$A = (B \wedge A) \vee (B^e \wedge A) = B \vee C.$$

Also , $0 = 0 \wedge A = (B \wedge B^e) \wedge A = B \wedge (B^e \wedge A) = B \wedge C$, which proves (5.31) .

We now state the final asiom :

(B6) If $A, B \in \Omega$ are not separate , then the greatest common part $A \wedge B \in \Omega$ exists.

It follows from (B6) that in the extended universe (5.3) , $A \wedge B$ exists for all $A, B \in \Omega'$. By (5.15), the least envelope $A \vee B = (A^e \wedge B^e)^e \in \Omega'$ exists always also. Equations (5.28) and (5.29) state that the operations $\wedge$ and $\vee$ are distributive with respect to one another. This fact, together with axiom (B4), can be expressed by saying that Ω' is a <u>complemented distributive lattice.</u>

Suppose that a body B is given. A finite collection $\mathcal{P} = \{P_i \mid i = 1, \ldots, n\}$ is called a <u>partition</u> of B if

$$B = \bigvee_{i=1}^{n} P_i \quad \text{and} \quad P_i \wedge P_j = 0 \text{ if } i \neq j . \tag{5.33}$$

We say that the partition $\mathcal{P}' = \{P_i \mid i = 1, \ldots, n'\}$ is a <u>refinement</u> of $\mathcal{P}$ if for each $P_i' \in \mathcal{P}'$ there is a $P_j \in \mathcal{P}$ such that $P_i' < P_j$.

<u>Theorem</u> : Any two partitions $\mathcal{P} = \{P_i \mid i = 1, \ldots, n\}$ $\mathcal{Q} = \{Q_j \mid j = 1, \ldots, m\}$ have a common refinement ; i.e., there is a partition which is a refinement of both $\mathcal{P}$ and $\mathcal{Q}$.

<u>Proof</u> : The required refinement consists of the nm parts $P_i \wedge Q_j$, $i = 1, \ldots, n$, $j = 1, \ldots, m$. The fact that these parts do indeed form a refinement of both $\mathcal{P}$ and $\mathcal{Q}$ follows easily from the distributive laws (5.28), (5.29), and from $P_i \wedge Q_j < \{P_i, Q_j\}$. Q.E.D.

The axioms (B1)-(B6) and the theorems derived from them reflect our common sense experiences with physical bodies. It is necessary here to make the mathematical description of bodies independent of any imbedding in "space", simply because there is no such thing as "space" in our developement.

The following purely mathematical special examples illustrate the concept of a material universe :

(α) Let Ω' consist of all subsets of an arbitrary set X and let $<$ represent set inclusion. We have $A \wedge B = A \cap B$ (intersection), $A \vee B = A \cup B$ (union) and $A^e = A^c$ (complement). The Newtonian mechanics of particle systems can be based on such a universe when X is a finite set.

(β) Let Ω' consist of all closures of open sets in an arbitrary topological space and let $<$ represent set inclusion. In this case the greatest common part of two bodies A, B is not always their intersection, but rather $A \wedge B = \overline{A^{\circ} \cap B^{\circ}}$ (a superimposed ° denotes the interior and a superimposed bar denotes the closure). The least envelope of a collection $\{A_i \mid i \in I\}$ is given by $\bigvee_{i \in I} A_i = \overline{\bigcup_{i \in I} A_i^{\circ}}$, i.e. by the closure of the union of the interiors of the sets in the collection. This least envelope is equal to the union $\bigcup_{i \in I} A_i$ if the collection is finite, but not necessarily if it is infinite. The exterior of a body is $A^e = \overline{A^c}$, i.e. the closure of the complement.

(γ) Let Ω consist of all finite unions of closed polyhedra and closed exteriors of polyhedra in a Euclidean space. Ω satisfies the axioms of a material universe when $<$ represents set inclusion. Greatest common parts, least envelopes, and exteriors are given by the same formulas as in example (β).

(δ) Let Ω consist of all closed regions with piecewise smooth boundaries, in a Euclidean space. If $<$ represents set inclusion, then the axioms (B1)-(B5) are satisfied, but (B6) is not. Since this example is of importance in conventional continuum mechanics, it may be desirable to develop the mechanics of material universes without postulating (B6). However, I have not yet been able to do so.

6. Systems of forces.

We assume that an event world $\mathcal{W}$ and a material universe Ω are given. A force system f is then defined by the following conditions :

(F1) With each instant $\mathcal{J} \in \Gamma$ and each pair (B, A) of separate bodies in Ω is associated an instantaneous vector $\underline{f}_{\mathcal{J}}(B, A) \in \mathcal{V}_{\mathcal{J}}$, called the force exerted by A on B at the instant $\mathcal{J}$.

(F2) If A, B_1, B_2 are mutually separate then

$$(6.1) \qquad \underline{f}_{\mathcal{J}}(B_1 \vee B_2, A) = \underline{f}_{\mathcal{J}}(B_1, A) + \underline{f}_{\mathcal{J}}(B_2, A)$$

holds for all instants $\mathcal{J} \in I$.

(F3) If A_1, A_2, B are mutually separate then

$$(6.2) \qquad \underline{f}_{\mathcal{J}}(B, A_1 \vee A_2) = \underline{f}_{\mathcal{J}}(B, A_1) + \underline{f}_{\mathcal{J}}(B, A_2)$$

holds for all $\mathcal{J} \in \Gamma$.

(F4) With each instant $\mathcal{J} \in \Gamma$ we can associate a positive number $\varkappa$, such that

$$(6.3) \qquad \sum_{i=1}^{n} | \underline{f}_{\mathcal{J}}(P_i, A) | \le \varkappa$$

holds for every pair (B, A) of separate bodies and every partition $\{P_i \mid i = 1, \ldots, n\}$ of B.

If the material universe Ω is a class of sets, as it is in the examples given in the previous section, then the conditions (F1)-(F4) may roughly be described as follows ; $\underline{f}_{\mathcal{J}}$ is a vector-valued function of two set variables, finitely additive in each variable, and of bounded variation in the first variable.

Since every body $B \in \Omega$ is separate from its exterior $B^e \in \Omega$, it is meaningful to consider $\underline{f}_{\mathcal{J}}(B, B^e)$, which is called the resultant force exerted on B at the instant $\mathcal{J}$. We say that the force system $\underline{f}$ is balanced at the instant $\mathcal{J}$ if this resultant force vanishes for all bodies, i.e. if

(6.4) $$\underline{f}_{\mathcal{J}}(B, B^e) = \underline{0} \in \mathcal{V}_{\mathcal{J}}$$

holds for all $B \in \Omega$.

<u>Theorem</u> : If the force system $\underline{f}$ is balanced at the instant $\mathcal{J}$ then

(6.5) $$\underline{f}_{\mathcal{J}}(B, A) = -\underline{f}_{\mathcal{J}}(A, B)$$

holds for all pairs (A, B) of separate bodies.

<u>Proof</u>: If $A = B^e$, then both sides of (6.5) vanish and hence are equal. If $A \neq B^e$, then $A \vee B = C \in \Omega$ exists. Of course, we have $A \wedge B = 0$ because A and B are separate. It follows from proposition (D) of Section 5 that

$$B = C \wedge A^e, A = C \wedge B^e$$

and hence, by (5.15),

(6.6) $$B^e = C^e \vee A, \quad A^e = C^e \vee B.$$

We have

$$C^e \wedge A = (B \vee A)^e \wedge A = B^e \wedge (A^e \wedge A) = B^e \wedge 0 = 0,$$

which means that C^e and A are separate. Similarly, it follows that C^e and B are separate. Therefore, in view of (6.6), the condition (F3) gives

(6.7) $$\begin{aligned} \underline{f}_{\mathcal{J}}(B, B^e) &= \underline{f}_{\mathcal{J}}(B, C^e \vee A) = \underline{f}_{\mathcal{J}}(B, C^e) + \underline{f}_{\mathcal{J}}(B, A), \\ \underline{f}_{\mathcal{J}}(A, A^e) &= \underline{f}_{\mathcal{J}}(A, C^e \vee B) = \underline{f}_{\mathcal{J}}(A, C^e) + \underline{f}_{\mathcal{J}}(A, B). \end{aligned}$$

Since $C = A \vee B$ and $A \wedge B = 0$, condition (F2) yields

(6.8) $$\underline{f}_{\mathcal{J}}(B, C^e) = + \underline{f}_{\mathcal{J}}(A, C^e) = \underline{f}_{\mathcal{J}}(C, C^e).$$

Adding the two equations (6.7) and observing (6.8) we obtain

(6.9) $$\underline{f}_{\mathcal{J}}(B, B^e) + \underline{f}_{\mathcal{J}}(A, A^e) - \underline{f}_{\mathcal{J}}(C, C^e) = \underline{f}_{\mathcal{J}}(B, A) + \underline{f}_{\mathcal{J}}(A, B).$$

Now, if the force system is balanced, the left side of (6.9) vanishes and hence (6.5) must hold. Q.E.D.

In words, the conclusion (6.5) of the theorem states that the force exrted by B on A is opposite to the force exerted by A on B. This statement may be called the law of action and reaction. Therefore, when the force system is balanced, then the law of action and reaction holds.

Consider a pair (B, A) of separate bodies, Let $\Omega_B =$ $\{P \mid P \in \Omega, P \prec B\}$ be the set of all parts P of B. We wish to define integration of real valued functions

(6.10) $$\phi : \Omega_B \to \mathcal{R}$$

with respect to a force system.

Consider first the special case when there is a partition $\{P_i \mid i = 1, \ldots, n\}$ of B and numbers $(\phi_1, \ldots, \phi_n)$ such that

(6.11) $$\phi(P) = \max\{\phi_i \mid P_i \wedge P \neq 0\}.$$

In particular, we have

(6.12) $$\phi(P) = \phi_i \quad \text{when} \quad P \prec P_i$$

We define the integral of ϕ with respect to the forces exerted by A by

(6.13) $$\int_B \phi d\underline{f} (A) = \sum_{i=1}^{n} \phi_i \underline{f}_{\mathcal{J}} (P_i, A) .$$

Consider now an arbitrary function of the type (6.10). With each partition $\mathcal{R} = \{P_i \mid i = 1, \ldots, n\}$ of B, we associate the sum

(6.14) $$\underline{u}_{\mathcal{R}} = \sum_{i=1}^{n} (P_i)\underline{f}_{\mathcal{J}} (P_i, A) \in \mathcal{V}_{\mathcal{J}}$$

We say that ϕ is integrable with respect to the forces exerted by

A if there is an instantaneous vector $\underline{u}$ $\mathcal{V}_{\mathcal{J}}$ with the following property: For every $\epsilon > 0$ we can find a partition $\mathcal{R}$ such that

(6.15) $$| \underline{u}_{\mathcal{R}} - \underline{u} | < \epsilon$$

whenever $\mathcal{R}'$ is a refinement of $\mathcal{R}$. If it exists, the vector $\underline{u} \in \mathcal{V}_{\mathcal{J}}$ is denoted by

(6.16) $$\underline{u} = \int_B \phi d\underline{f}_{\mathcal{J}}(A) \in \mathcal{V}_{\mathcal{J}}$$

and is called the <u>integral of ϕ with respect to the forces exerted by</u> A.

The integral thus defined is a generalized version of the familiar Riemann-Stieltjès integral.

Let

(6.17) $$\underline{w} : \Omega_B \rightarrow \mathcal{V}_{\mathcal{J}}$$

be a function whose values are instantaneous vectors. If $\underline{a} \in \mathcal{V}_{\mathcal{J}}$, then the function $\phi = \underline{a} \cdot \underline{w}$ defined by $\phi(P) = a \cdot (\underline{w}(P))$ for $P \prec B$ is of the type (6.10). Assuming that the integral $\int_B (\underline{a} \cdot \underline{w}) d\underline{f}_{\mathcal{J}}(A)$ exists for each $\underline{a} \in \mathcal{V}_{\mathcal{J}}$ it is easily seen to depend linearly on $\underline{a}$. Hence there is a linear transformation on $\mathcal{V}_{\mathcal{J}}$ whose value for $\underline{a}$ is this integral. The transpose of this linear transformation will be denoted by

(6.18) $$\int_B \underline{w} \otimes d\underline{f}_{\mathcal{J}}(A),$$

so that

(6.19) $$[\int_B \underline{w} \otimes d\underline{f}_{\mathcal{J}}(A)]^T \underline{a} = \int_B (\underline{a} \cdot \underline{w}) d\underline{f}_{\mathcal{J}}(A)$$

holds for all $\underline{a} \in \mathcal{V}_{\mathcal{J}}$. The trace of the transformation (6.18) will be denoted by

$$\text{tr} \left[\int_B \underline{w} \otimes d\underline{f}_{\mathcal{J}}(A) \right] = \int_B \underline{w} \cdot d\underline{f}_{\mathcal{J}}(A). \tag{6.20}$$

7. Dynamical processes.

As before, we assume that an event world $\mathcal{W}$ and a material universe Ω are given. A kinematical process π of a class C^n (C^∞, C^ω) is defined by the following conditions:

(K1) With each body $B \in \Omega$ is associated a non-empty world motion $\pi(B)$ of class $C^n(C^\infty, C^\omega)$.

(K2) π is monotone in the sense that $A \prec B$ implies $\pi(A) \subset \subset \pi(B)$. (7.1)

(K3) Given any event $e \in \mathcal{D}_{\pi(B)}(\mathcal{J})$ and any neighborhood $\mathcal{N}$ of e in the instantaneous space $\mathcal{J}$, there is a part $P \prec B$ such that

$$e \in \mathcal{D}_{\pi(P)}(\mathcal{J}) \subset \mathcal{N}. \tag{7.2}$$

Roughly, (K2) states that a body has always occupies more "space" than any of its parts, and (K3) states that a body has arbitrarily "small" parts every where.

The following three propositions are immediate consequences of (K2): (i) If A and B are not separate, then

$$\pi(A \wedge B) \subset \pi(A) \cap \pi(B). \tag{7.3}$$

(ii) If

$$\pi(A) \cap \pi(B) = \emptyset \tag{7.4}$$

then A and B are separate. (iii) If $A \neq B^e$ then

$$\pi(A) \cup \pi(B) \subset \pi(A \vee B). \tag{7.5}$$

Theorem: If $A \neq B^e$ then for all instants

$$\mathcal{D}_{\pi(A)}(\mathcal{J}) \cup \mathcal{D}_{\pi(B)}(\mathcal{J}) = \mathcal{D}_{\pi(A \vee B)}(\mathcal{J}) \tag{7.6}$$

Proof: It is clear from (7.5) that

$$(7.7)\qquad \mathcal{D}_{\pi(A)}(\mathcal{J}) \cup \mathcal{D}_{\pi(B)}(\mathcal{J}) \subset \mathcal{D}_{\pi(A \vee B)}(\mathcal{J}).$$

Suppose now that there is an event e such that

$$(7.8)\qquad e \in \mathcal{D}_{\pi(A \vee B)}(\mathcal{J})$$

but

$$(7.9)\qquad e \notin \mathcal{D}_{\pi(A)}(\mathcal{J}) .$$

Since $\mathcal{D}_{\pi(A)}(\mathcal{J})$ is a closed subset of the instantaneous space $\mathcal{J}$ it follows from (7.9) that there is a neighborhood $\mathcal{N}$ of e such that

$$(7.10)\qquad \mathcal{N} \cap \mathcal{D}_{\pi(A)}(\mathcal{J}) = \emptyset$$

By (K3) and (7.8) we can find a part

$$(7.11)\qquad P < A \vee B$$

such that

$$(7.12)\qquad e \in \mathcal{D}_{\pi(P)}(\mathcal{J}) \subset \mathcal{N} .$$

Taking the intersection of (7.12) with $\mathcal{D}_{\pi(a)}(\mathcal{J})$, we infer from (7.10) that $\mathcal{D}_{\pi(P)}(\mathcal{J}) \cap \mathcal{D}_{\pi(A)}(\mathcal{J}) = \emptyset$ and hence that

$$\pi(P) \cap \pi(A) = \emptyset .$$

It follows that p and A must be separate : $P < A^e$. This result and (7.11) show that $P < B$. Therefore, using (K2) and (7.12) we obtain

$$(7.13)\qquad e \in \mathcal{D}_{\pi(P)}(\mathcal{J}) \subset \mathcal{D}_{\pi(B)}(\mathcal{J}) .$$

We conclude that if (7.8) holds, then also

$$e \in \mathcal{D}_{\pi(A)}(\mathcal{J}) \cup \mathcal{D}_{\pi(B)}(\mathcal{J}) ,$$

which proves (7.7). Q.E.D.

It is not legitimate to replace $\bigcup$ by $\bigcap$ and $\vee$ by $\wedge$ in (7.6). In fact it is possible that A and B are separate and yet

$$\mathcal{P}_{\pi(A)}(\mathcal{J}) = \mathcal{P}_{\pi(B)}(\mathcal{J}). \tag{7.14}$$

If (7.14) holds we say that A and B are coincident at the instant $\mathcal{J}$. If there are parts $P < A$ and $Q < B$ such that P and Q are coincident we say that A and B interpenetrate at the instant $\mathcal{J}$. If there are no such parts and yet

$$\mathcal{P}_{\pi(A)}(\mathcal{J}) \cap \mathcal{P}_{\pi(B)}(\mathcal{J}) \neq \emptyset \tag{7.15}$$

we say that A and B are in contact at the instant $\mathcal{J}$.

We assume now that not only a kinematical process π but also a force system $\underline{f}$ is given. Let $\mathcal{J}$ be some instant and suppose that $\underline{w}$ is a function which associates with each worldline ℓ an instantaneous vector $\underline{w}(\ell) \in \mathcal{V}_{\mathcal{J}}^{\ell}$. Consider a body $B \in \Omega$. For each part $P < B$ we select a worldline $\ell \in \pi(P)$. Having done so, we can consider the function which associates with $P \in \Omega_B$ the instantaneous vector $\underline{w}(\ell) \in \mathcal{V}_{\mathcal{J}}^{\ell}$. This is a function of the type (6.17) and hence may be integrable with respect to the forces exerted by some body A separate from B. If this is the case no matter how the $\ell \in \pi(P)$ are selected, and if the resulting integral is independednt of this selection, we say that $\underline{w}$ is integrable for the process π with respect to the forces exerted by A and we denote its integral by (6.18), even though $\underline{w}$ now has a meaning different from that of Sect. 6.

For example, suppose that ϕ is a frame of reference and that every worldline $\ell \in \pi(B)$ is of class C^1 relative to ϕ (cf. Sect. 4).

Then we can associate with each $\mathcal{L} \in \pi(B)$ not only the velocity (4.12) but also the corresponding instantaneous velocity

$$(7.16) \qquad \underline{v}_{\mathcal{J}}(\mathcal{L}) = \frac{d}{ds}\left[\bar{x}(\mathcal{J}+s)(\mathcal{J})\right]_{s=0} \in \mathcal{V}_{\mathcal{J}}.$$

The integral

$$(7.17) \qquad \pi_{\mathcal{J}}(B) = \int_B \underline{v}_{\mathcal{J}} \cdot d\underline{f}_{\mathcal{J}}(B^e) ,$$

defined in accordance with (6.20), is called the <u>rate of working</u> at at the instant $\mathcal{J}$ of the forces exerted on B relative to the frame of reference ϕ. I have not yet been able to devise a good proof of existence of the integral (7.17) under suitable conditions, and we here <u>assume</u> that the kinematical process π, the force system $\underline{f}$, and the frame of reference ϕ are such that the integral (7.17) exists for all bodies $B \in \Omega$.

<u>Theorem</u>: The rate of working (7.17) is independent of the frame of reference if and only if the following two conditions are satisfied:

(i) The resultant force acting on B at the instant $\mathcal{J}$ vanishes:

$$(7.18) \qquad \underline{f}_{\mathcal{J}}(B, B^e) = \underline{0}$$

(ii) Let $g \in \mathcal{J}$ be chosen arbitrarily and define $\underline{p}$ by $\underline{p}(\mathcal{L}) = \mathcal{L}(\mathcal{J}) - g \in \mathcal{V}_{\mathcal{J}}$.

Then

$$(7.19) \qquad \int_B \underline{p} \otimes d\underline{f}_{\mathcal{J}}(B^e) = K_{\mathcal{J}}(B, g)$$

is a symmetric tensor over $\mathcal{V}_{\mathcal{J}}$.

The skew part of the integral $\int_B \underline{p} \otimes d\underline{f}_{\mathcal{J}}(A)$ is called the <u>moment</u> about $g \in \mathcal{J}$ of the forces exerted on B by A at the instant $\mathcal{J}$. If $A = B^e$ then this moment is called the <u>resultant moment</u> on B. Thus, c ndition (ii) states that the resultant moment on B vanishes at the instant $\mathcal{J}$

Proof: Let Φ and Φ^* be two frames of reference. The instantaneous velocity relative to Φ^* of the worldline ℓ is given by

(7.20) $$\underline{v}^*_{\mathcal{J}}(\mathcal{J}) = \frac{d}{ds}[\bar{x}^*(\mathcal{J}+s)(\mathcal{J})]\Big|_{s=0} \in \mathcal{V}_{\mathcal{J}},$$

where $\bar{x}^*(\mathcal{J}) \in \Phi^*$ satisfies

(7.21) $$\bar{x}^*(\mathcal{J})(\mathcal{J}) = \ell(\mathcal{J}),$$

in accordance with (4.11).

Now, $e^* = \bar{x}^*(\mathcal{J}+s)(\mathcal{J})$ $e = \bar{x}(\mathcal{J}+s)(\mathcal{J})$ are related by (4.9), i.e.

(7.22) $$\bar{x}^*(\mathcal{J}+s)(\mathcal{J}) = f(s) + Q(s)[\bar{x}(\mathcal{J}+s)(\mathcal{J}) - g].$$

Taking into account (7.20), (7.16), (7.21), and (4.10), we find that differentiating (7.22) with respect to s and then putting s=0 gives

(7.23) $$\underline{v}^*_{\mathcal{J}}(\ell) = \underline{v}_{\mathcal{J}}(\ell) + \underline{u} + W\,\underline{p}(s),$$

where

(7.24) $$\underline{u} = \frac{d}{ds} f(s)\Big|_{s=0}, \quad W = \frac{d}{ds} Q(s)\Big|_{s=0}.$$

If we substitute (7.23) into the formula

(7.25) $$\pi^*_{\mathcal{J}}(B) = \int_B \underline{v}^*_{\mathcal{J}} \cdot d\underline{f}_{\mathcal{J}}(B^e)$$

for the rate of working relative to the frame Φ^*, we find

(7.26) $$\pi^*_{\mathcal{J}}(B) = \pi_{\mathcal{J}}(B) + \underline{u} \cdot \int_B d\underline{f}_{\mathcal{J}}(B^e) + \int_B (W\,\underline{p}) \cdot d\underline{f}_{\mathcal{J}}(B^e).$$

Since $(W\,\underline{p}) \cdot \underline{w} = \mathrm{tr}\,[W\,(p \otimes w)]$ and $\int_B d\underline{f}_{\mathcal{J}}(B^e) = \underline{f}_{\mathcal{J}}(B, B^e)$, we see that (7.26) may be written

(7.27) $$\pi^*_{\mathcal{J}}(B) = \pi_{\mathcal{J}}(B) + \underline{u} \cdot \underline{f}_{\mathcal{J}}(B, B^e) + \mathrm{tr}\,[W\,K_{\mathcal{J}}(B, g)]$$

Now, since Q(s) is orthogonal, the tensor W given by (7.24) is skew.

Moreover, the frame ϕ^* can be chosen such that $\underline{u} \in \mathcal{V}_{\mathcal{J}}$ is any prescribed instantaneous vector and W any prescribed skew tensor over $\mathcal{V}_{\mathcal{J}}$ Thus, it follows from (7/27) that $\pi^*_{\mathcal{J}}(B) = \pi_{\mathcal{J}}(B)$ for every frame ϕ^* if and only if $\underline{f}_{\mathcal{J}}(B, B^e)$ vanishes and $K_{\mathcal{J}}(B, g)$ is symmetric. Q.E.D.

A pair $(\pi, \underline{f})$, consisting of a kinematical process and a force system $\underline{f}$ will be called a dynamical process if the rate of working (7.17), for all instants $\mathcal{J} \in \Gamma$ and all bodies $B \in \Omega$, is independent of the frame of reference. The theorem just proved implies that $(\pi, \underline{f})$ is a dynamical process if and only if the force system $\underline{f}$ is always balanced and if the resultant moments vanish for all bodies and all instants. The first of these conditions restricts the force system $\underline{f}$ alone, but the second involves both π and $\underline{f}$.

An automorphism α of the event-world $\mathcal{W}$ as defined in Sect. 3 induces a transformation

(7.28) $$(\pi, \underline{f}) \rightarrow (\pi^{\alpha}, \underline{f}^{\alpha})$$

on dynamical processes as follows; π^{α} is defined by

(7.29) $$\pi^{\alpha}(B) = \{ \ell^{\alpha} \mid \ell \in \pi(B) \},$$

where ℓ^{α} is defined according to (3.8). $\underline{f}^{\alpha}_{\mathcal{J}\alpha}(A, B)$ is defined to be the instantaneous vector in $\mathcal{V}_{\mathcal{J}\alpha}$ obtained from $\underline{f}_{\mathcal{J}}(A, B) \in \mathcal{V}_{\mathcal{J}}$ by means of the inner product space isomorphism from $\mathcal{V}_{\mathcal{J}}$ onto $\mathcal{V}_{\mathcal{J}\alpha}$ induced by α. It is easily seen that the transformation (7.28) does in fact carry dynamical processes into dynamical processes.

8. Constitutive classes

In a very general way, the mechanical properties of bodies can be described by means of classes of dynamical processes. A certain mechanical property of a body is then expresses by saying that the body can participate only in processes that belong to the class defining this property.

A constitutive class for a pair (A, B) of separate bodies is a class $\mathcal{L}$ (A, B) of dynamical processes satisfying the following two conditions :

(C1) Let $(\pi, \underline{f})$ and $(\pi', \underline{f}')$ be two dynamical processes such that

(8.1) $$\underline{f}_{\mathcal{J}}(A, B) = \underline{f}'_{\mathcal{J}}(A, B)$$

for all instants $\mathcal{J} \in \Gamma$ and such that

(8.2) $$\pi(P) = \pi'(P)$$

for all parts P of A or B Then

(8.3) $(\pi, \underline{f}) \in \mathcal{L}(A, B)$ if and only if $(\pi', \underline{f}') \in \mathcal{L}(A, B)$.

(C2) If $(\pi, \underline{f}) \in \mathcal{L}(A, B)$ then $(\pi^{\alpha}, \underline{f}^{\alpha})\ \mathcal{L}(A, B)$ for every automorphism α of the event-world $\mathcal{U}^{\mathfrak{C}}$.

Condition (C1) expresses the requirement that $\mathcal{L}$ (A, B) should describe the machanical interaction between A and B . This interaction concerns only A and B ; what happens to the remainder of the universe should be irrelevant . The condition (C2) expresses the idea that events have no individuality beyond that conferred to them by their role in a dynamical precess.

In a complete mechanical thory, a constitutive class $\mathcal{L}$ (A, B) should be prescribed for every pair (A, B) of separate bodies. The object of the theory is then to determine the precesses that belong to all of these classes. Usually, however, it is only a particular body A that is taken as the center of interest. The classes $\mathcal{L}(B, A^e)$, $B \prec A$, describe the interaction between the parts B of A and the exterior world A^e, while the classes $\mathcal{L}(B, C)$, $\{B, C\} \prec A$, $B \wedge C = 0$, describe the internal properties of A.

We shall now describe how Newtonian particle mechanics fits into the the framework developed here: It is assumed that the material universe Ω contains a distinguished body A, called the particle system. The particle system A is assumed to have a partition $\{P_i \mid i = 1, \ldots, n\}$ whose elements P_i have no parts. The P_i are called the particles of the system. Each part of A is the least envelope of finitely many of these particles.

For each P_i the constitutive class $\mathcal{L}(P_i, A^e)$ consists of all dynamical processes $(\pi, \underline{f})$ with the following properties: (i) $\pi(A^e)$ is a frame of reference; (ii) $(P_i) = \{\ell_i\}$ consists only of one world-line ℓ_i, which is assumed to be of class C^2 relative to the frame $\pi(A^e)$; (ii) the force exerted by A^e on P_i is given by

(8.4) $$\underline{f}_{\mathcal{J}}(P_i, A^e) = - m_i \underline{a}_{i\mathcal{J}},$$

where m_i is a prescribed positive number, called the inertial mass of P_i, and where $\underline{a}_{i\mathcal{J}}$ is the instantaneous acceleration of ℓ_i at the instant $\mathcal{J}$ relative to the frame $\pi(A^e)$. The class $\mathcal{L}(P_i, A^e)$ describes the inertial interaction between the particle P_i and the remainder of the universe A^e.

If P_i and P_j are two different particles, then the constitutive class $\mathcal{L}(P_i, P_j)$ consists of all dynamical processes such that $\pi(P_i) = \{\ell_i\}$ and $\pi(P_i) = \{\ell_i\}$ are singletons and such that

$$(8.5)\qquad \underline{f}_{\mathcal{J}}(P_i;P_j) = [\ell_i(\mathcal{J}) - \ell_j(\mathcal{J})]\ F_{ij}(\delta(\ell_i(\mathcal{J}), \ell_j(\mathcal{J}))),$$

Where F_{ij} is a prescribed real-valued function of a real variable. This function describes the "forces-law" for the interaction between P_i and P_j. In the the case of the Newtonian theory of gravitation, F_{ij} is given by

$$(8.6)\qquad F_{ij}(\delta) = -\frac{m_i' m_j'}{\delta},$$

where the m_i' are prescribed positive numbers, called the gravitational masses of the P_i. The theory fits the observed data when the gravitational masses are taken to be proportional to the inertial masses, a fact for which there is no good explanation within classical mechanics .

In conventional continuum mechanics material properties are described by means of constitutive equations. These involve processes that are described in terms of classical space-time , which makes use of an absolute space $\mathcal{E}$ and a numerical time-scale, represented by the set $\mathcal{R}$ of real numbers. A neo-classical event-world $\mathcal{W}^{\ell}$ can be related to $\mathcal{E}$ and $\mathcal{R}$ by mappings $\mathcal{W}^{\ell} \to \mathcal{E} \times \mathcal{R}$ that have the following properties : if $(x_1, \tau_1) \in \mathcal{E} \times \mathcal{R}$ corresponds to $e_1 \in \mathcal{W}^{\ell}$ and $(x_2, \tau_2) \in \mathcal{E} \times \mathcal{R}$ to $e_2 \in \mathcal{W}^{\ell}$, then

$$(8.7)\qquad \tau(e_1, e_2) = \tau_1 - \tau_2,$$

and

$$(8.8)\qquad \delta(e_1, e_2) = d(x_1, x_2)$$

when e_1 and e_2 are simultaneous, i.e. when $\tau_1 = \tau_2$. Any such mapping will be called a representation of the event-world $\mathcal{W}^{\ell}$.

Suppose that two representations are given and that (x, τ) corresponds to the event e in the first representation, while (x^*, τ^*) corresponds to e in the second representation. It follows easily from the representation theorem of Sect. 2 that

$$(8.9) \qquad \begin{aligned} x^* &= c(\tau) + Q(\tau)(x - q) , \\ \tau^* &= \tau - \gamma , \end{aligned}$$

where $c(\tau)$ and q are points in $\mathcal{E}$, where $Q(\tau)$ is an orthogonal transformation of the translation space of $\mathcal{E}$, and where γ is a number. If constitutive equations are to determine constitutive classes in the sense described above, they should be conditions on dynamical processes, independent of the representations used to describe these processes. Hence constitutive equations should be invariant under transformations of the form (8.9). This requirement is called "principle of material frame - indifference" or "principle of material objectivity" in conventional mechanics.

The condition (C1) on constitutive classes can be employed to justify another principle of conventional continuum mechanics: the <u>principle of local action.</u> This principle states that the stress tensor at a point should depend only on the motion of an arbitrarily small neighborhoof of the point. Now the stress tensor at an event is determined by the forces $\underline{f}_{\mathcal{J}}(B, C)$, where B and C are bodies which occupy regions $\mathcal{B}_{\pi(B)}(\mathcal{J})$ and $\mathcal{B}_{\pi(C)}(\mathcal{J})$ that are contained in an aribitrarily small neighborhood of the event in question in the instantaneous space $\mathcal{J}$. The constitutive classes $\mathcal{L}(B, C)$ for these bodies should give rise to a constitutive equation for the stress tensor. By (C1) the motion of bodies that occupy regions outside the arbitrarily small neighborhood of the event under consideration does not affect membership or non-memership of processes in the class $\mathcal{L}$ (B, C) and hence may be disregarded in the constitutive equation for the stress tensor. A more precise formulation of this idea must await a precise developement of the theory of stress within the framework presented in these lectures.

CENTRO INTERNAZIONALE MATEMATICO ESTIVO

(C. I. M. E.)

R. A. Toupin

"ELASTICITY AND ELECTRO-MAGNETIC"

Corso tenuto a Bressanone dal 31 maggio al 9 giugno 1965

ELASTICITY AND ELECTRO-MAGNETISM

by

R. A. Toupin
(I. B. M. Research Center, Yorktown Heights, N. Y.)

The purpose of these preliminaries is to introduce the notation, terminology, and principal mathematical results to be used in the course of the lectures. These will allow an easy, uninterrupted development of the physical theory. The results presented here may be found in greater detail in the following sources :

Geometric Integration Theory
H. Whintney
Princeton University Press (1957)

Ricci-Calculus
J. A. Schouten
Springer-Verlag (1954)

Finite Dimensional Vector Spaces
P. Halmos

Tensor Fields
J. L. Ericksen
Appendix, Classical Field Theories
Handbuch der Physik, III/1(1960)

1. COTENSORS

Let V^n denote an n-dimensional real vector space. Elements of V^n will be denoted by boldface, lower case Latin letters, v, u,... and will be called vectors. A real valued multilinear (i. e., linear in each argument) function of r vectors is called an r-cotensor. Thus, denoting the real numbers by R, an r-cotensor . Thus, denoting the real numbers by R an r-cotensor is a multilinear mapping

$$a : V^{rn} \to R ,$$

where

$$W^{rn} = V^n \times V^n \times \ldots \times V^n$$

is the r-fold Cartesian product of V^n. Cotensors will be denoted by lower case, Greek boldface letters.

The sum of any two cotensors and the product of a cotensor by a real number are defined by the relations

$$(\underset{\sim}{a}+\underset{\sim}{\beta})(v, u) = \underset{\sim}{a}(v, u) + \underset{\sim}{\beta}(v, u)$$

(1.1) $$(\lambda \underset{\sim}{a})(u) = \lambda \underset{\sim}{a}(u), \quad \lambda \in R, \quad \underset{\sim}{u} \in V^n .$$

As in (1.1), we always denote real numbers by lightface, lower case Greek or Latin letters. With these definitions, the set of all r-cotensors is a certain linear space which we denote by V_r^n, and $\dim (V_r) = n^r$. We call V^n the carrier space of V_r^n. For $r = 1$, we write V_n and call the elements of V_n^n, covectors. The space of covectors is called the conjugate of V^n. More generally, the space of real valued linear functions of the elements of any linear space L is called the conjugate space L^c.

Let $\underset{\sim}{e}_i$, $i = 1, 2, \ldots, n$ denote a basis (linearly independent set of vectors) in the carrier space V^n . Then an arbitrary vector $v \in V^n$ has the representation

$$(1.2) \qquad \underset{\sim}{v} = v^i \underset{\sim}{e}_i, \quad v^i \in R, \quad i = 1, 2, \ldots, n$$

where the <u>components</u> v^i of $\underset{\sim}{v}$ with respect to the basis e_i are uniquely determined by $\underset{\sim}{v}$ and $\underset{\sim}{e}_i$. As in (1.2) we use the summation convention wherein, if the same letter appears in a given term of an expression in both a superior and inferior position (not at the same level, however) , summation over the corresponding index set is implied without writing the summation sign. When the index set is not clear from the context, the summation sign will be used .

It is evident that an r-cotensor $\underset{\sim}{\alpha}$ is uniquely determined by its set of values

$$(1.3) \qquad \alpha_{i_1 i_2 \ldots i_r} = \alpha_{(i)} = \alpha(\underset{\sim}{e}_{i_2}, \underset{\sim}{e}_{i_2}, \ldots, \underset{\sim}{e}_{i_r}) .$$

For $r = 1$, the set of convectors $\underset{\sim}{\epsilon}^i$, $i = 1, 2, \ldots, n$ defined by

$$(1.4) \qquad \underset{\sim}{\epsilon}^i(\underset{\sim}{e}_j) = \delta^i_j = \begin{cases} 1 & \text{if } i = j \\ 0 & \text{if } i \neq j \end{cases}$$

are linearly independent and constitute a basis in the space V_n of covectors. The sets $\underset{\sim}{e}_i$ and $\underset{\sim}{\epsilon}^j$ so related are called reciprocal bases for V^n and the conjugate space V_n.

Every covector has the representation

$$(1.5) \qquad \alpha = \alpha_i \underset{\sim}{\varepsilon}^i ,$$

and if v is represented as in (1.2), then

(1.6) $$\underset{\sim}{\alpha}(\underset{\sim}{v}) = v^i \alpha_i \ .$$

More generally, now, for $r > 1$, the n^r r-cotensors defined by

(1.7) $$\underset{\sim}{\varepsilon}^{i_1} \otimes \underset{\sim}{\varepsilon}^{i_2} \otimes \dots \otimes \underset{\sim}{\varepsilon}^{i_r}(\underset{\sim}{e}_{j_1}, \underset{\sim}{e}_{j_2}, \dots, \underset{\sim}{e}_{j_r}) = \delta^{i_1}_{j_1} \delta^{i_2}_{j_2} \cdots \delta^{i_r}_{j_r}$$

are linearly independent and constitute a basis in V_{n^r} called the <u>tensor basis</u> of V_{n^r} corresponding to the basis $\underset{\sim}{e}_i$ in the carrier space V^n of V_{n^r}. Every r-cotensor $\underset{\sim}{\alpha}$ has the representation

(1.8) $$\underset{\sim}{\alpha} = \alpha_{i_1 i_2 \dots i_r} \underset{\sim}{\varepsilon}^{i_1} \otimes \underset{\sim}{\varepsilon}^{i_2} \otimes \dots \otimes \underset{\sim}{\varepsilon}^{i_r}$$

where the tensor components $\alpha_{(i)}$ of $\underset{\sim}{\alpha}$ are given by (1.3).

The tensor product of an r-cotensor $\underset{\sim}{\alpha}$ and an s-cotensor $\underset{\sim}{\beta}$ is the (r + s)-cotensor $\underset{\sim}{\alpha} \otimes \underset{\sim}{\beta}$ defined by

(1.9) $$(\underset{\sim}{\alpha} \otimes \underset{\sim}{\beta})(\underset{\sim}{v}_1, \underset{\sim}{v}_2, \dots, \underset{\sim}{v}_r, \underset{\sim}{u}_1, \underset{\sim}{u}_2, \dots, \underset{\sim}{u}_s) = \underset{\sim}{\alpha}(\underset{\sim}{v}_1, \underset{\sim}{v}_2, \dots, \underset{\sim}{v}_r)\, \underset{\sim}{\beta}(\underset{\sim}{u}_1, \underset{\sim}{u}_2, \dots, \underset{\sim}{u}_s).$$

The tensor components of $\underset{\sim}{\alpha} \otimes \underset{\sim}{\beta}$ are given in terms of the tensor components of $\underset{\sim}{\alpha}$ and $\underset{\sim}{\beta}$ by

(1.10) $$(\underset{\sim}{\alpha} \otimes \underset{\sim}{\beta})_{i_1 i_2 \dots i_r j_1 j_2 \dots j_s} = \alpha_{i_1 i_2 \dots i_r} \beta_{j_1 j_2 \dots j_s} \ .$$

2. TENSORS AND MIXED TENSORS

A multilinear real valued function

$$T : V_{rn} \to R,$$

where

$$V_{rn} = V_n \times V_n \times \dots V_n$$

is the r-fold Cartesian product of V_n, is called an r-tensor*. All that has been said in § 1 can now be repeated with the roles of V^n and V_n interchanged and with obvious changes in the terminology. In particular, every r-tensor T has the representation

$$T = T^{i_1 i_2 \dots i_r} \underset{\sim}{e}_{i_1} \otimes \underset{\sim}{e}_{i_2} \otimes \dots \otimes \underset{\sim}{e}_{i_r}, \tag{2.1}$$

where the tensor components $T^{(i)}$ of $\underset{\sim}{T}$ are given by

$$T^{i_1 i_2 \dots i_r} = \underset{\sim}{T}(\underset{\sim}{e}^{i_1}, \underset{\sim}{e}^{i_2}, \dots \underset{\sim}{e}^{i_r}). \tag{2.2}$$

The tensor product $\underset{\sim}{T} \otimes \underset{\sim}{S}$ of an r-tensor and an s-tensor is defined in obvious analogy to the tensor product of r-cotensors.

The set of all r-tensors with addition and multiplication by scalars defined by

---

*In Tensor analysis, it is common to call an r-tensor a tensor of rank r. But we shall use the term rank of a tensor in an entirely different sense below; hence, we avoid the common terminology here.

$$(\underset{\sim}{T}+\underset{\sim}{S})(\underset{\sim}{\alpha}_1, \underset{\sim}{\alpha}_2, \ldots, \underset{\sim}{\alpha}_r) = \underset{\sim}{T}(\underset{\sim}{\alpha}_1, \underset{\sim}{\alpha}_2, \ldots, \underset{\sim}{\alpha}_r) + \underset{\sim}{S}(\underset{\sim}{\alpha}_1, \underset{\sim}{\alpha}_2, \ldots, \underset{\sim}{\alpha}_r)$$

(2.3)

$$(\lambda \underset{\sim}{T})(\underset{\sim}{\alpha}_1, \underset{\sim}{\alpha}_2, \ldots, \underset{\sim}{\alpha}_r) = \lambda \underset{\sim}{T}(\underset{\sim}{\alpha}_1, \underset{\sim}{\alpha}_2, \ldots, \underset{\sim}{\alpha}_r)$$

is a linear space V^{n^r} of dimension n^r.

By definition, $(V^n)^c = V_n$; i.e., the space of covectors is the conjugate of the space of vectors. The conjugate $(V_n)^c$ of the space of covectors is, by definition, what we have called the space of 1-tensors V^{n^1}. Now V^n, the carrier space, has the same dimension as V^{n^1}; hence, they are isomorphic. The <u>natural isomorphism</u> $\phi : V^n \rightarrow V^{n^1}$ is defined by

(2.4) $$\underset{\sim}{\phi}(\underset{\sim}{v})(\underset{\sim}{\alpha}) = \underset{\sim}{\alpha}(\underset{\sim}{v}) = \alpha_i v^i \quad .$$

It is customary to denote $\underset{\sim}{\phi}(\underset{\sim}{v})$ and $\underset{\sim}{v}$ by the same letter and not to distinguish between vectors and 1-tensors whose carrier space V^n is the corresponding space of vectors. We adopt this convention here, but the natural isomorphism and all these agreements should be kept in mind. More generally now, the conjugate space $(V_{n^r})^c$ of the space of r-cotensors has the same dimension as the space V^{n^r} of r-tensors and, hence, they are isomorphic, but distinct. The <u>natural isomorphism</u> $\Phi : V^{n^r} - (W_{n^r})^c$ is defined as follows. Call an r-tensor $\underset{\sim}{T}$ simple if it is the tensor product of r-vectors ; $\underset{\sim}{T} = \underset{\sim}{v}_1 \otimes \underset{\sim}{v}_2 \otimes \cdots \otimes \underset{\sim}{v}_r$. Define $\underset{\sim}{\alpha}(\underset{\sim}{T})$ for every simple r-covector by

(2.5) $$\underset{\sim}{\alpha}(\underset{\sim}{T}) = \underset{\sim}{\alpha}(\underset{\sim}{v}_1 \otimes \underset{\sim}{v}_2 \otimes \cdots \otimes \underset{\sim}{v}_r) = \alpha(\underset{\sim}{v}_1, \underset{\sim}{v}_2, \ldots, \underset{\sim}{v}_r)$$

and set

(2.6) $$\underset{\sim}{\alpha}(\lambda \underset{\sim}{T} + \eta \underset{\sim}{S}) = \lambda \underset{\sim}{\alpha}(\underset{\sim}{T}) + \eta \underset{\sim}{\alpha}(\underset{\sim}{S})$$

if $\underset{\sim}{T}$ and $\underset{\sim}{S}$ are simple, but $T + S$ not necessarily simple. But every r-tensor $\underset{\sim}{T}$ is the sum of a finite number of simple r-tensors (cf., the representation (2.1)). Hence, (2.5) and (2.6) define $\underset{\sim}{\alpha}(\underset{\sim}{T})$ for arbitrary $\underset{\sim}{T} \in V^{n^r}$. The natural isomorphism between V^{n^r} and $(V_{n^r})^c$ is then defined by

$$(2.7) \qquad \underset{\sim}{\Phi}(\underset{\sim}{T})(\underset{\sim}{\alpha}) = \underset{\sim}{\alpha}(\underset{\sim}{T}) \; .$$

Here, as in the case of vectors and 1-tensors, it is conventional not to distinguish between "co-cotensors" $\underset{\sim}{\Phi}(\underset{\sim}{T})$ and the tensor $\underset{\sim}{T}$ related by the natural isomorphism established by (2.7). Thus, $\underset{\sim}{\Phi}(\underset{\sim}{T})$ and $\underset{\sim}{T}$ are denoted by the common symbol $\underset{\sim}{T}$, and

$$(2.8) \qquad \underset{\sim}{T}(\underset{\sim}{\alpha}) = \underset{\sim}{\alpha}(\underset{\sim}{T}) = a_{i_1 i_2 \ldots i_r} T^{i_1 i_2 \ldots i_r} \; ,$$

where $a_{(i)}$ and $T^{(i)}$ are the components of $\underset{\sim}{\alpha}$ and $\underset{\sim}{T}$ with respect to arbitrary tensor bases in V_{n^r} and V^{n^r}.

More generally, now, a real valued multilinear function

$$(2.9) \qquad \underset{\sim}{M} : V_1^{n_1} \times V_2^{n_2} \times \ldots \times V_r^{n_r} \to R$$

of r vector arguments drawn from an arbitrary collection of r vector spaces is called a <u>mixed</u> r-tensor unless $V_1 = V_2 = \ldots V_r$.

It suffices to illustrate the general case by the case $r = 2$. Then

$$\underset{\sim}{M} : V^n \times U^m \to R \; ,$$

say. Let $\underset{\sim}{e}_i$ and $\underset{\sim}{E}_\alpha$, $i = 1, 2, \ldots, n$, $\alpha = 1, 2, \ldots, m$ be bases in V^n and U^m, respectively. Then, $\underset{\sim}{M}$ is uniquely determined by its components

$$(2.10) \qquad M_{i\alpha} = \underset{\sim}{M}(\underset{\sim}{e}_i, \underset{\sim}{E}_\alpha).$$

Let $\underset{\sim}{M} + \underset{\sim}{N}$ and $\lambda\underset{\sim}{M}$ be defined by

$$(\underset{\sim}{M} + \underset{\sim}{N})(\underset{\sim}{v}, \underset{\sim}{u}) = \underset{\sim}{M}(\underset{\sim}{v}, \underset{\sim}{u}) + \underset{\sim}{N}(\underset{\sim}{v}, \underset{\sim}{u}),$$

$$(\lambda\underset{\sim}{M})(\underset{\sim}{v}, \underset{\sim}{u}) = (\underset{\sim}{M}(\underset{\sim}{v}; \underset{\sim}{u}).$$

Then, the set of all such mixed tensors is a linear space W_{mn} of dimension mn. A basis in W_{mn} consists in the mnon elements defined by

$$(2.12) \qquad (\underset{\sim}{\varepsilon}^i \otimes \underset{\sim}{\mathcal{E}}^\alpha)(\underset{\sim}{e}_j, \underset{\sim}{e}_\beta) = \delta^i_j \delta^\alpha_b,$$

and an arbitrary $M \in W_{mn}$ has the representation

$$(2.23) \qquad \underset{\sim}{M} = M_{i\alpha} \underset{\sim}{\varepsilon}^i \otimes \underset{\sim}{\mathcal{E}}^\alpha .$$

where the components of $\underset{\sim}{M}$, the $M_{i\alpha}$, are given by (2.10).

Consider the special case of the above where $V^n = U_m$, the conjugate space of the second argument U^m. Then,

$$\underset{\sim}{M} : U_m \times U^m \to R$$

and we set

$$(2.14) \qquad M^\alpha_\beta = M(\underset{\sim}{\mathcal{E}}^\alpha, \underset{\sim}{E}_\beta)$$

where the $\underset{\sim}{\mathcal{E}}^\alpha$ and $\underset{\sim}{E}_\beta$ are reciprocal, $\underset{\sim}{E}_\alpha(\underset{\sim}{\mathcal{E}}^\beta) = \delta^\beta_\alpha$. Thus,

$$(2.15) \qquad \underset{\sim}{M} = M^\alpha_\beta \underset{\sim}{E}_\alpha \otimes \underset{\sim}{\mathcal{E}}^\beta .$$

Every such mixed tensor $\underset{\sim}{M}$ determines a unique linear transformation

$$\underset{\sim}{M}^{*} : U^{m} \to U^{m}$$

defined as follows :

$$\underset{\sim}{M}^{*}(\underset{\sim}{u})(\underset{\sim}{\beta}) = \underset{\sim}{M}(\underset{\sim}{\alpha}, \underset{\sim}{u})$$

where $\underset{\sim}{M}^{*}(\underset{\sim}{u})(U_m)^c$ is that unique element of U^m determined by the natural isomorphism $\underset{\sim}{\theta}$ (cf. 2.4) . With respect to tensor bases in U_m and U^m

$$M^{*\alpha}_{\beta} = M^{\alpha}_{\beta},$$

where

$$M^{*\alpha}_{\beta} = \underset{\sim}{\xi}(\underset{\sim}{M}^{*}(\underset{\sim}{E}_{\beta}))$$

are the components of $\underset{\sim}{M}^{*}$. In a similar way, the mixed tensor $\underset{\sim}{M}$ determines a linear transformation $\underset{\sim}{M}^{**} : U_m \to U_m$ defined $\underset{\sim}{M}^{**}(\underset{\sim}{\alpha})(\underset{\sim}{v}) = \underset{\sim}{M}(\underset{\sim}{\alpha}, \underset{\sim}{v})$ and the tensor components of $\underset{\sim}{M}^{**}$ defined by $M^{**\alpha}_{\beta} = \underset{\sim}{E}_{\beta}(\underset{\sim}{M}^{**}(\underset{\sim}{\xi}^{\alpha}))$ are also equal, respectively, to the components M^{α}_{β} of $\underset{\sim}{M}$. These definitions justify and are consistent with the usual rules and conventions of tensor algebra. There, it is traditional not to distinguish between $\underset{\sim}{M}$, M^{*}, and $\underset{\sim}{M}^{**}$ and to denote all three by the common symbol $\underset{\sim}{M}$. In the following we shall use absolute notations or the kernel index notations of tensor algebra interchangeably according to whichever seems most efficient and expressive in a given context. By the components of tensors, cotensors, or mixed tensors we shall always mean <u>tensor components</u> as these have been defined above . In the physical applications, a mixed tensor, r-tensor, or r-cotensor is generally introduced into the physical theory with a specific logical meaning; e.g. as a 1 - tensor and not a, vector, or as a linear transformation of some vector space, and not as a mixed tensor, but then the natural isomorphism established above is used freely to define other operations in which the mixed tensor, r-tensor, or r-cotensor plays a different logical role.

3. THE SYMMETRY PARTS OF TENSORS AND COTENSORS

Let $\underset{\sim}{L}$ be a nonsingular linear transformation

(3.1) $$\underset{\sim}{L}: V^n \to V^n$$

of an n-dimensional vector space V^n. Then $\underset{\sim}{L}$ induces a linear tensor transformation

(3.2) $$\underset{\sim}{L}_r : V_{n}^{r} \to V_{n}^{r}$$

in the space V_{n}^{r} of r-cotensors having V^n as carrier space. The transformation $\underset{\sim}{L}_r$ is defined by

(3.3) $$(\underset{\sim}{L}_r \alpha)(\underset{\sim}{L}v_1, \underset{\sim}{L}v_2, \ldots, \underset{\sim}{L}v_r) = \underset{\sim}{\alpha}(\underset{\sim}{v}_1, \underset{\sim}{v}_2, \ldots, \underset{\sim}{v}_r) .$$

If

(3.4) $$\underset{\sim}{L}_r(\underset{\sim}{\alpha}) = \underset{\sim}{\alpha} ,$$

then $\underset{\sim}{\alpha}$ is said to be <u>invariant under the tensor transformation</u> $\underset{\sim}{L}_r$. More generally, let $U \subset V_{n}^{r}$ be a proper subspace of V_{n}^{r}. Then if

(3.5) $$\underset{\sim}{L}_r(U) \subset U$$

for every $\underset{\sim}{L}_r$ in some set $\{\underset{\sim}{L}_r\}$ of tensor transformations, then U is an <u>invariant subspace</u> under the set of transformations $\{\underset{\sim}{L}_r\}$.

Consider the symmetric (permutation) group S^r on the first r integers and let $\underset{\sim}{\Pi} = \begin{pmatrix} 1 & 2 & 3 \ldots r \\ \pi_1 & \pi_2 & \pi_3 \cdots \pi_r \end{pmatrix} S^r$. Then, in terms of an arbitrary r-cotensor $\underset{\sim}{\alpha}$ we define the r-cotensor $H(\underset{\sim}{\alpha})$ by

(3.6) $$\underset{\sim}{\Pi}\underset{\sim}{\alpha}(\underset{\sim}{v}_1, \underset{\sim}{v}_2, \ldots, \underset{\sim}{v}_r) = \underset{\sim}{\alpha}(\underset{\sim}{v}_{\pi_1}, \underset{\sim}{v}_{\pi_2}, \ldots, \underset{\sim}{v}_{\pi_r}) .$$

The r-cotensor $\tilde{H}\tilde{\alpha}$ is called an <u>isomer</u> of $\tilde{\alpha}$. Defining

$$\tilde{\Pi}(\tilde{\alpha}+\tilde{\beta}) = \tilde{\Pi}(\tilde{\alpha}) + \tilde{\Pi}(\tilde{\beta}), \quad \tilde{\Pi}(\lambda\tilde{\alpha}) = \lambda\tilde{\Pi}(\tilde{\alpha}), \tag{3.7}$$

every permution $\tilde{\Pi} \in S^r$ determines a linear transformation

$$\tilde{\Pi}: V_{n^r} \rightarrow V_{n^r}. \tag{3.8}$$

It is easy to see that every linear transformation $\tilde{\Pi}$ defined in this way commutes with every tensor transformation $\tilde{L}_r$:

$$\tilde{\Pi}\tilde{L}_r(\tilde{\alpha}) = \tilde{L}_r\tilde{\Pi}(\tilde{\alpha}). \tag{3.9}$$

More generally, now, consider the enveloping algebra $A^{r!}$ of the symmetric group S^r. Define the transformations $\tilde{\Pi}+\tilde{\Omega}$, and $\lambda\tilde{\Pi}$ in V_{n^r} by

$$(\tilde{\Pi}+\tilde{\Omega})\tilde{\alpha} = \tilde{\Pi}\tilde{\alpha} + \tilde{\Omega}\beta, \quad (\lambda\tilde{\Pi})\tilde{\alpha} = \lambda\tilde{\Pi}\tilde{\alpha}. \tag{3.10}$$

Then, each element of $A^{r!}$, say,

$$\tilde{A} = \sum_{r=1}^{r!} a^s \tilde{\Pi}_s, \quad \tilde{\Pi}_s \in S^r \tag{3.11}$$

determines a linear transformation of V_{n^r}, defined by the above and

$$\tilde{\ }(\tilde{\alpha}) = \sum_{s=1}^{r!} a^s \tilde{\Pi}(\tilde{\alpha}). \tag{3.12}$$

The algebra $A^{r!}$ possesses a resolution of its identity element $\tilde{I}$,

$$\tilde{I} = \tilde{J}_1 + \tilde{J}_2 + \ldots + \tilde{J}_p, \tag{3.13}$$

where each $\tilde{J}_k$, $k = 1, 2, \ldots, p$ is idempotent and irreducible and such that

(3.14) $\quad \underset{\sim}{J}_k^2 = \underset{\sim}{J}_k, \quad \underset{\sim}{J}_k \underset{\sim}{J}_h = 0 = \underset{\sim}{J}_h \underset{\sim}{J}_k, \quad h \neq k.$

Irreducible means that there exists no decomposition of any $\underset{\sim}{J}_h$ into a sum $\underset{\sim}{J}_h = \underset{\sim}{A} + \underset{\sim}{B}$ such that $\underset{\sim}{A}$ and $\underset{\sim}{B}$ are idempotent and $\underset{\sim}{A}\underset{\sim}{B} = \underset{\sim}{B}\underset{\sim}{A} = 0$. From the existence of the resolution of the identity (3. 13) it follows that every r-cotensor α can be resolved into symmetry parts $\underset{\sim}{\alpha}_k = \underset{\sim}{J}_k(\underset{\sim}{\alpha})$ as follows :

(3.15) $\quad \underset{\sim}{\alpha} = \underset{\sim}{I}\underset{\sim}{\alpha} = \sum_{k=1}^{p} \underset{\sim}{J}_k \underset{\sim}{\alpha} = \sum_{k=1}^{p} \underset{\sim}{\alpha}_k ,$

and one has

(3.16) $\quad \underset{\sim}{J}_k(\underset{\sim}{\alpha}_h) = \delta_{kh} \, \underset{\sim}{\alpha}_h .$

If $\underset{\sim}{J}_k \underset{\sim}{\alpha} = \underset{\sim}{\alpha}$, α is said to have symmetry $\{k\}$. The sum $\underset{\sim}{\alpha} + \underset{\sim}{\beta}$ and product $\lambda \underset{\sim}{\alpha}$ of r-cotensors of symmetry $\{k\}$ are r-cotensors of the same symmetry class $\{k\}$. Hence, $\overset{*}{V}{}_n{}_r$ is resolved as follows :

(3.17) $\quad {}_nV_r = {}_nV^1_r \oplus {}_nV^2_r \oplus \cdots \oplus {}_nV^p_r$

into a direct sum of subspaces of r-cotensors of given symmetry. (Some of these subspaces may be empty ; i.e., may have dimension 0.)

If follows from the commutativity property (3.9) that the subspaces ${}_nV^k_r$ of r-cotensors of given symmetry are invariant subspaces of every tensor transformation $\underset{\sim}{L}_r$ of ${}_nV_r$;

$$\underset{\sim}{L}_r({}_nV^k_r) = \underset{\sim}{L}_r \underset{\sim}{J}_k({}_nV_r) = \underset{\sim}{J}_k \underset{\sim}{L}_r({}_nV_r)$$

$$\subset \underset{\sim}{J}_k({}_nV_r) ,$$

or

$$\underset{\sim}{L}(V^r_{n^r}) \subset V^k_{n^r} .$$

Theorem: (Weyl) The resolution (3.17) of V_{n^r} into a direct sum of invariant subspaces of r-cotensors of given symmetry is a maximal decomposition of V_{n^r} under the set of all tensor transformations $\{\underset{\sim}{L}_r\}$ induced in V_{n^r} by the set of all non singular linear transformations $\{\underset{\sim}{L}\}$ of V^n. In other words, if $\{\underset{\sim}{L}\}$ is the set of all nonsingular linear transformations of V^n and $\{\underset{\sim}{L}_r\}$ the corresponding set of tensor transformations induced in V_{nr}, then no proper subspace of any $V^k_{n^r}$, k 1, 2, ..., p is invariant under the set $\{\underset{\sim}{L}_r\}$.

If $r > 1$, the number p(r) of symmetry classes $\{k\}$ is always ≥ 2. Amongst these p symmetry classes for every value of $r > 1$ is the class of symmetric r-cotensors for which

$$\underset{\sim}{\Pi}\,\underset{\sim}{a} = \underset{\sim}{a}, \quad \text{for every permutation } \underset{\sim}{\Pi} \in S^r$$

and the class of antisymmetric r-cotensors for which

$$\underset{\sim}{\Pi}\,\underset{\sim}{a} = \underset{\sim}{a} \qquad \text{for every even permutation } \underset{\sim}{\Pi}$$

$$\underset{\sim}{\Pi}\,\underset{\sim}{a} = -\underset{\sim}{a} \qquad \text{for every odd permutation } \underset{\sim}{\Pi} .$$

For brevity, antisymmetric r-cotensors are called r-covectors. The subspaces of symmetric and antisymmetric r-cotensors are denoted by the special symbols $V_{(n^r)}$ and $V_{[n^r]}$, respectively. For $r = 2$, the resolution (3.17) reduces to the familiar decomposition

(3.18) $$V_{n^2} = V_{(n^2)} \oplus V_{[n^2]}$$

of 2-cotensors into their symmetric and antisymmetric parts. But for $r > 2$,

$$V_{n^r} = V_{(n^r)} \oplus V_{[n^r]} \oplus U,$$

where, in general, U is not empty.

If $\alpha_{i_1 i_2 \ldots i_r}$ are the components of an r-cotensors with respect to some basis e^i, we denote the components of the symmetric and antisymmetric parts of $\underset{\sim}{\alpha}$ by $\alpha_{(i_1 i_2 \ldots i_r)}$ and $\alpha_{[i_1 i_2 \ldots i_r]}$, respectively.

$$\alpha_{(i_1 i_2 \ldots i_r)} = (1/r!) \sum_{\Pi \in S^r} \alpha_{i_{\pi_1} i_{\pi_2} \ldots i_{\pi_r}}$$

$$\alpha_{[i_1 i_2 \ldots i_r]} = (1/r!) \sum_{\Pi \in S^r} (-)^{\sigma_\pi} \alpha_{i_{\pi_1} i_{\pi_2} \ldots i_{\pi_r}},$$

where $\sigma_\pi = 0, 1$ for even and odd permutations, respectively.

All that has been said above for r-cotensors holds with minor changes for r-tensors, perhaps with one exception. If L is a linear transformation of the carrier space V^n, then the linear <u>tensor</u> transformation of the space of r-tensors V^{n^r} induced by L is defined by

$$(3.19) \qquad \underset{\sim}{L}^r(\underset{\sim}{T}) = T^{i_1 i_2 \ldots i_r} (\underset{\sim}{L}\, \underset{\sim}{e}_{i_1}) \otimes (\underset{\sim}{L}\, \underset{\sim}{e}_{i_2}) \ldots \otimes (\underset{\sim}{L}\, \underset{\sim}{e}_{i_r})$$

where

$$T = T^{i_1 i_2 \ldots i_r} \underset{\sim}{e}_{i_1} \otimes e_{i_2} \otimes \cdots \otimes \underset{\sim}{e}_{i_r}.$$

The definitions of $\underset{\sim}{L}_r$ (3.3) and $\underset{\sim}{L}^r$ (3.19) imply that

$$(3.20) \qquad \overline{\overline{\underset{\sim}{T}(\underset{\sim}{\alpha}}}\,) = \underset{\sim}{T}(\underset{\sim}{\alpha})$$

where $\overline{\underset{\sim}{T}} = \underset{\sim}{L}^{r}(\underset{\sim}{T})$ and $\overline{\underset{\sim}{\alpha}} = \underset{\sim}{L}_{r}(\underset{\sim}{\alpha})$, provided that $\underset{\sim}{L}$ is non-singular so that $\underset{\sim}{L}_{r}$ is defined. The subspaces of antisymmetric and symmetric r-tensors are denoted by $V^{[n^r]}$ and $V^{(n^r)}$, respectively . For brevity, we call antisymmetric r-tensors, r-vectors.

If $\underset{\sim}{T}^{k}$ is a tensor of symmetry class $\{k\}$ and $\underset{\sim}{\alpha}_{h}$ a cotensor of symmetry class $\{h\}$, then

$$\underset{\sim}{T}^{k}(\underset{\sim}{\alpha}_{h}) = 0 \quad \text{if} \quad \{k\} \neq \{h\}. \tag{3.21}$$

It follows that

$$\underset{\sim}{T}(\underset{\sim}{\alpha}) = \underset{\sim}{\alpha}(\underset{\sim}{T}) = \sum_{\{k\}} \underset{\sim}{T}^{k}(\underset{\sim}{\alpha}_{k}) \tag{3.22}$$

if $\underset{\sim}{T} = \sum_{\{k\}} \underset{\sim}{T}^{k}$, $\quad \underset{\sim}{\alpha} = \sum_{\{h\}} \underset{\sim}{\alpha}_{h}$.

4. THE GRASSMAN ALGEBRA

Let V^n be an n-dimensional vector space and consider the direct sum (Grassman space)

(4.1) $$G = V_{[1]} \oplus V_{[n]} \oplus V_{[n^2]} \oplus \cdots \oplus V_{[n^n]}$$

of the spaces of r-covectors, $r = 0, 1, 2, \ldots, n$ having the common carrier space V^n. The space $V_{[1]}$ is the space of scalars, and $V_{[n]} = V_n$ is the space of covectors. Then

(4.2) $$\dim (G) = 1 + n + \binom{n}{2} + \binom{n}{3} = (1+1)^n = 2^n .$$

Let $\underset{\sim}{\alpha}_r$ and $\underset{\sim}{\beta}_s$ be an r-covector and an s-covector, respectively, and let $\underset{\sim}{J}_{[r]}$ be the idempotent linear transformation (antisymm etrizer) that projects V_{n^r} into $V_{[n^r]}$:

(4.3) $$\underset{\sim}{J}_{[r]} (V_{n^r}) = V_{[n^r]} .$$

The exterior (Grassman) product $\underset{\sim}{\alpha}_r \vee \underset{\sim}{\beta}_s$ is the (r+s)-covector defined by

(4.4) $$\underset{\sim}{\alpha}_r \quad \underset{\sim}{\beta}_s = \frac{(r+s)!}{r!\,s!} \underset{\sim}{J}_{[r+s]} (\underset{\sim}{\alpha}_r \quad \underset{\sim}{\beta}_s) .$$

In words, the exterior product of an r-covector and an s-covector is the antisymmetric part of their tensor product times the numeral factor $(r+s)/r!\,s!$. From the associativity of the tensor product and the property

$$\underset{\sim}{J}_{[r+s]} (\underset{\sim}{\alpha}_r \otimes \underset{\sim}{\beta}_s) = \underset{\sim}{J}_{[r+s]} (\underset{\sim}{J}_{[r]} \underset{\sim}{\alpha} \otimes \underset{\sim}{J}_{[s]} \underset{\sim}{\beta})$$

it follows that

$$(4.5)\qquad \underset{\sim}{\alpha}_r \vee (\underset{\sim}{\beta}_s \vee \underset{\sim}{\chi}_t) = (\underset{\sim}{\alpha}_r \vee \underset{\sim}{\beta}_s) \vee \underset{\sim}{\chi}_t$$

so that the exterior product defined by (4.4) is associative, but

$$(4.6)\qquad \underset{\sim}{\alpha}_r \vee \underset{\sim}{\beta}_s = (-)^{rs}\ \underset{\sim}{\beta}_s \vee \underset{\sim}{\alpha}_r \ .$$

Also, $\underset{\sim}{\alpha}_r \vee (\underset{\sim}{\beta}_s + \underset{\sim}{\gamma}_s) = \underset{\sim}{\alpha}_r \vee \underset{\sim}{\beta}_s + \underset{\sim}{\alpha}_r \vee \underset{\sim}{\gamma}_s$, and $\underset{\sim}{\alpha}_r \vee (\lambda \underset{\sim}{\beta}_s) = (\lambda \underset{\sim}{\alpha}_r) \vee \underset{\sim}{\beta}_s = \lambda \underset{\sim}{\alpha}_r \vee \underset{\sim}{\beta}_s$. Thus, if $\underset{\sim}{\alpha} = (\underset{\sim}{\alpha}_0, \underset{\sim}{\alpha}_1, \ldots, \underset{\sim}{\alpha}_n)$ and $\underset{\sim}{\beta} = \underset{\sim}{\beta}_0, \underset{\sim}{\beta}_1, \ldots, \underset{\sim}{\beta}_n$ are any two elements of the Grassman space G and we define their product by

$$(4.7)\qquad \underset{\sim}{\alpha} \vee \underset{\sim}{\beta} = \underset{\sim}{\alpha}_0 \underset{\sim}{\beta}_0, \sum_{r+s=1} \underset{\sim}{\alpha}_r \vee \underset{\sim}{\beta}_s, \sum_{r+s=2} \underset{\sim}{\alpha}_r \vee \underset{\sim}{\beta}_s, \ldots, \sum_{r+s=n} \underset{\sim}{\alpha}_r \vee \underset{\sim}{\beta}_s \ ,$$

then the bilinear mapping $\underset{\sim}{\Gamma} : G \times G \to G$ with $\underset{\sim}{\Gamma}(\underset{\sim}{\alpha}, \underset{\sim}{\beta}) = \underset{\sim}{\alpha} \vee \underset{\sim}{\beta}$ determines tha linear associative <u>Grassman algebra</u> $(\underset{\sim}{\Gamma}, G)$. A corresponding algebra is defined in the same way in the space

$$(4.8)\qquad G^c = V^{[1]} + V^{[n]} + V^{[n^r]} + \ldots + V^{[n^n]}$$

which we denote by $(\underset{\sim}{\Gamma}, G^c)$. Since G^c is the conjugate of G, if $\underset{\sim}{\alpha} \in G$ and $\underset{\sim}{v} \in G^c$, $\underset{\sim}{v}(\underset{\sim}{\alpha})$ is defined and given by

$$(4.9)\qquad \underset{\sim}{v}(\underset{\sim}{\alpha}) = \underset{\sim}{\alpha}(\underset{\sim}{v}) = \sum_{r=0}^{n} \underset{\sim}{\alpha}_r (\underset{\sim}{v}^r) \ ,$$

where

$$\underset{\sim}{\alpha} = \underset{\sim}{\alpha}_0 \oplus \underset{\sim}{\alpha}_1 \oplus \cdots \oplus \underset{\sim}{\alpha}_n, \quad \underset{\sim}{v} = \underset{\sim}{v}^1 \oplus \cdots \oplus \underset{\sim}{v}^n \ .$$

A set of vectors (covectors) $\underset{\sim}{v}_1, \underset{\sim}{v}_2, \ldots, \underset{\sim}{v}_r$ in V^n is linearly independent if and only if the r-vector (covector) $\underset{\sim}{v}_1 \vee \underset{\sim}{v}_2 \vee \ldots \vee \underset{\sim}{v}_r$ is different from zero.

An r-vector $\underset{\sim}{w}_r \in V^{[n,r]}$ is <u>simple</u> if and only if there exists a set of r-vectors $\underset{\sim}{v}_1, \underset{\sim}{v}_2, \ldots, \underset{\sim}{v}_r$ such that $\underset{\sim}{w}_r = \underset{\sim}{v}_1 \vee \underset{\sim}{v}_2 \vee \ldots \vee \underset{\sim}{v}_r$.

If $\underset{\sim}{w}$ is an r-vector and $\underset{\sim}{v}$ a vector, then $\underset{\sim}{v}$ is a <u>divisor</u> of $\underset{\sim}{w}$ if and only if $\underset{\sim}{w} \vee \underset{\sim}{v} = \underset{\sim}{v} \vee \underset{\sim}{w} = 0$. It is known that $\underset{\sim}{v}$ is divisor of $\underset{\sim}{w}$ if and only if there exists an $(r-1)$-vector $\underset{\sim}{u}$ such that $\underset{\sim}{w} = \underset{\sim}{u} \vee \underset{\sim}{v}$.

Let $\underset{\sim}{\alpha}$ be any r-octensor. Then with respect to each argument of $\underset{\sim}{\alpha}$, say the pth, there is a set of $n^{(r-1)}$ covectors $\underset{\sim}{\alpha}^{(p)}_{i_1 i_2 \ldots i_{r-1}}$ defined by

$$(4.10) \quad \underset{\sim}{\alpha}^{(p)}_{i_1 i_2 \ldots i_r}(\underset{\sim}{u}) = \underset{\sim}{\alpha}(\underset{\sim}{e}_{i_1}; \underset{\sim}{e}_{i_2}, \ldots, \underset{\sim}{e}_{i_{p-1}}, \underset{\sim}{u}, \underset{\sim}{e}_{i_{p+1}}, \ldots, \underset{\sim}{e}_{i_r}),$$

$\underset{\sim}{e}_i$ a basis. The number of linearly independent covectors in this set is called the <u>p^{th} rank of $\underset{\sim}{\alpha}$</u>. The pth rank of r-tensors and of general mixed tensors is defined in the obvious analogous way. Every rank of an r-covector or of an r-vector has one and the same value, which is called simply its <u>rank</u>. The rank of a 2-covector (2-tensor) is always an even number. If $2s$ is the rank of the r-covector $\underset{\sim}{\alpha}$, then $\underset{\sim}{\alpha}$ is expressible as the sum of s simple 2-covectors:

$$(4.11) \quad \underset{\sim}{\alpha} = \underset{\sim}{\beta}_1 \vee \underset{\sim}{\gamma}_1 + \underset{\sim}{\beta}_2 \vee \underset{\sim}{\gamma}_2 + \ldots + \underset{\sim}{\beta}_s \vee \underset{\sim}{\gamma}_s, \text{ if rank}(\underset{\sim}{\alpha}) = 2s,$$

where the $\underset{\sim}{\beta}_p$, $\underset{\sim}{\gamma}_p$, $p = 1, 2, \ldots, s$ are linearly independent. Thus, if the $\underset{\sim}{\beta}_p$, $\underset{\sim}{\gamma}_p$ are the first 2s elements of a basis in V^n, the matrix of components of $\underset{\sim}{\alpha}$ with respect to such a basis has the values

(4.12) $$\|\alpha_{ij}\| = \mathrm{diag}\,(Q, \ldots Q, 0, 0, \ldots 0),$$

where Q is the 2×2 matrix $\begin{Vmatrix} 0 & 1 \\ -1 & 0 \end{Vmatrix}$.

If $\underset{\sim}{\alpha}_r$ is an r-covector and $\underset{\sim}{v}^r$ an r-vector, their scalar product $\underset{\sim}{\alpha}_r \cdot v^r = v^r \cdot \underset{\sim}{\alpha}_r$ is defined by

(4.13) $$\underset{\sim}{\alpha}_r \cdot \underset{\sim}{v}^r = (1/r!)\underset{\sim}{\alpha}_r(\underset{\sim}{v}^r)\,.$$

The interior product of an (r+s)-covector and an s-vector is then defined by

(4.14) $$(\underset{\sim}{\alpha}_{r+s} \wedge \underset{\sim}{v}^s) \cdot \underset{\sim}{w}^r = \underset{\sim}{\alpha}_{r+s} \cdot (\underset{\sim}{v}^s \vee \underset{\sim}{w}^s)\,, \quad \text{for all} \quad \underset{\sim}{w}^r\,.$$

The interior product of an (r+s)-vector and an s-covector is defined by

(4.15) $$(\underset{\sim}{v}^{r+s} \wedge \underset{\sim}{\alpha}_s) \cdot \underset{\sim}{\beta}_r = \underset{\sim}{v}^{r+s} \cdot (\underset{\sim}{\beta}_r \vee \underset{\sim}{\alpha}_s)\,, \quad \text{for all} \quad \underset{\sim}{\beta}_r\,.$$

5. DUALITY

Let $\underset{\sim}{e} \neq 0$ be an arbitrary n-covector with carrier space V^n. Every n-covector with an n-dimensional carrier space is simple ; therefore , there exists a linearly independent set of covectors $\underset{\sim}{e}^i$, $i = 1, 2, \ldots, n$ such that

(5.1) $$\underset{\sim}{e} = \underset{\sim}{e}^1 \vee \underset{\sim}{e}^2 \vee \ldots \vee \underset{\sim}{e}^n .$$

Let $\underset{\sim}{E}$ denote the corresponding n-vector defined by the reciprocal set $\underset{\sim}{e}_i$ of vectors :

(5.2) $$\underset{\sim}{E} = \underset{\sim}{e}_1 \vee \underset{\sim}{e}_2 \vee \ldots \vee \underset{\sim}{e}_n .$$

If the $\underset{\sim}{e}_i$ are the basis vectors in V^n , then the corresponding tensor components of $\underset{\sim}{e}$ and $\underset{\sim}{E}$ given by

(5.3) $$E^{i_1 i_2 \ldots i_n} = \underset{\sim}{E}(\underset{\sim}{e}^{i_1}, \underset{\sim}{e}^{i_2}, \ldots, \underset{\sim}{e}^{i_n}),$$
$$e_{i_1 i_2 \ldots i_n} = \underset{\sim}{e}(\underset{\sim}{e}_{i_1}, \underset{\sim}{e}_{i_2}, \ldots, \underset{\sim}{e}_{i_n})$$

have the values $E^{12 \ldots n} = +1$, $e_{12 \ldots n} = +1$. The value of every other component of $\underset{\sim}{E}$ and $\underset{\sim}{e}$ is determined by the values of these two components and the antisymmetry of $\underset{\sim}{E}$ and $\underset{\sim}{e}$. The components $E^{i_1 i_2 \ldots i_n}$ and $e_{i_1 i_2 \ldots i_n}$ are called the <u>permutation symbols.</u> Note that every linearly independent set of vectors $\underset{\sim}{e}_i$ (i.e., every basis) determines a corresponding $\underset{\sim}{E}$ and $\underset{\sim}{e}$. These should perhaps be distinguished by writing $\underset{\sim}{E}(i)$ and $\underset{\sim}{e}(i)$ to indicate their dependence on the basis $\underset{\sim}{e}_i$. The components of a given $\underset{\sim}{E}(i)$, $\underset{\sim}{e}(i)$ with respect to another basis $\underset{\sim}{e}_{i'}$, say, do not, in general, have the values given by the permutation symbols (5.3). Rather, they are determined by the general relations

$$(5.4)\qquad \underset{\sim}{E}(i') = (\det \underset{\sim}{S})\, \underset{\sim}{E}(i)\,, \qquad \underset{\sim}{e}(i') = (\det \underset{\sim}{S})^{-1}\, \underset{\sim}{e}(i)\,,$$

where $S^{i}_{j'}$ is the matrix which defines the $\underset{\sim}{e}_{i'}$ as a linear combination of the set $\underset{\sim}{e}_i$:

$$(5.5)\qquad \underset{\sim}{e}_{j'} = S^{i}_{j'}\, \underset{\sim}{e}_i.$$

It follows that the mixed tensor $\underset{\sim}{\delta} = \underset{\sim}{E}(i) \otimes \underset{\sim}{e}(i)$ is independent of i. The mixed tensor $\frac{1}{r!}\underset{\sim}{\delta}$ so defined is the identity transformation $\frac{1}{r!}\underset{\sim}{\delta} : V^{[n^n]} - V^{[n^n]}$. More generally, the identity transformation $(1/r!)\delta_r : V^{n^r} - V^{n^r}$ has tensor components with respect to an arbitrary basis $\underset{\sim}{e}_i$ in V^n given by

$$(5.6)\qquad \delta^{i_1 i_2 \ldots i_r}_{j_1 j_2 \ldots j_r} = (1/n-r)!)\, E^{i_1 i_2 \ldots i_r k_1 \ldots k_{n-r}}\, e_{j_1 j_2 \ldots j_r k_1 \ldots k_{n-r}}$$

where $\underset{\sim}{E}$ and $\underset{\sim}{e}$ are the n-vector and n-covector defined in terms of an arbitrary linearly independent set of vectors.

Formulas like (5.6) point up the need for a more efficient and condensed notation when dealing with components of r-vectors and r-covectors. For

$$\alpha^{\cdots}_{\cdots\, i_1 i_2 \ldots i_r}$$

where α is antisymmetric in the indices $i_1 \ldots i_r$, let us write

$\alpha^{\cdots}_{\cdots(i)}$, and for the contracted product such as occurs in (5.6), let us write

(5.7) $$(1/r!)\alpha^{\cdots}_{\cdots i_1 i_2 \cdots i_r}\beta^{\cdots i_1 i_2 \cdots i_r}_{\cdots} = \alpha^{\cdots}_{\cdots(i)}\beta^{\cdots(i)}_{\cdots}$$

where $\underset{\sim}{\alpha}$ and $\underset{\sim}{\beta}$ are general mixed tensors. Thus, for examples, the scalar product defined in (4.13) is given in the condensed notation, in terms of components by

$$\alpha_r \cdot v^r = \alpha_{r(i)} v^{r(i)} ,$$

and, more briefly, by $\alpha_{(i)}v^{(i)}$, when the value of r is clear from the context or unimportant for the meaning of the term.

For each choice of $\underset{\sim}{e}$ and $\underset{\sim}{E}$ we can show that the mappings

(5.8) $$\underset{\sim}{D}: V^{[{}_n^r]} \to V_{[{}_n^{n-r}]}$$
$$\underset{\sim}{D'}: V_{[{}_n^r]} \to V^{[{}_n^{n-r}]},$$

defined by

(5.9) $$\underset{\sim}{D}(\underset{\sim}{v}^r) = \underset{\sim}{e} \wedge \underset{\sim}{v}^r ,$$
$$\underset{\sim}{D'}(\underset{\sim}{\alpha}_r) = \underset{\sim}{E} \quad \underset{\sim}{\alpha} ,$$

have the property

(5.10) $$\underset{\sim}{D}\ \underset{\sim}{D'} = \underset{\sim}{\delta}^r = \underset{\sim}{D'}\ \underset{\sim}{D}.$$

Thus, $\underset{\sim}{D}$ and $\underset{\sim}{D'}$ are 1-1 and onto. $\underset{\sim}{D}\,\underset{\sim}{v}^r$ is called the <u>dual</u>

of $\underset{\sim}{v}^r$ and $\underset{\sim}{D}'\underset{\sim}{\alpha}_r$ is called the <u>dual</u> of $\underset{\sim}{\alpha}_r$. It must be kept in mind that the duality isomorphism between the spaces of r-vectors and (n-r)-covectors established by D and D' depends on the choice of basis used to define $\underset{\sim}{E}$ and $\underset{\sim}{e}$. Because of the relations (5.4), two bases $\underset{\sim}{e}_i$ and $\underset{\sim}{e}_{i'}$ determine different isomorphisms unless $\underset{\sim}{e}_i$ and $\underset{\sim}{e}_{i'}$ are related by a transformation with determinant +1 ; i.e., by a unimodular transformation.

The interior and exterior products of r-covectors and r-vectors introduced above and the duality mappings based on an n-vector and n-covector are related to the classical cross product of Gibbs' vector analysis in the following way :

<u>Definition of cross product</u> :

$$\underset{\sim}{\alpha} \times \underset{\sim}{\beta} = \underset{\sim}{D}'(\underset{\sim}{\alpha} \vee \underset{\sim}{\beta})$$

$$\underset{\sim}{u} \times \underset{\sim}{v} = \underset{\sim}{D}(\underset{\sim}{u} \vee \underset{\sim}{v}) .$$

6. QUADRATIC FORMS

A. 2-cotensor

(6.1) $\underset{\sim}{g} : V^n \times V^n \to R$

is <u>symmetric</u> if

(6.2) $\underset{\sim}{g}(\underset{\sim}{v}, \underset{\sim}{u}) = \underset{\sim}{g}(\underset{\sim}{u}, \underset{\sim}{v})$.

A symmetric 2-cotensor is called a <u>quadratic form</u> . Its rank with respect to either argument has a common value called the <u>rank</u> of g. If the rank is n, g is <u>nonsingular</u>; otherwise, g is <u>singular.</u> If $\underset{\sim}{g}(\underset{\sim}{v}, \underset{\sim}{v}) > 0$ (<0) for all $\underset{\sim}{v} \neq 0 \in V^n$, $\underset{\sim}{g}$ is <u>positive (negative) definite</u> ; otherwise, g is <u>indefinite.</u> A basis $\underset{\sim}{e}_i$ can always be found such that $g_{ij} = g(e_i, e_j) =$ diag (1, 1, . , . , -1, -1, . . . 0, 0) . The diagonal matrix defined by $\underset{\sim}{g}$ in this way is called the <u>signature</u> of g. It is also called the <u>canonical form</u> of the matrix of components g_{ij} .

If $\underset{\sim}{g}$ is nonsingular, then the mapping (cf. the discussion in 2)

(6.3) $\underset{\sim}{g}^* : V^n \to V_n$

defined by

(6.4) $\underset{\sim}{g}^*(\underset{\sim}{v})(\underset{\sim}{w}) = \underset{\sim}{g}(\underset{\sim}{v}, \underset{\sim}{w})$

is 1-1 and onto . In physical theories and in Riemannian geometry where a particular nonsingular quadratic form $\underset{\sim}{g}$ plays a central and dominant role, it is customary to identify the elements $\underset{\sim}{v} \in V^n$ and their images $\underset{\sim}{g}(\underset{\sim}{v})$ under $\underset{\sim}{g}$, and to regard them merely as different representations of the "same" vector. Here we shall denote the covector $\underset{\sim}{g}(\underset{\sim}{v})$ by $\underset{\sim}{v}^{\dagger}$ and the vector $\underset{\sim}{g}^{*\,-1}(\underset{\sim}{a})$ by $a^{\dagger}$. Also, while it is customary to denote the components of the inverse of $\underset{\sim}{g}^{*}$ by $g^{\alpha\beta}$, we shall denote $\underset{\sim}{g}^{*\,-1}$ by $\underset{\sim}{g}^{\dagger}$ and its components by $g^{\dagger\alpha\beta}$.

These are special case of the general rules of tensor notation concerning the "raising and lowering" of indices by transvection with a fundamental symmetric, nonsingular 2-cotensor and its inverse. More generally now, "raising" all the indices of an r-covector defines an isomorphism

(6.5) $$\underset{\sim}{g}^{*} : V_{[n^r]} \rightarrow V^{[n^r]}$$

defined by

(6.6) $$a^{\dagger\, i_1 i_2 \ldots i_r} = g^{\dagger\, i_1 j_1} g^{\, i_2 j_2} \ldots g^{\dagger\, i_r j_r} a_{j_1 j_2 \ldots j_r}$$

$$\underset{\sim}{a}^{\dagger} = \underset{\sim}{g}^{*}(\underset{\sim}{a}) ,$$

and "lowering" all the indices of an r-vector defines the inverse transformation

(6.7) $$g^{*\,-1} : v^{[n^r]} \rightarrow V_{[n^r]} ,$$

and one has

(6.8) $$\underset{\sim}{a}^{\dagger\dagger} = \underset{\sim}{g}^{*\,-1}(\underset{\sim}{a}^{\dagger}) = \underset{\sim}{a} .$$

Every non-singular quadratic form in V^n determines such an isomorphism between the spaces $V^{[n^r]}$ and $V_{[n_r]}$.

The isomorphisms g_r^* determined in this way by different quadratic forms are distinct, as are the duality transformations $\underset{\sim}{D}(i)$ and $\underset{\sim}{D}(i')$ determined by linearly independent sets $\underset{\sim}{e}_i$ and $\underset{\sim}{e}_{i'}$ not related by a unimodular transformation.

Let

(6.9) $$g = (1/n!)\, E^{i_1 i_2 \dots i_n} E^{j_1 j_2 \dots j_n} g_{i_1 j_1} \dots g_{i_n j_n}$$

denote the determinant of the quadratic form $\underset{\sim}{g}$. Note that the value of the determinant depends on the basis used to define $\underset{\sim}{E}$.

Two ordered linearly independent sets of vectors $\underset{\sim}{v}_i$ and $\underset{\sim}{v}_{i'}$ in V^n are said to have the same orientation if they are related by a transformation with positive determinant : otherwise, they are said to have opposite orientation. A V^n together with an ordered linearly independent set of vectors in it is an oriented n-dimensional vector space $\vec{V}^n$. (V^n, $\underset{\sim}{v}_i$) and (V^n, $\underset{\sim}{v}_{i'}$) determine the same $\vec{V}^n$ provided $\underset{\sim}{v}_i$ and $\underset{\sim}{v}_{i'}$ have the same orientation ; otherwise, they are regarded as different oriented V^n, say $\vec{V}^n$ and $\overleftarrow{V}^n$.

Consider now a $\vec{V}^n$ and set

(6.10) $$\underset{\approx}{D} = (\epsilon\sqrt{|g|}\,)\ \underset{\sim}{D}(i)$$

where $\epsilon = +1$ if the set $\underset{\sim}{e}_i$ used to define the dual transformation has the same orientation as $\vec{V}^n$ and $\epsilon = -1$ otherwise. This definition of $\underset{\approx}{D}$ is independent of the basis used to define $\underset{\sim}{D}(i)$. One then has

(6.11) $$g_r^{*-1}\, \underset{\approx}{D} = (-)^{r(n-r)}\ \frac{g}{|g|}\ \underset{\approx}{D}^{-1}\, g_r^* .$$

7. CHAINS AND COCHAINS

Let A^n denote an n-dimensional affine space with translations V^n. We call elements of V^n vectors and the elements of A^n points. Points are denoted by lightface, lower-case Latin letters p, q, ..., etc. We write

$$\underset{\sim}{v}(p) = p + \underset{\sim}{v}$$

for the image of p under the translation $\underset{\sim}{v}$, and

$$\underset{\sim}{v} = p - q$$

for the unique element of V^n defined by the pair of points (p.q) . We say that the vector $\underset{\sim}{v}$ points from q to p.

An oriented r-simplex $s_r \subset A^n$ is determined by giving an ordered set of r+1 points $p_0 p_1 p_2 \ldots p_r$ called the vertices of s_r. The simplex s_r consists in the set of points p given by

$$(7.1) \qquad p = p_0 + \sum_{i=1}^{r} a^i \underset{\sim}{v}_i, \quad 0 < a^i < 1 , \quad \underset{\sim}{v}_i = p_i - p_0 ,$$

and it is assumed that the vectors $\{ \underset{\sim}{v}_i \}$ are linearly independent. The orientation of s_r is given by the orientation of the set of r-vectors $\{ \underset{\sim}{v}_i \}$.

The r-vector $\underset{\sim}{s}_r$ of s_r is defined by

$$(7.2) \qquad \underset{\sim}{s}_r = \frac{1}{r!} \underset{\sim}{v}_1 \vee \underset{\sim}{v}_2 \vee \cdots \vee \underset{\sim}{v}_r .$$

A Euclidean space E^n is an affine space with a positive definite quadratic form $\underset{\sim}{q}(\underset{\sim}{u}, \underset{\sim}{v})$ in the translation space V^n of E^n.

The <u>r-direction</u> $\underset{\sim}{d}_r$ of an r-simplex $s_r \subset E^n$ is defined by

(7.3) $$\underset{\sim}{d}_r = \frac{\underset{\sim}{s}_r}{|\underset{\sim}{s}_r|_q}, \qquad |\underset{\sim}{s}_r|_q = q_r(\underset{\sim}{s}_r, \underset{\sim}{s}_r),$$

where q_r is the quadratic form in $V^{[r]}$ whose definition has been given in §5. We call $|\underset{\sim}{s}_r|_q$, the <u>r-volume</u> of s_r.

The <u>center</u> p_c of an r-simplex (6.1) is the point defined by

(7.4) $$p_c = p_0 + \left(\frac{1}{r+1}\right) \sum_{i=1}^{r} \underset{\sim}{v}_i .$$

An <u>r-cube</u> $t_r \subset A^n$ is the set of points given by

(7.5) $$p = p_0 + \sum_{i=1}^{r} a^i \underset{\sim}{v}_i, \quad 0 < a^i < 1,$$

where the r vectors $\underset{\sim}{v}_i$ are linearly independent, and the orientation of t_r is the orientation determined by the set of r vectors $\{\underset{\sim}{v}_i\}$. The point $p_e = p_0 + \frac{1}{2}\Sigma \underset{\sim}{v}_i$ in (7.5) is the <u>center</u> of t_r. The r-vector of t_r is $\underset{\sim}{t}_r = \underset{\sim}{v}_1 \vee \underset{\sim}{v}_2 \vee \ldots \vee \underset{\sim}{v}_r$, its r-direction is $\underset{\sim}{d}_r = \underset{\sim}{t}_r / |\underset{\sim}{t}_r|$, and $|\underset{\sim}{t}_r|_q$ is its r-volume.

An <u>r-chain</u> $c_r \subset A^n$ is a linear combination of a finite number of nonoverlapping r-simplexes $s_{r\alpha}$, $\alpha = 1, 2, \ldots$ with real coefficients. Two r-simplexes c_r and c'_r are nonoverlapping if $c_r \cap c'_r$ is the empty set of points in A^n, or the points of an s-simplex, $s < r$. By $0\, s_r$ is meant the empty set of points; by $-1\, s_r$ is meant the simplex comprising the same set of points as the simplex s_r, but with opposite orientation. If $c_r = \sum_\alpha c^\alpha s_{r\alpha}$, and $c'_r = \sum_\alpha c'^\alpha s_{r\alpha}$ are two r-chains in

A^n, we define $\lambda c_r + \eta c'_r$ as the r-chain $\sum_{\alpha} (\lambda c^{\alpha} + \eta c'^{\alpha}) s_{r\alpha}$ so that the set C_r of all r-chains in A^n is a linear space.

An r-cochain F is a real-valued linear function, $F: C_r \to R$, of r-chains.

8. COCHAINS DEFINED BY INTEGRATION OF r-COVECTOR FIELDS

Let A^n be an affine space with translations V^n. A tensor field in A^n is a mapping $T: A^n \to V$ which assigns to each point $p \in A^n$ a tensor $\underset{\sim}{T}(p)$ in a tensor space V with carrier space V^n.

Let $\underset{\sim}{q}$ be an arbitrary positive definite quadratic form in V^n so that $|\underset{\sim}{T}|_q$ and $|p-q|_q = \underset{\sim}{q}(\underset{\sim}{v}, \underset{\sim}{v})$, $\underset{\sim}{v} = p-q$ define certain norms in V and A^n, respectively, in terms of which we may define the continuity of a tensor field T. If T is continuous with respect to the norm $|\ |_q$, then it is continuous with respect to any other norm $|\ |_{q'}$, defined in this way, We say this to emphasize that the considerations of this section are independent of the choice of the positive definite quadratic form $\underset{\sim}{q}$.

Set

$$\nabla \underset{\sim}{T}(p, v) = \lim_{t \to 0^+} \frac{\underset{\sim}{T}(p+tv) - \underset{\sim}{T}(p)}{t} ;$$

then, if the limit exists, ∇T is linear in $\underset{\sim}{v}$ and hence is a tensor with carrier space V^n; we denote it by $\nabla \underset{\sim}{T}(p)$ and call it the gradient of T at the point p. If $\nabla \underset{\sim}{T}(p)$ is defined for each $p \in A^n$, then ∇T is a tensor field in A^n. Let $R \subset A^n$ be a region of A^n. If T is continuous at each point $p \in R$, we say that the field T is continuous in R. If ∇T exists and is continuous at each point $p \in R$, we say that T is 1-smooth or smooth in R. If $\nabla \nabla T$ exists and is continuous in R, we say that ∇T is 2-smooth in R. Proceeding in this way, we define r-smooth tensor fields in regions R of A^n.

An <u>r-covector field</u> in A^n is a special case

(8.1) $$\varphi : A^n \to V_{[n^r]}$$

of a tensor field defined more generally above. For brevity, we shall call an r-covector field in A^n, an <u>r-form</u> in A^n.

The integral of an r-form over an r-chain,

(8.2) $$F(c_r , \varphi) = \int_{c_r} \varphi$$

is defined as follows. First of all, we set

(8.3) $$F(c_r , \varphi) = F(\sum_\alpha a^\alpha s_{r\alpha}, \varphi) = \sum_\alpha a^\alpha \int_{s_{r\alpha}} \varphi ,$$

so that it suffices to define the integral of φ over an r-simplex s_r. Now for every value of $\epsilon > 0$ an r-simplex s_r can be (subdivided) expressed as an r-chain of the form

$$s_r = \sum_{\beta=1}^{N(\epsilon)} s_{r\beta} ,$$

where each $s_{r\beta}$ has the same r-direction as s_r, and $\operatorname{diam}_q(s_{r\beta}) < \epsilon \operatorname{diam}_q(s_r)$, (q arbitrary). Let $S_1, S_2, \ldots$ be any sequence of such subdivisions of s_r such that $\epsilon_k \to 0$. For the subdivision S_k, let $p_{k\beta}$ be the center of the simplex $s_{r\beta}$. Set

(8.4) $$\int_{s_r} \varphi = \lim_{k \to \infty} \sum_{\beta=1}^{N_k} \varphi(p_{k\beta}) \cdot s_{r\beta} .$$

An elegant proof that the limit (8.4) exists and is independent of the sequence $S_1, S_2, \ldots$ if φ is continuous in p is given by Whitney . It follows from the definitions (8.2) and (8.4) that every continuous r-form in A^n determines a unique r-cochain F and that $F(a\varphi + b\varphi') = aF(\varphi) + bF(\varphi')$. A cochain defined in this way is called a <u>continuous r-cochain.</u> If the r-form φ is s-smooth, then F is <u>s-smooth</u>.

A continuous r-form φ in $R \subset A^n$ is called <u>regular</u> in R if there exists a continuous (r+1)-form rot φ such that, for every simplex $s_{r+1} \subset R$

$$(8.5) \qquad \int_{\partial s_{r+1}} \varphi = \int_{s_{r+1}} \text{rot } \varphi ,$$

where ∂s_{r+1} denotes the <u>boundary</u> of the simplex s_{r+1} <u>oriented</u> as follows : if s_{r+1} has vertices $[p_0 p_1 \ldots p_{r+1}]$, then s_{r+1} is the r-chain given by the sum of r-simplexes

$$(8.6) \qquad \partial s_{r+1} = \sum_{t=0}^{r+1} (-)^t s_r(p_0 0_1 \ldots \hat{p}_{r+1-t}, \ldots, p_{r+1}) ,$$

where $s_r(p_0 p_1 \ldots p_{r+1-t}, \ldots, p_{r+1})$ denotes the simplex with ordered vertices $p_0 p_1 \ldots p_{r+1}$ with p_{r+1-t} omitted.

With these definitions, the famous theorems of Gauss, Stokes, Kelvin, Poincaré, and others may be viewed as a special case of the

<u>Divergence theorem; Every smooth r-form</u> φ <u>in</u> $R \subset A^n$ is <u>a regular r-form in R; moreover, for smooth</u> φ,

$$(8.7) \qquad \text{rot } \varphi = \partial \nabla \varphi = (r+1) \; J_{r+1}(\Delta\varphi) .$$

A proof of the divergence theorem is not difficult fo simplexes and chains. Later, we shall consider r-forms in a smooth manifold, and the definition of $\int \varphi$ will be extended to smooth manifolds and smooth manifolds embedded in a smooth manifold . In this way we get a quick proof of the divergence theorem for a much wider class of regions in A^n.

A regular r-form φ is closed (irrotational) in R if rot $\varphi = 0$. Thus , in other words, an r-form φ is irrotational in R if and only if the continuous r-chain $F(\varphi)$ has the property

$$F(\partial s_{r+1}, \varphi) = \int_{\partial s_{r+1}} \varphi = 0 \quad \text{every} \quad s_{r+1} \subset R.$$

An r-form φ is circulation free in R if there exists a regular (r-1)-form π such that, throughout R, $\varphi = \text{rot } \pi$. The r-form π is called a potential of the circulation free r-form φ. Every circulation free r-form in an arbitrary region $R \subset A^n$ is irrational,

$$\int_{\partial c_r} \varphi = \int_{\partial c_r} (\text{rot } \pi) = \int_{\partial\partial c_r} \pi = 0 . \tag{8.8}$$

where we have used the property $\partial\partial c_r = 0$ of the boundary operator ∂. But it is not true that every irrotational r-form φ in an arbitrary region R is circulation free. (Let R be the annulus $p = p_0 + \underset{\sim}{v}$, $a \leq |\underset{\sim}{v}| \leq b$, $a > 0$, $b > 0$, in E^2 and let φ be the 1-form with components (0, 1) in every polar coordinate system for which p_0 is the origin.) Whitney has shown , however, that every irrotational r-form φ in a starshaped region $R \subset A^n$ is circulation free in A^n, and he has given an explicit construction of a potential π for φ in R.

The potential π of a circulation free r-form φ is not unique. Clearly, if the r-form $\varphi = \text{rot } \pi$ in R, then $\varphi = \text{rot } \pi'$ in R also where $\pi' =$ $= \pi + \text{rot } \gamma$ where γ is any regular (r-2)-form.

9. CONTINUOUS r-COVECTOR FIELDS DEFINED BY CERTAIN r-COCHAINS

In the previous section, the continuous r-cochain $F(\varphi)$ was defined for every continuous r-form φ in A^n. In this section, it will be shown how every r-cochain F of a certain class determines a unique r-form $\varphi(F)$. The characterization of this class of cochains and the proof of the existence of $\varphi(F)$ are due to Whitney. I shall sketch here in some detail Whitney's work, for I feel that it has wide applications in continuum mechanics. A very special case of Whitney's theorem to be discussed below will be recognized by experts in continuum mechanics as a new and novel approach to the concept of stress and the existence of a stress tensor. In classical field theories, r-forms represent the most basic and primitive physical quantities. The electromagnetic field, the gravitional field, the charge and current fields, and the stress tensor (a vector-valued 2-form) are familiar examples. But the concept of a field (r-form) is sophisticated indeed (except, perhaps, a 0-form) for it carries with it the rather complicated notion of its r-direction at each point. I believe that the concept of a real valued linear function of r-simplexes or chains lies closer to physical intuition than the concept of an r-form. Therefore, in the part of these lectures which concerns physical theory, the definitions of the basic physical quantities to occur will be given in terms of the values of r-cochains. To make contact with the more traditional view which introduces the gravitational field of force or the electromagnetic field as primitives we shall need the following results.

An r-cochain $F: C_r \to R$ is semi-sharp if

(a) For each bounded region $R \subset A^n$, there exists an N_R such that

$$|F(s_r)| < N_R \, |\underset{\sim}{s}_r|_q .$$

(b) For each point $p \in A^n$ and $\epsilon > 0$ there exists a ζ such that for any (r+1)-simplex s_{r+1} contained in the r-cube $U_\zeta(p)$ of diameter ζ and center p,

$$|F(\partial s_{r+1})| \leq |\underset{\sim}{s}_{r+1}|_q .$$

(c) One may choose ζ in (b) such that for any r-simplex s_r and vector $\underset{\sim}{v} \in V^n$ (the translation space of A^n)

$$|F(T_{\underset{\sim}{v}} s_r - s_r)| \leq \epsilon |\underset{\sim}{s}_r| \text{ if } s_r \subset U_\zeta(p), \ |\underset{\sim}{v}| \leq \zeta.$$

In (c) $T_{\underset{\sim}{v}} s_r$ is the simplex consisting in the set of points of s_r translated by $\underset{\sim}{v}$ and having the same orientation as s_r.

Theorem (Whitney): If F is a semi-sharp r-cochain, there exists a continuous r-covector field φ such that

$$F(c_r) = \int_{c_r} \varphi .$$

In other words, every semi-sharp r-cochain is a continuous r-cochain.

The proof of the theorem rests in part on the following

Lemma (Whitney); Let φ be a real valued function of simple r-vectors such that

(1) φ is homogeneous of degree one:

$$\varphi(a\underset{\sim}{v}_r) = a\varphi(\underset{\sim}{v}_r), \quad \underset{\sim}{v}_r = \underset{\sim}{v}_1 \vee \underset{\sim}{v}_2 \vee \dots \vee \underset{\sim}{v}_r$$

(2) $\sum_{\beta=0}^{r+1} \varphi(\underset{\sim}{s}_{r\beta}) = 0$, <u>for every (r+1)-simplex</u> s_{r+1}, <u>where the boundary of</u> $\underset{\sim}{s}_{r+1}$ <u>is given by</u> $\sum_{\beta=0}^{r+1} \underset{\sim}{s}_{r\beta}$.

(In words, the last condition reads, the sum of the values of φ on the r+1 - oriented r-vectors of the faces of every (r+1)-simplex is zero.) <u>Then there exists a unique r-covector</u> $\underset{\sim}{\varphi}$ <u>such that</u> $\underset{\sim}{\varphi}\,\underset{\sim}{v}_r = \varphi(\underset{\sim}{v}_r)$.

In other words, every φ with properties (1) and (2) is <u>linear</u> in $\underset{\sim}{v}_r$ and defines a unique r-covector $\underset{\sim}{\varphi}$. The uniqueness of $\underset{\sim}{\varphi}$ is immediate. I present a somewhat simpler proof of the linearity of φ than given by Whitney.

Set

$$F(\underset{\sim}{v}_1, \underset{\sim}{v}_2, \ldots, \underset{\sim}{v}_r) = \varphi(\underset{\sim}{v}_1 \vee \underset{\sim}{v}_2 \vee \ldots \vee \underset{\sim}{v}_r) .$$

Then, by (1), F is homogeneous of degree 1 in each argument. We show that it must be linear. Suppose r=1, and consider a 2-simplex with faces having 1-directions, $\underset{\sim}{u}, -\underset{\sim}{v}$, and $\underset{\sim}{v}-\underset{\sim}{u}$. Then, by (2),

$$F(\underset{\sim}{u}) + F(\underset{\sim}{v}-\underset{\sim}{u}) + F(-\underset{\sim}{v}) = 0 ,$$

and using (1),

$$F(\underset{\sim}{v}-\underset{\sim}{u}) = F(\underset{\sim}{v}) + F(-\underset{\sim}{u}),$$

or, setting $\underset{\sim}{w} = -\underset{\sim}{u}$,

$$F(\underset{\sim}{v}+\underset{\sim}{w}) = F(\underset{\sim}{v}) + F(\underset{\sim}{w}) ,$$

which proves that F is linear if r = 1. The general case r > 1 is illustrated sufficiently be case r = 2. Consider the 3-simplex s with $\underset{\sim}{v}_i = p_i - p_0$, i = 1, 2, 3, $p_0p_1p_2p_3$ the vertices of s. It follows from (2) that

$$(9.1)\qquad F(\underset{\sim}{v}_2-\underset{\sim}{v}_3,\ \underset{\sim}{v}_1-\underset{\sim}{v}_3)+F(\underset{\sim}{v}_1,\underset{\sim}{v}_2)+F(\underset{\sim}{v}_2,\underset{\sim}{v}_3)+F(\underset{\sim}{v}_3,\underset{\sim}{v}_1)=0$$

for all linearly independent $\underset{\sim}{v}_1, \underset{\sim}{v}_2, \underset{\sim}{v}_3$. In (9.1) replace the arguments $\underset{\sim}{v}_1, \underset{\sim}{v}_2, \underset{\sim}{v}_3$ by $\underset{\sim}{v}_1$, $a\underset{\sim}{v}_2$, and $a\underset{\sim}{v}_3$, respectively. Then, for every $a > 0$,

$$(9.2)\qquad F(a(\underset{\sim}{v}_2-\underset{\sim}{v}_3),\ \underset{\sim}{v}_1-a\underset{\sim}{v}_3)+F(\underset{\sim}{v}_1,a\underset{\sim}{v}_2)+F(a\underset{\sim}{v}_2,a\underset{\sim}{v}_3)=0\ .$$

Now use (1) to obtain

$$(9.3)\qquad aF(\underset{\sim}{v}_2-\underset{\sim}{v}_3,\ \underset{\sim}{v}_1-a\underset{\sim}{v}_3)+aF(\underset{\sim}{v}_1,\underset{\sim}{v}_2)+a^2F(\underset{\sim}{v}_2,\underset{\sim}{v}_3)+aF(\underset{\sim}{v}_3,\underset{\sim}{v}_1)=0.$$

Divide (9.3) by a. The limit of the resulting expression for $a \to 0$ is

$$(9.4)\qquad F(\underset{\sim}{v}_2-\underset{\sim}{v}_3,\underset{\sim}{v}_1)+F(\underset{\sim}{v}_1,\underset{\sim}{v}_2)+F(\underset{\sim}{v}_3,\underset{\sim}{v}_1)=0.$$

Now using (1), the antisymmetry of F, and setting $-\underset{\sim}{v}_3=\underset{\sim}{w}$, one gets

$$F(\underset{\sim}{v}_2+\underset{\sim}{w},\ \underset{\sim}{v}_1)=F(\underset{\sim}{v}_2,\underset{\sim}{v}_1)+F(\underset{\sim}{w},\underset{\sim}{v}_1),$$

which proves the linearity of F in its first argument. By antisymmetry, it is linear also in the second argument. Hence, F is an antisymmetric, multilinear function of two vectors i.e., φ is a 2-covector.

<u>Proof of the main theorem</u> : Let F be a semi-sharp cochain so that (a), (b), and (c) hold. Choose any point p and an r-direction $\underset{\sim}{d}$. Let $s_1, s_2, \ldots$ be a sequence of r-simplexes containing p and having the common r-direction $\underset{\sim}{d}$; $\underset{\sim}{d}=\underset{\sim}{s}_i/|\underset{\sim}{s}_i|_q$, i=1, 2,... . Suppose that diam $(s_i) \to 0$. Set

(9.5) $$\varphi(p,\underset{\sim}{d}) = \lim_{i\to\infty} F(s_i)/|\underset{\sim}{s}_i|_q .$$

Existence and uniqueness of the limit is proved as follows : Let s be any r-simplex with r-direction $\underset{\sim}{d}$ such that $s \subset U_\zeta(p)$. Then one can choose an r-cube τ containing p such that $T_{\underset{\sim}{v}_k}\tau \subset s$, $k = 1, 2, \ldots, n$ and

(9.6) $$\left| s - \sum_{k=1}^{n} T_{\underset{\sim}{v}_k} \tau \right| \leq \epsilon_1 ,$$

for every $\epsilon_1 > 0$. Then , by property (a) of F,

(9.7) $$\left| F\left(s - \sum_{k=1}^{n} T_{\underset{\sim}{v}_k} \tau\right)\right| \leq N \left| s - T_{\underset{\sim}{v}_k} \right|_q \leq N\epsilon_1 .$$

By property (c) of F, it follows that, for each value of k,

(9.8) $$\left| F\left(T_{\underset{\sim}{v}_k} \tau - \tau\right)\right| < \epsilon\, |\tau|_q ,$$

so that by summing (9.8) over the s values of k,

(9.9) $$F\left|\left(\sum_{k=1}^{n} T_{\underset{\sim}{v}_k} \tau - s\,\tau\right)\right| \leq \sum_{k=1}^{n}\left| F\left(T_{\underset{\sim}{v}_k}\tau - \tau\right)\right| \leq n\,\epsilon\, |\tau|_q\; |s|_q .$$

Therefore ,

$$|F(s)|\tau| - F(\tau)|s|| = \left| F(s) - F\left(\sum_{k=1}^{n} T_{\underset{\sim}{v}_k} \tau\right) |\tau| + F\left(\sum_{k=1}^{n} T_{\underset{\sim}{v}_k} \tau\right)|\tau| - F(\tau)|s|\right|$$

$$\leq N\epsilon_1|\tau| + \left| F\left(\sum_{k=1}^{n} T_{\underset{\sim}{v}_k} \tau\right)|\tau| - F(\tau)|s|\right|$$

$$\leq N\epsilon_1|\tau| + |\tau|\left|F(\Sigma T_{\underset{\sim}{v}_k} \tau) - nF(\tau)\right| + |nF(\tau)|\tau| - F(\tau)|s||$$

$$\leq N\epsilon_1|\tau| + \epsilon|\tau||s| + |\tau|N(n|\tau| - |s|)$$

$$\leq N\epsilon_1|\tau| + |\tau||s|\epsilon + N\epsilon_1|\tau|.$$

Dividing by $|\tau|\ |s|$ one gets

$$\left|\frac{F(s)}{|s|} - \frac{F(\tau)}{|\tau|}\right| \leq 2\frac{\epsilon_1 N}{|s|} + \epsilon.$$

For each s , choose $\epsilon_1 = |s|\epsilon/(2N)$. Then

(9.10) $$\left|\frac{F(s)}{|s|} - \frac{F(\tau)}{|\tau|}\right| \leq 2\epsilon.$$

The inequality (9.10) holds for every $s \subset U_\zeta(p)$; hence, for any pair s and $s' \subset U_\zeta(p)$

(9.11) $$\left|\frac{F(s)}{|s|} - \frac{F(s')}{|s'|}\right| \leq 4\epsilon .$$

This proves the existence and uniqueness of the limit in (9.5) . Using (9.11) it is also easy to see that $\varphi(p, \underset{\sim}{d})$ is continuous in the argument p :

$$(9.12)\quad |\varphi(p,\underset{\sim}{d}) - (p',\underset{\sim}{d}) = \left| \lim_{i\to\infty} \frac{F(s_i)}{|s_i|} - \lim_{j\to\infty} \frac{F(s'_j)}{|s'_j|} \right| ,$$

where s_i, $i = 1, 2, \ldots$ is a sequence of r-simplexes with r-direction $\underset{\sim}{d}$ each of which contains the point p and s'_j, $j = 1, 2, \ldots$ is a sequence with r-direction $\underset{\sim}{d}$, each of which contains the point p'. We may choose $s_i = T_{\underset{\sim}{v}} s'_i$, where $\underset{\sim}{v} = p - p'$. Then (9.12) becomes

$$(9.13)\quad |\varphi(p,\underset{\sim}{d}) - \varphi(p',\underset{\sim}{d})| = \lim_{i=\infty} \left| \frac{F(T_{\underset{\sim}{v}} s'_i) - F(s'_i)}{|s'_i|} \right| .$$

Using property (c) of F we now get

$$(9.14)\quad |\varphi(p,\underset{\sim}{d}) - \varphi(p',\underset{\sim}{d})| < \epsilon ,\ \text{for all}\ |\underset{\sim}{v}| = |p-p'| \leq \zeta ,\ s'_i \subset U_\zeta(p) .$$

From the definition of the Riemann integral $\int_s f(p)dp$ of a continuous function $f: A^n \to R$, it now follows that ,

$$(9.14)\qquad F(s) = \int_s \varphi(p,\underset{\sim}{d})\, dp$$

for every simplex s with r-direction $\underset{\sim}{d}$. Cut s into simplexes $s_1, s_2, \ldots, s_n$ of diam $< \zeta$. For each s_i let $p_i \in s_1$.
Then

$$\left| F(s_i) - \int_{s_i} \varphi(p,\underset{\sim}{d})dp \right| \leq \left| F(s_i) - \varphi(p_i,\underset{\sim}{d}) |s_i| \right| + \left| \int_{s_i} [\varphi(p_i,\underset{\sim}{d}) - \varphi(p,\underset{\sim}{d})]\, dp \right|$$

$$\leq 5\ \epsilon\ |s_i| \ ;$$

hence,

$$\left| F(s) - \int_s \varphi(p, \underset{\sim}{d})dp \right| \leq 5\epsilon \quad |s| \; ;$$

which proves (8.14).

To this point, only the properties (a) and (c) of F have been used. Property (c) is now invoked to prove that φ has the properties (1) and (2) of the Lemma. Now $\varphi(p, \underset{\sim}{d})$ has been defined only for r-directions $\underset{\sim}{d}$; i.e., only for simple r-vectors such that $|\underset{\sim}{d}| = 1$. Define $\varphi(p, \underset{\sim}{v})$ for all simple r-vectors $\underset{\sim}{v}$ by setting

$$\varphi(p, \underset{\sim}{v}) = |\underset{\sim}{v}| \, \varphi(p, v / |\underset{\sim}{v}|) .$$

Since, by definition, $\varphi(p, \underset{\sim}{d}) = - \varphi(p, -\underset{\sim}{d})$,

$$\varphi(p, a\underset{\sim}{v}) = a\,\varphi(p, \underset{\sim}{v}) \tag{9.15}$$

for all real a. Hence, $\varphi(p, \underset{\sim}{v})$ satisfies the first condition of the lemma.

Let $p \in s$ be an interior point of an (r+1)-simplex and let s_λ be s contracted towards p by the factor λ and set $\partial s_\lambda = \sum_i s_{\lambda i}$, where the $s_{\lambda i}$ are the oriented faces of s_λ. Then

$$|\underset{\sim}{s}_\lambda| = \lambda^{r+1} |\underset{\sim}{s}| , |\underset{\sim}{s}_{\lambda i}| = \lambda^r |\underset{\sim}{s}_i| .$$

Let $s_\lambda \subset U_\zeta(p)$ for $\lambda \leq \lambda_0$ and let $\underset{\sim}{d}_i$ be the r-direction of s_i. Then

$$\left| \sum_i \varphi(p, \underset{\sim}{s}_{\lambda i}) - \int_{s_\lambda} \varphi(p, \underset{\sim}{d})dp \right| = \left| \sum_i \int_{s_{\lambda i}} [\varphi(p, \underset{\sim}{d}_i) - \varphi(q, \underset{\sim}{d}_i)]dq \right|$$

$$\sum_{\le i} \epsilon |\underset{\sim}{s}_{\lambda i}| = \epsilon \sum_i \lambda^r |\underset{\sim}{s}_i|$$

$$\le \epsilon \lambda^r |\partial \underset{\sim}{s}| .$$

Also, by property (b),

$$\left| \int_{\partial s_\lambda} \varphi(p, \underset{\sim}{d}) dp \right| = |F(\partial s_\lambda)| \le \epsilon |\partial \underset{\sim}{s}_\lambda| \le \epsilon \lambda^r |\partial \underset{\sim}{s}| .$$

Hence,

$$|\sum_i \varphi(p, \underset{\sim}{s}_{\lambda i})| \le 2\epsilon \lambda^r |\partial \underset{\sim}{s}|$$

and dividing by λ^r we get

$$|\sum_i \varphi(p, \underset{\sim}{s}_i) \le 2\epsilon \; |\partial \underset{\sim}{s}|.$$

Since ϵ is arbitrary, $\varphi(p, \underset{\sim}{v})$ has property (2) of the lemma. If follows that ψ is an r-covector for each value of p. Define $\underset{\sim}{\varphi}(p)$ by

(9.16) $$\varphi(p, \underset{\sim}{v}_r) = \underset{\sim}{\varphi}(p) \cdot \underset{\sim}{v}_r ,$$

and denote the corresponding r-covector field by φ (i.e., φ is a continuous r-form). Then

(9.17) $$F(s) = \int_s \varphi(p, \underset{\sim}{d}) \, dp = \int_s \varphi$$

which is the assertion of the theorem.

Remark on notation. Let $d\underset{\sim}{s}(p)$ be an r-vector with r-direction $\underset{\sim}{d}_r$ such that the r-volume of any $s' \subset s$ with r-direction $\underset{\sim}{d}_r$ is given by $\int_s |d\underset{\sim}{s}(p)|$. Then, the integral of the r-form φ is also denoted by

(9.18) $$F(s) = \int_s \varphi = \int_s \underset{\sim}{\varphi}(p) \; D\underset{\sim}{s}(p) = \int_s \underset{\sim}{\varphi}(p) d\underset{\sim}{s}(p) =$$

Other expressions for F(s) are as follows : Let $\underset{\sim}{D}(i)$ and $\underset{\approx}{D}$ be the duality transformations defined with respect to an arbitrary n-covector and quadratic form $\underset{\sim}{q}$. The first of these is independent of q and the second depends only on the orientation of the n-covector used to define $\underset{\sim}{D}(i)$. Set

$$\underset{\approx}{D}\, \underset{\sim}{\varphi} = \underset{\sim}{\tilde{\varphi}} \;, \quad \underset{\approx}{D}' \; d\underset{\sim}{s} = d\underset{\sim}{\tilde{s}} \;,$$

$$\underset{\sim}{D}(i) \underset{\sim}{\varphi} = \underset{\sim}{\hat{\varphi}}, \; \underset{\sim}{D}'(i) \; d\underset{\sim}{s} = d\underset{\sim}{\hat{s}} \quad .$$

Then ,

$$F(s) = \int_s \varphi = \int_s \underset{\sim}{\varphi} \; d\underset{\sim}{s} = (-)^{n(n-r)} \int_s \underset{\sim}{\tilde{\varphi}} \cdot d\underset{\sim}{\tilde{s}}$$

(9.19) $$= (-)^{n(n-r)} \int_s \underset{\sim}{\hat{\varphi}} \cdot d\underset{\sim}{\hat{s}}$$

The quantities $\underset{\sim}{\hat{\varphi}}$, $\underset{\sim}{\tilde{\varphi}}$,$d\underset{\sim}{\hat{s}}$, and $d\underset{\sim}{\tilde{s}}$ depend on a quadratic form and an n-covector. These could even be chosen as continuous functions of p. But the integral $\int_s \varphi$ is independent of any quadratic form or n-form D(i, p) ; its value depends only on the r-form $\varphi : A^n \to V_{\left[{}^{r}_{n}\right]}$ and on the oriented r-simplex s. To use the representations (9.19) masks this independence and the simplicity of the definition of $\int_s \varphi$. Nevertheless, many of the standard and traditional formulas of classical vector analysis rest upon the possibility of these alternative representations, and to exhibit the relation between some of the results in the physical theory involving r-forms and known classical results, it is necessary to introduce expressions like (9.19) . In particular , in classical vector analysis, the (n-r-1) - forms defined by

$$\text{(9.20)} \qquad \begin{aligned} \operatorname{div} \hat{\varphi} &= D \ \operatorname{rot} \varphi \ , \\ \operatorname{div} \tilde{\varphi} &= D \ \operatorname{rot} \varphi \ , \end{aligned}$$

are called the <u>natural</u> and <u>absolute divergence</u> of φ , respectively. The divergence theorem or the definition of a regular r-form Eq. (7.5) then appears in the following guises :

$$\text{(9.21)} \qquad \int_S \underset{\sim}{\varphi} \, d\underset{\sim}{s} = \int_S \operatorname{rot} \underset{\sim}{\varphi} \, d\underset{\sim}{s} \ ,$$

$$\text{(9.22)} \qquad = (-)^{n(n-r)} \int_S (\operatorname{div} \underset{\sim}{\hat{\varphi}}) \, . \, d\underset{\sim}{\hat{s}} \ ,$$

$$\text{(9.23)} \qquad \int_S \underset{\sim}{\hat{\varphi}} \, d\underset{\sim}{\hat{s}} = (-)^{n} \int_S (\operatorname{div} \underset{\sim}{\hat{\varphi}}) \, . \, d\underset{\sim}{\hat{s}} \ ,$$

and in similar guises with $\underset{\sim}{D}$ replaced by $\underset{\approx}{D}$.

10. SMOOTH MANIFOLDS

In the physical theory considered in the lectures, it is not assumed that space or space-time is an affine space, and we require a definition of r-covector fields and a theory of integration of r-forms in a smooth manifold. We sketch here the theory as presented by Whithney.

An n-dimensional smooth manifold M^n is a mathematical system of the following sort. M^n is a connected topological space with open sets U, U', ... together with a collection of <u>coordinate systems</u> x_i, i in some index set . Each coordinate system is a homeomorphism

$$x_1 : O_i \quad M^n, \quad O_i \subset A^n \tag{10.1}$$

of an open set in A^n into M^n, A^n a fixed n-dimensional affine space. A finite or denumerable set of the <u>coordinate patches</u> $U_i = \chi_i(O_i)$ cover M^n. If $U_i \cap U_j \neq 0$, then $\chi_i^{-1} \chi_j = \chi_{ij}$ is defined in some $O_{ij} \subset O_i$ and $\chi_{ij} : O_{ij} \to A^n$ is a mapping from an open set in A^n into some other open set in A^n. We require that the <u>gradient</u> of χ_{ij} defined by

$$\underset{\sim}{\nabla} \chi_{ij}(p, \underset{\sim}{v}) = \lim_{t \to 0^+} \frac{\chi_{ij}(p+t\underset{\sim}{v}) - \varepsilon_{ij}(p)}{t} \tag{10.2}$$

exist and be continuous in p throughout its domain . It can be shown that $\chi_{ij}(p, \underset{\sim}{v})$ is linear in $\underset{\sim}{v}$: hence, $\underset{\sim}{\nabla} \chi_{ij}(p, \underset{\sim}{v}) = \underset{\sim}{\nabla} \chi_{ij}(p) \cdot \underset{\sim}{v}$ and $\underset{\sim}{\nabla} \chi_{ij}(p)$ is a linear transformation in V^n, the translation space of A^n, The coordinate systems of a smooth manifold have the property that $\nabla \chi_{ij}$ has rank n at each p in the domain of χ_{ij} . The manifold M^n is <u>s-smooth</u>

if each χ_{ij} is smooth (i.e., if $\nabla^s \chi_{ij}$ for all i and j, where defined, exists and is continuous).

If a subset of the U_i exists such that the U_i cover M^n and such that the Jacobians det $|\nabla\chi_{ij}|$ are all positive, then M^n is orientable and the set of coordinate systems χ_i orient M^n. Any coordinate system χ'_j related to one of the χ_i by a transformation with positive Jacobian is a preferred coordinate system of the oriented M^n.

In an s-smooth manifold M^n is defined in terms of coordinate systems χ_i, call any mapping $\chi = \chi_i \circ \varphi : O \to M^n$ obtained by composing any χ_i with an r-smooth, $r \geq s$, homeomorphism $\varphi: O \to O_i$, $O \subset A^n$, an admissible coordinate system for M^n. Henceforth, by a coordinate system of M^n, is meant any admissible coordinate system for M^n.

A mapping $f: M \to M'$ of one k-smooth manifold M into another M' is s-smooth, $s \leq k$, if $\chi_i^{-1} f \chi_j$, where defined, is s-smooth. When M' = R (real line) f is a real-valued s-smooth function in M; when M = R, or a connected open set in $O \subset R$, f is an s-smooth parametrized curve in M'.

Consider all the smooth curves in M^n defined by f, f', ... which contain a given point $x \in M^n$. We may assume, without loss in generality, that the domain of each f, f', ... contains the point $0 \in R$ and that $f(0) = x$. Let χ_1 be a coordinate system of M^n such that $x \in U_i$. (We say that χ_1 is a coordinate system about x.) Then $f_i = \chi_i^{-1} f$ is defined near 0.

Let

$$\dot{f}_i = \frac{df_1}{dt}(0) \quad . \tag{10.3}$$

Then, by definition, $\dot{f}_i$ is a vector in V^n, the translation space of A^n. Call two smooth curves through x *equivalent* (in particular, *tangent*) if $\dot{f}_i = \dot{f}'_i$. Since $\dot{f}_i = \nabla \chi_{ij}(\dot{f}'_j)$, this definition of equivalence is independent of the coordinate system. By a vector $v(x)$ at x in M^n we mean an equivalence class of smooth parametrized curves in M^n which pass through x. We call the vector $\dot{f}_i \in V^n$ a *representation* of the corresponding set of equivalent curves $v(x)$. The sum $v(x) + u(x)$ and the multiple $\lambda v(x)$ of vectors at x in M^n is defined by addition and multiplication of their representations in V^n. From the linearity of the law of transformation of the representations corresponding to two coordinate systems, we see that the definition of $v(x) + u(x)$, and of $\lambda v(x)$ is independent of coordinate system. The set of all vectors $v(x)$ at x in M^n forms an n-dimensional vector space $V^n(x)$, the *tangent space of* M^n *at* x. Each coordinate system χ_i of M^n about x defines an isomorphism

$$(10.4) \qquad \nabla \chi_i(p) : V^n \to V^n(x), \quad p = \chi_i^{-1}(x),$$

where $\nabla \chi_i(p, \dot{f}_i) = v(x)$, $v(x)$ the equivalence class of smooth curves having the representation $\dot{f}_i$ in the coordinate system χ_i. The n-dimensional vector spaces $V^n(x)$ and $V^n(x')$, $x \neq x'$, thouth isomorphic, are distinct. In general, $v(x) + u(x')$ is not defined. One could define $v(x) + u(x')$ by adding their representations in some coordinate system χ_i, provided x and x' were both in U_i; but such a definition is not independent of χ_i.

A mapping

$$(10.5) \qquad f: M^m \to M^n, \quad m < n$$

of one smooth manifold into another is <u>regular</u> if the following holds. Let $x = f(X)$, and χ_i be a coordinate system of M^m about X, χ_α a coordinate system of M^n about x. Then $f_{\alpha i} = \chi^{-1}_{\alpha} \chi_i$ maps some $R \subset A^m$ into A^n. Let

$$(10.6) \qquad \nabla f_{\alpha i}(p, \underset{\sim}{v}) = \nabla f_{\alpha i}(p) \cdot \underset{\sim}{v} = \lim_{t \to 0^+} \frac{f_{\alpha i}(p + \underset{\sim}{v} t) - f_{\alpha i}(p)}{t},$$

$p \in A^m$, $\underset{\sim}{v} \in V^m$. Then, by definition, $\underset{\sim}{\nabla} f_{\alpha i}(p)$ is a linear transformation of V^m into V^n. The mapping f is regular at X if $\nabla f_{\alpha i}$ is continuous at $p = \chi_i^{-1}(X)$ and has rank m. This definition of regularity at X is independent of the coordinate systems χ_i and χ_α f is regular in an open set $Q \subset M^m$ if it is regular at each $X \subset\!\!\epsilon\, Q$; it is <u>regular</u> if it is regular for all $X \in M^m$. Now in M^m and M^n, the coordinate systems χ_i and χ_α establish the isomorphisms $\nabla \chi_i$ and $\nabla \chi_\alpha$ between V^n and $V^n(x)$, and between V^m and $V^m(X)$, respectively. The gradient of the mapping ∇f is then defined by

$$(10.7) \qquad \nabla f = \nabla \chi_\alpha \, \nabla f_{\alpha i} \, (\nabla \chi_i)^{-1}.$$

The gradient of f is defined for every $X \in M^n$, and its definition is independent of coordinate systems. Its value at a point X of M^n defines a linear transformation $\nabla f(X) : V^m(x) \to V^n(x)$, $x = f(X)$, of the tangent space of M^m at X into the tangent space of M^n at the point $f(X)$.

11. INTEGRATION IN MANIFOLDS

A tensor whose carrier space is the tangent space $V^n(x)$ at a point x of a smooth manifold M^n is called a <u>tensor at x</u>. A <u>tensor field</u> in a manifold is a function

$$T: M^n \to \left\{ V^{n\,k}_{\;n}(x); \quad x \in M^n \right\}$$

such that $T(x) \in V^{n\,k}_{\;n}(x)$. (The values of k and are the same all x; and $T(x)$ is a tensor at x.) The field T is continuous or s-smooth if its <u>representation</u> $\chi_i(T)$ defined by

$$\chi_i(T)(p, \underset{\sim}{v}_1, \underset{\sim}{v}_2, \ldots, \overset{1}{\underset{\sim}{\alpha}}, \overset{2}{\underset{\sim}{\alpha}}, \ldots) =$$

(11.1)
$$T(x, \underset{\sim}{\nabla} \chi_i \cdot \underset{\sim}{v}_1, \underset{\sim}{\nabla} \chi_i \cdot \underset{\sim}{v}_2, \ldots, \underset{\sim}{\nabla}_{\chi i}^{-1} \cdot \overset{1}{\underset{\sim}{\alpha}}_2, \ldots),$$

$p = \chi_i^{-1}(x)$, $v_k \in V^n$, $\overset{k}{\alpha} \in V_n$, in any coordinate system χ_i of M^n is continuous or s-smooth. This definition of smooth-ness is independent of the coordinate system in an r-smooth manifold provided $s \leq r$.

The gradient ∇f of a regular mapping $f: M^m \to M^n$ considered in the previous section is an example of the more general concept of a <u>multi-point tensor.</u> We may view ∇f as a bilinear function

$$\nabla f: V^m(X) \times V_n(x) \to R.$$

It is a <u>mixed tensor field over M^m.</u> In general, it is not a field in M^n because only in special subcases of $m = n$ is ∇f defined at every point of M^n.

Let φ be an r-cotensor field in M^n and let : $M^m \to M^n$ be regular. Then we define a corresponding r-cotensor field $f^*(\varphi)$ in M^m by

(11.2) $$(f^*\varphi)(X, \underset{\sim}{U}_1, \underset{\sim}{U}_2, \ldots, \underset{\sim}{U}_r) = \varphi(x, \underset{\sim}{\nabla} f \underset{\sim}{U}_1, \underset{\sim}{\nabla} f \underset{\sim}{U}_2, \ldots, \underset{\sim}{\nabla} f \underset{\sim}{U}_r) ,$$

where each $U_k \in V^m(X)$, $x = f(X)$, and $\underset{\sim}{\nabla} f \underset{\sim}{U}_k = \nabla f\,(X, \underset{\sim}{U}_k) \in V^n$.

Now $\int_s \varphi$ has been defined only for simplexes and linear combinations of simplexes in an affine space A^n . We wish to extend this definition to curvilinear simplexes and certain open sets in a manifold. This can be done following Whitney. First of all, consider an n-form φ in A^n . Say that φ is summable over the open set $R \subset A^n$ if there exists a polyhedron P with the property that for every $\epsilon > 0$

(11.3) $$\left| \int_Q \varphi \right| < \epsilon \text{ for all polyhedra } Q \subset R - P .$$

Then if φ is summable over R, there exists a number $\int_R \varphi$ such that

(11.4) $$\left| \int_Q \varphi - \int_R \varphi \right| < \epsilon \text{ if } P \subset Q \subset R.$$

This defines $\int_R \varphi$ for certain φ and open sets $R \subset A^n$. If, with respect to an arbitrary positive definite (metric) in the V^n or A^n, $R = \sup\ \{ Q : Q \subset R,\ Q \text{ a polyhedron} \} < \infty$ and $|\varphi| < \infty$, then φ is summable over R and, therefore, $\int_R \varphi$ is defined . We need also the fact that if φ is summable over R, then so also is $\pi\varphi$ where π is any bounded continuous realvalued function in R.

The next result we need is the transformation formula. Let f be any one - one regular mapping of the open set $R \subset A^n$ onto the open set $R' \subset A'^n$ with $\bar{J}_f(p) > 0$ in R. Let φ be a continuous n-form in R' summable over R'. Then $f^* \varphi$ is summable over R, and

$$\int_R f^* \varphi = \int_{R'} \varphi \, . \tag{11.5}$$

Next we define $\int_{M^n} \varphi$ for any compact smooth oriented manifold M^n. Let spt (φ) denote the closure of the set of points $x \in M^n$ where $\varphi(x) \varphi \neq 0$. If spt $(\varphi) \subset U_i$ where U_i is a coordinate patch of M_n, χ_i a preferred coordinate system of M^n, then we set

$$\int_{M^n} \varphi = \int_{O_i} \chi_i \varphi, \quad \text{spt}\,(\varphi) \subset \chi_i(O_i) \, . \tag{11.6}$$

That this definition of $\int_{M^n} \varphi$ is independent of the coordinate system follows from the transformation formula (11.5).

Suppose next that spt(φ) is not contained in any U_i. In this case, let $\sum_i \pi_i$ be a partition of unity with the following property. Let the finite set of coordinate patches U_i, i = 1, 2, ..., N cover M^n, where the χ_i orient M^n. Then one can construct a set of real-valued smooth functions π_i in M^n such that $\pi_i(x) = 0$, $x \in U_i$, and $\sum_i \pi_i(x) = 1$, $x \in M^n$. Express $\varphi(x)$ as $\sum_i \varphi_i(x)$, $\varphi_i(x) = \pi_i(x) \varphi(x)$. Then $\int_{U_i} \varphi_i$ is defined as above, and we set

$$\int_{M^n} \varphi = \sum_i \int_{U_i} \varphi_i \, . \tag{11.7}$$

The definition (11.7) can be shown to be independent of the partition of unity $\Sigma\, \pi_i$ and the coordinate systems χ_i .

Now let R be any open subset of an oriented manifold M^n such that $\bar{R}$ is compact. Let φ be any continuous n-form defined in a neighborhood of R. Then there exists a finite number of coordinate patches U_i which cover $\bar{R}$, χ_i preferred coordinate systems of M^n . Define the π_i as above. Then , for some neighborhood U of $\bar{R}$, $\sum_i \pi_i(x) = 1,\ x \in U$. Set $\varphi_i = \pi_i\, \varphi$ and define

$$\text{(11.8)} \qquad \int_R \varphi_i = \int_{\chi_1^{-1}(R)} \chi_i\, \varphi_i = \int_R \varphi = \sum_i \int_R \varphi_i \ .$$

Again, it can be shown that this definition of $\int_R \varphi$ is independent of the coordinate systems χ_i and of the partition of unity $\Sigma\, \pi_i$.

We record here Whitney's version of the divergence theorem in manifolds. For the complete definition of a <u>standard n-manifold,</u> I refer the reader to Whitney's book. It can be remarked that every smooth image of a polyhedron in A^n is a standard n-manifold, and every regular region in the sense of Kellogg is a standard n-manifold . To state the theorem we need at least the following partial description of a standard n-manifold. There is a connected compact topological space, $\bar{M}$, a closed subset $\partial\bar{M}$ of $\bar{M}$ (call it the boundary of M) and a closed subset $\partial_0\bar{M}$ of $\partial\,\bar{M}$ (call it the edges and vertices of $\partial\bar{M}$) . $\bar{M} - \partial\,\bar{M}$ is a smooth oriented n-dimensional manifold, and smooth manifolds. At each point of $x \in (\partial M - \partial_0 M)$ there is defined an <u>outward normal vector</u> $\underset{\sim}{\nu}(x) \neq 0$, which, by continuity, is defined at points of $\bar{M} - \partial\bar{M}$ near x. The orientation of the smooth manifold $\bar{M} - \partial\,\bar{M}$ is fixed by the ordered set $\{v(x)\,.\ v_1(x), \ldots, v_{n-1}(x)\}$ of vectors, where the last n-1 elements determine the orientation of the part of $\partial\overline{M} - \partial_0\overline{M}$ containing the point x. Let φ be an n-form such that

(a) φ is defined, continuous, and bounded in $\partial\overline{M} - \partial_0\overline{M}$, and is regular in $\overline{M} - \partial\overline{M}$,

(b) φ is summable over $\partial\overline{M} - \partial_0\overline{M}$,

(c) rot φ is summable over $\overline{M} - \partial\overline{M}$.

Then

$$\int_{\partial\overline{M} - \partial_0\overline{M}} \varphi = \int_{\overline{M} - \partial\overline{M}} \operatorname{rot} \varphi \,. \tag{11.9}$$

Hencefort, when the meaning is clear from the context we shall abbreviate (11.9) to read

$$\int_{\partial\overline{M}} \varphi = \int_{\overline{M}} \operatorname{rot} \varphi \,. \tag{11.10}$$

When φ is smooth and M is 2-smooth, rot φ is given by rot $\varphi(x) = x_i^* \, \partial\vee\varphi$, where x_i is any coordinate system about x and $\partial\vee\varphi$ is defined in (7.7) .

Let f: $M^r \to M^n$ be any regular mapping of an oriented r-dimensional smooth manifold into an n-dimensional smooth manifold and let φ be an r-form in M^n. Then $f^*\varphi$ is an r-form in M^r and we set

$$\int_{R'} \varphi = \int_R f^* \varphi, \qquad R' = f(R) \,. \tag{11.11}$$

This defines the integral of an r-form φ in M^n, $r \leq n$ over every r-dimensional smooth manifold or piece R' of a smooth manifold in M^n . If $\overline{R'} = f(\overline{R})$, and $\overline{R} \subset M^r$

is a standard manifold, then

$$\int_{\partial\overline{R'}} \varphi = \int_{\overline{R}} \operatorname{rot} \varphi \tag{11.12}$$

provided the (r-1)-form φ is continuous in some neighborhood of R', $f^* \varphi$ is regular in R, $f^* \varphi$ is summable over $\partial\overline{R} - \partial_0\overline{R}$, and f^* rot φ is summable over R.

12. CURVILINEAR r-COCHAINS

Let M^n be an n-dimensional smooth manifold with coordinate systems $\chi_i : O_i \to U_i \subset M^n$. If $s \subset O_i$ is an r-simplex in A^n, we call $\chi_i(s)$ a <u>curvilinear r-simplex</u> in M^n. A <u>curvilinear r-chain</u> in M^n is an expression of the form $c = \Sigma a^i s_i$, where each s_i is a curvilinear simplex in M^n, $s_i \cap s_j$ = sum of chains of lower dimension, a^i are positive or negative integers or zero. We drop the adjective "curvilinear" when the meaning is clear without it. The sum of r-chains in M^n is defined in the same way as the sum of r-chains in A^n. An <u>r-cochain</u> in M^n is a linear function of r-chains in M^n. If $c = \Sigma a^i s_i$, then $\int_{s_i} \varphi$ is defined as in the last section and we set

$$(12.1) \qquad \int_c \varphi = \Sigma a^i \int_{s_i} \varphi.$$

The value of $\int_c \varphi$ is independent of its representation as a sum of r-simplexes. Thus, every r-form φ in M^n, summable over every curvilinear r-simplex in M^n, determines a unique r-cochain in M^n by (12.1). If the (r-1)-form φ is regular in a neighborhood of c, then

$$(12.2) \qquad \int_{\partial c} \varphi = \int_c \operatorname{rot} \varphi,$$

and if φ is smooth, then $\operatorname{rot} \varphi = \partial \vee \varphi = \chi_i (\partial \vee \chi_i^{-1} \varphi)$. (cf., § 8 for the definition of $\partial \vee \gamma$, γ an r-form in A^n). By the triangulation theorem for smooth manifolds, every compact oriented manifold can be expressed as a curvilinear r-chain of the form $M^r = \Sigma s_i$. Thus, for any compact r-dimensional manifold

M^r in M^n and any regular (r-1)-form φ in M^n, we have a relation like (12.2) with c replaced by M^r. We say that an r-cochain defined by integration of an r-form φ in M^n is continuous or s-smooth accordingly as the r-form φ is continuous or s-smooth.

The three properties (a), (b), and (c) of § 9 which characterize a semi-sharp r-cochain in an affine space A^n do not have an immediate invariant significance for r-cochains in a manifold. But since they are purely local conditions, using the coordinate systems of a manifold, the properties can be staded in terms of the inverse images $\chi_i^{-1}(s) = s'$ of curvilinear r-simplexes contained in some $U_i \subset M^n$ and of the corresponding values of the r-cochain. Thus, if F is an r-cochain in M^n and we set $F(s) = F'(s')$, $s' = \chi_i^{-1} s$, then F is called semisharp if and only if F' is semi-sharp. If follows that semi-sharp r-cochains and continuous r-cochains in a manifold are in 1-1 correspondence. Every semi-sharp r-cochain in M^n determine a unique continuous r-form φ in M^n such that $F(c) = \int_c \varphi$ and, conversely, every continuous r-form φ in M^n defines a unique semi-sharp r-cochain by this same rule of association.

13. MOTIONS AND INVARIANT COCHAINS

Let M^m and M^n denote smooth manifolds of the dimensions indicated. Let $O \subset R$ denote a connected open set of real numbers. Then a regular mapping

$$(13.1) \qquad f: M^n \times O \to M^m, \quad m \geq n+1$$

is called a parametrized motion of M^n in M^m. The <u>trajectory</u> of a point $X \in M^n$ is the smooth parametrized curve $f_X: O \to M^m$, $f_X(t) = f(X,t)$. The <u>orbit</u> of X is the set of points $\{x;\ x = f(X,t),\ t \in O\}$.

When $m = n+1$, define a <u>motion</u> of M^n as a smooth mapping

$$(13.2) \qquad f' : M^{n+1} \to M^n$$

such that $\nabla f'$ has rank n in M^{n+1}. The <u>orbit</u> of a point $X \in M^n$ is the set of points $\{x;\ f'(x) = X\}$. Using the implicit function theorem, it can be shown that every connected piece of an orbit is a smooth parametrized curve in M. Thus, when $m = n+1$, there is no distinction <u>locally</u> between motions and parametrized motions of M^n in M^{n+1}. But the orbits of a motion may not be connected so that motions and parametrized motions in the large are not in 1 - 1 correspondence. Moreover, the existence of motions depends on the character of the manifolds M^n and M^m. In the following we shall be concerned only with local properties of motions and parametrized motions only.

Assume now that $m = n+1$. Then (12.1) has a unique smooth inverse $f^{-1}: M^{n+1} \to M^n \times O$ which consists in two smooth mappings

$f' : M^{n+1} \to M^n$ and $T : M^{n+1} \to O$, where $\nabla f'$ has rank n throughout its domain and ∇T has rank 1. Thus, every parametrized motion $(m = n+1)$ determines a unique motion f', but not conversely. Two parametrized motions with the same orbits determine one and the same motion by this construction. Through each point $x \in U = f(M^n \times O)$ in the range of f there passes one and only one trajectory. Let $\underset{\sim}{v}(x) \in V^n(x)$ denote the tangent vector of the trajectory through x. From the regularity assumption, the vector field v with these values in U is continuous. We call this vector field in U, the <u>velocity field of the parametrized motion</u> f. If $f' : M^{n+1} \to M^n$ is a motion, then at each point x in the domain of f', $\Delta f'$ has one and only one linearly independent proper vector $\underset{\sim}{w}(x) \neq 0$ such that

(13.3) $$\underset{\sim}{\Delta} f'(x) \cdot \underset{\sim}{w}(x) = 0 .$$

The velocity field of a motion is the equivalence class of smooth vector fields in the domain of f' such that $\underset{\sim}{v}(x) \neq 0$ is a proper vector of $\nabla f'(x)$ with proper value zero at each point.

Let $f_t : M^n \quad M^{n+1}$ be defined by $f_t(X) = f(X, t)$. Then

(13.4) $$T_\tau = f_{t+\tau} \cdot f_t^{-1} , \quad t, \quad t+\tau \in O$$

is a 1 - 1 regular mapping

(13.5) $$T_\tau : U_t \to U_{t+\tau}$$

of an open set $U_t \subset M^{n+1}$ onto an open set $U_{t+\tau} \subset M^{n+1}$. We have

(13.6) $$T_0 = 1 \text{ (the identity)} , \quad T_\tau \cdot T_{\tau'} = T_{\tau+\tau'} .$$

Let s_t be any curvilinear r-simplex in U_t. Then $s_{t+\tau} = T_\tau(s_t)$ is a curvilinear r-simplex in $U_{t+\tau}$. Let φ be an r-form in M^{n+1}. Then

$$F(s_t,\tau) = \int_{s_{t+\tau}} \varphi \tag{13.7}$$

is defined for all $s_t \subset U_t$. If $F(s_t, \tau)$ is independent of τ for all such s_t, we say that φ is an <u>invariant r-form under the motion f</u>, or that the corresponding r-cochain $F(s_t)$, defined by integration of φ is an <u>invariant r-cochain under the motion</u> f.

<u>Theorem</u> : Let $F(s_t, \tau)$ be defined by (13.7). Let v be the velocity field of the parametrized motion f. Then, if the r-form φ and the (r-1)-form $v \wedge \varphi$ are regular in the range of f, $F(s_t,\tau)$ is differentiable in τ for each s_t and

$$\frac{dF(s_t, \tau)}{d\tau} = \int_{s_{t+\tau}} [\, v \wedge \text{rot}\, \varphi + \text{rot}\, (v \wedge \varphi)\,] . \tag{13.8}$$

To prove (13.8) consider the value of

$$\nabla F = F(s_t, \tau + \tau') - F(s_t,\tau) = \int_{s_{t+\tau+\tau'}} \varphi - \int_{s_{t+\tau}} \varphi .$$

The quantity ∇F differs from the integral of φ over the boundary $\partial(s_{t+\tau} \times I)$, $I = \{t;\ \tau \leq t \leq \tau + \tau'\}$, by an integral of φ over $\partial s_{t+\tau}\ I$;

$$\Delta F = \int_{\partial(s_{t+\tau}\ I)} \varphi + \int_{(\partial s_{t+\tau})\times I} \varphi$$

$$= \int_{s_{t+\tau} I} \text{rot}\ \varphi + \int_{(\ s_{t+\tau})\times I} \varphi ,$$

where we have used the regularity of φ. But the integrals on the right can be expressed as iterated integrals,

$$\begin{aligned} \nabla F &= \int_I d\tau'' \int_{S_{t+\tau}} v \wedge (\mathrm{rot}\,\varphi) + \int_I d\tau'' \int_{\partial S_{t+\tau}} v \wedge \varphi \\ &= \int_I d\tau'' \int_{S_{t+\tau}} v \wedge \mathrm{rot}\,\varphi + \int_I d\tau'' \int_{S_{t+\tau}} \mathrm{rot}\,(v \wedge \varphi), \end{aligned} \tag{13.9}$$

where we have used the regularity of $v \wedge \varphi$. Thus, dividing (13.9) by τ'' and taking the $\lim \tau'' \to 0$ yields the formula (13.8).

It follows that if φ and $v \wedge \varphi$ are regular, then φ and the r-cochain $F = \int \varphi$ are invariant under the parametrized motion f with velocity field v if and only if

$$\mathcal{L}_v \varphi = v \wedge \mathrm{rot}\,\varphi(v \wedge \varphi) = 0, \tag{13.10}$$

at each point in the range of f. The continuous r-form $\mathcal{L}_v \varphi$ is called the Lie derivative of with respect to the vector field v. Let $\underset{\sim}{D}(x)$ be the duality transformation defined by any linearly independent set of vectors at x. Set $\hat{\varphi} = \underset{\sim}{D}(\varphi)$ and div $\hat{} = \underset{\sim}{D}(\mathrm{rot}\,\varphi)$. The dual of $\mathcal{L}_v \varphi$ is then given by

$$\underset{\sim}{D}(\mathcal{L}_v \varphi) = (\mathrm{div}\,\hat{\varphi}) \times v + \mathrm{div}\,(\hat{\varphi} \times v). \tag{13.11}$$

If the vectors used to define $\underset{\sim}{D}$ are the tangent vectors to the coordinate curves of a coordinate system, then rot φ and div $\hat{\varphi}$ are given by the differentiation formulas $\partial \wedge \varphi$ and $\partial \wedge \hat{\varphi}$ provided that φ is smooth. But the formulas (13.8) and (13.11) hold under the weaker hypotheses made in deriving them.

14. PHYSICAL UNITS. PHYSICAL DIMENSIONS AND PHYSICAL QUANTITIES.

Weyl has remarked:

"However, not only points are required to be represented by reproducible symbols, but also every other kind of geometric entity, and when passing to physics all sorts of physical quantities like velocities, forces, field strengths, wave functions, and what not, expect a similar sympolic treatment. One often acts as though once the points have been submitted to it by fixing a frame of refe rence for them, all these other things will follow suit without necessitating further provisions. This is certainly not true; at least further units of measurement have to be fixed at random so as to make the scheme of reference complete."

For the purposes of these lectures, the following formal definitions will be adopted. For a discussion of the historical use and development of the concepts and mathematics of physical units and dimensions, see Ericksen's Appendix on Tensor Fields in The Classical Field Theories, Vol. III/1, Handbuch der Physik.

A physical unit $\underset{\approx}{U}$ is vector different from zero in a one-dimensional vector space U called a <u>unit space</u>. Thus, any physical unit $\underset{\approx}{U} \in U$ is a basis for U. A function f; $U \to L$ with values in some linear space is said to have <u>physical dimension</u> $[\underset{\approx}{U}^n]$ if.

$$(14.1) \qquad f(a\,\underset{\approx}{U}) = a^{-n} f(\underset{\approx}{U}) ;$$

i.e., if f is a homogeneous function of degree -n. More generally now, if f; $U_1 \times U_2 \times \ldots \times U_n \to L$ is homogeneous of degree $-n_k$ in the argument U_k, we write

$$\text{(14.2)} \qquad \text{phys. dim. } f = \left[\underline{\underline{U}}_1^{\,n_1} \; \underline{\underline{U}}_2^{\,n_2} \ldots \underline{\underline{U}}_k^{\,n_k} \right] ,$$

In these lectures we restrict the use of the term <u>physical quantity</u> and use it only to mean the following . A physical quantity is an r-cochain in some manifold M^n with values which are homogeneous functions of a set $\underset{\approx}{U}_a$, $a = 1, 2, \ldots N$ of independent physical units. Thus, if F is a physical quantity, it has a definite physical dimension, and if F is semi-sharp or continuous, there exists an r-form in M^n such that

$$\text{(14.3)} \qquad F(c, \underset{\approx}{U}_1, \ldots, \underset{\approx}{N}_N) = \int_c \varphi(\underset{\approx}{U}_1, \ldots, \underset{\approx}{U}_N) ,$$

and we write

$$\text{(14.4)} \qquad \text{phys.dim. } F = \text{phys. dim. } \varphi = \left[\underline{\underline{U}}_1^{\,n_1} \ldots \underline{\underline{U}}_N^{\,n_N} \right] ,$$

with suitable values for the exponents $n_1, \ldots, n_N$.
In particular , if phys. dim. $F = [\underset{\approx}{U}]$, we write,

$$\text{(14.5)} \qquad F(c, \underset{\approx}{U}) = \int_c \varphi\, (\underset{\approx}{U}) .$$

The real number $F(c, \underset{\approx}{U})$ is called the <u>amount of the physical quantity</u> F <u>in</u> c <u>measured in units</u> $\underset{\approx}{U}$; $\varphi(\underset{\approx}{U})$ is called the <u>density of the physical quantity</u> F <u>measured in units</u> $\underset{\approx}{U}$.

Our use of the term " physical quantity" in these lectures lies closest to the concept and definition of an "extensive" quantity as that term is used in the literature of thermostatics.

LECTURE I

ELECTRIC CHARGE AND MAGNETIC FLUX

" In the application of mathematics to the calculation of electrical quantities, I shall endeavour in the first place to deduce the most general conclusions from the data at our disposal, and in the next place to apply the results to the simplest cases that can be chosen."

J. C. Maxwell

I.1. INTRODUCTORY REMARKS

The electrodynamics of elastic media finds important engineering applications in the phenomena of photoelasticity and piezoelectricity. The classical theories of these effects and of many others are embraced as special cases by the general theory of the electromagnetic field in material media to be presented in these lectures. The lectures will stress an orderly introduction of the physical concepts and laws of nature which are common to a broad class of special theories .

We begin with a mathematical theory of electricity and magnetism and only later introduce the mechanical and kinematical concepts of length, time, force, stress, energy, and momentum. It is common knowledge that modern concepts of lenth, time , and simultaneity had their origins in the theory of electricity and magnetism and that these concepts are not entirely consistent with the Newtonian viewpoint.

One objective of the treatment of the electromagnetic field given here is to trace again these origins.

I. 2. ELECTRIC CHARGE AND MAGNETIC FLUX

It is possible to understand the classical equations of Maxwell and Lorentz governing the electric, magnetic, charge, and current fields in terms of only three primitive concepts .

1) the distribution of magnetic flux
2) the distribution of electric charge
3) the universal relation between these two distributions

For this purpose we begin by taking the concept of an event as primitive and undefined, just as a point is primitive and undefined in Euclidean geometry. We call the set of all events, event-space, and denote if by $\mathcal{E}$. We lay down the first of some eighteen assumptions or principles to be introduced in the course of the lectures.

A1 : The set of all events is an orientable 4-dimensional smooth manifold.

Events will be denoted by lower case Greek letters ξ , ξ', ξ'' , etc. The coordinates of an event are denoted by ξ^{α} , where the index α ranges over the values 1, 2, 3, and 4. Greek lower case indices will always have this same range.

To each, oriented 2-dimensional submanifold $\mathcal{E}^2 \subset \mathcal{E}$ we assign a real number $F(\mathcal{E}^2 , \underset{\sim}{\Phi}')$ called the magnetic flux through $\mathcal{E}^2$ where $\underset{\sim}{\Phi}$ is a unit of magnetic flux.

A.2. Magnetic flux is a continuous 2-cochain in $\mathcal{E}$ with phys. dim . [$\underset{=}{\Phi}$].

It follows from A2 that there exists an electromagnetic field φ, a continuous 2-form in $\mathcal{E}$, such that

$$F(\mathcal{E}^2, \underset{\approx}{\Phi}) = \int_{\mathcal{E}^2} \varphi(\underset{\approx}{\Phi}) , \tag{2.1}$$

$$\text{phys. dim. } \varphi = [\underset{=}{\Phi}].$$

A3: (Faraday's Law of Magnetic Induction) The magnetic flux of every cycle $C^2 \subset \mathcal{E}$ vanishes,

$$\oint_{C^2} \varphi = 0 \tag{2.2}$$

From Faraday's law of magnetic induction follows the existence of electromagnetic potentials , regular 1-forms in $\mathcal{E}$, such that the magnetic flux through any $\mathcal{E}^2$ is given by

$$F(\mathcal{E}^2, \underset{\approx}{\Phi}) = \oint_{\partial \mathcal{E}^2} \alpha(\Phi) , \tag{2.3}$$

where

$$\text{phys.dim. } \alpha = [\underset{=}{\Phi}], \tag{2.4}$$

$$\text{rot } \alpha = \varphi . \tag{2.5}$$

The distribution of magnetic flux does not determine a unique electromagnetic potential for if α satisfies (2.3) for every $\mathcal{E}^2$, then so also does α' given by

$$\alpha' = \alpha + \text{rot } \beta , \tag{2.6}$$

where β is an arbitrary regular 0-form in $\mathcal{E}$ (scalar field).

Next we introduce the definition and certain properties of <u>electric charge</u> in precise analogy with magnetic flux. To each oriented 3-dimensional submanifold of events $\mathcal{E}^3 \subset \mathcal{E}$ we assign a real number $C(\mathcal{E}^3, \underset{\approx}{Q})$ called the electric charge of $\mathcal{E}^3$ measured in units of electric charge $\underset{\approx}{Q}$.

A4: Electric charge is a continuous 3-cochain in $\mathcal{E}$ and phys. dim. $C = [\underset{\approx}{Q}]$.

According to A4 there exists a 3-form χ in $\mathcal{E}$, the <u>charge-current field,</u> such that

$$C(\mathcal{E}^3, \underset{\approx}{Q}) = \int_{\mathcal{E}^3} \chi(\underset{\approx}{Q}) \tag{2.7}$$

$$\text{phys. dim. } \chi = [\underset{\approx}{Q}] .$$

A5. (Law of Conservation of Charge) The electric charge of every cycle C^3 is zero.

$$\oint_{C^3} \chi = 0 . \tag{2.8}$$

It follows from the law of conservation of charge that there exist charge-current potentials η, 2-forms in $\mathcal{E}$, such that

$$C(\mathcal{E}^3, \underset{\approx}{Q}) = \oint_{\partial\mathcal{E}^3} \eta(\underset{\approx}{Q}) , \tag{2.9}$$

$$\text{phys. dim } \eta = [\underset{\approx}{Q}], \tag{2.10}$$

$$\text{rot } \eta = \chi . \tag{2.11}$$

The charge distribution does not determine a unique charge current potential. If η satisfies (2.9) for every $\mathcal{E}^3$, then so also does η' given by

$$\eta' = \eta + \operatorname{rot} \beta , \tag{2.12}$$

where β is an arbitrary 1-form in $\mathcal{E}$.

When the electromagnetic field and the charge-current field are not only continuous but regular, then Faraday's law of induction and the law of conservation of electric charge are equivalent to the local conditions

$$\begin{aligned} \operatorname{rot} \varphi &= 0 , \\ \operatorname{rot} \chi &= 0 . \end{aligned} \tag{2.13}$$

Moreover, if φ and χ are smooth, these local conditions are expressible in terms of the differential equations,

$$\begin{aligned} \partial \vee \varphi &= 0 , \\ \partial \vee \chi &= 0 . \end{aligned} \tag{2.14}$$

But the smoothness of the distributions of magnetic flux and electric charge assumed in A2 and A4 is already so severe as to rule out electromagnetic fields and charge-current fields commonly considered in applications. Weaker assumptions sufficiently general to include all the applications are the following alternatives to A2 and A4. Call an r-form in $\mathcal{E}$ piecewise regular if $\mathcal{E}$ can be subdivided into a finite number of standard manifolds, $\mathcal{E} = \bigcup_i \mathcal{E}_i$, $\mathcal{E}_i \cap \mathcal{E}_j = 0$, such that the r-form is regular in each. Then replace A2 and A4 by

A2'. There exists a piecewise regular electromagnetic potential α such that the magnetic flux measured in units of $\underset{\approx}{\Phi}$ through any $\mathcal{E}^2$ is given by

$$(2.1') \qquad F(\mathcal{E}^2, \underset{\approx}{\Phi}) = \oint_{\partial \mathcal{E}^2} \alpha(\underset{\approx}{\Phi}) .$$

A4': There exists a piecewise regular charge-current potential η such that the electric charge in $\mathcal{E}^3$ measured in units $\underset{\sim}{Q}$ is given by

$$(2.4') \qquad C(\mathcal{E}^3, \underset{\sim}{Q}) = \oint_{\partial \mathcal{E}^3} \eta(\underset{\sim}{Q}) .$$

Faraday's law of magnetic induction and the law of conservation of electric charge now follow as consequence of A2' and A4' in each $\mathcal{E}_i$ the electromagnetic field and the charge-current field are given by

$$(2.15) \qquad \begin{aligned} \varphi &= \operatorname{rot} \alpha , \\ \chi &= \operatorname{rot} \eta . \end{aligned}$$

II.3. DISCUSSION

No definitions of "magnetic flux per unit area" or of "electric charge per unit volume" have been given . No concept of lenght, area, volume, or of time has been used nor can be inferred from the definitions and properties assigned thus far to magnetic flux and electric charge. The definitions and properties of magnetic flux and electric charge given here in this generality must be attributed to Bateman and Kottler.

If a coordinate system is introduced in about an event ξ, and a choice is made for the unit of electric charge, then the charge current field χ has a value at ξ represented by components $\chi_{\alpha\beta\mu}(\xi, \underset{\sim}{Q})$. If the coordinates and the unit of charge are transformed, these components of the charge - current field undergo the transformation

(3.1) $\chi_{\alpha'\beta\mu'}(\xi, \underset{\approx}{Q}') = Q^{-1} \; s^{\alpha}_{\alpha'} \, s^{\beta}_{\beta'} \, s^{\mu}_{\mu'} \, \chi_{\alpha\beta\mu}(\xi, \underset{\approx}{Q})$,

where $\underset{\approx}{Q}' = \underset{\approx}{Q}\underset{\approx}{Q}$, and $s^{\alpha}_{\alpha'}$ are the components of the gradient of the coordinate transformation. Similarly, the components of the electromagnetic field φ under the transformation

(3.2) $\varphi_{\alpha'\beta'}(\xi, \underset{\approx}{\Phi}') = \Phi^{-1} \, s^{\alpha}_{\alpha'} \, s^{\beta}_{\beta'} \, \varphi_{\alpha\beta}(\xi, \underset{\sim}{\Phi})$,

where = . These laws of transformation for the comonents of the lectromagnetic and charge-current fields were discovered, after some trial ' ' error, by Lorents. From the present viewpoint, these transformation formulas are consequences of the definitions of magnetic flux and electric charge. We see that these transformation formulas merely reflect the fact that electric charge and magnetic flux, relative to given units, are numbers assigned to certain sets of events, and that these numbers do not depend in any way upon coordinate systems, observers, measuring apparatus, clocks, rigid rods, or anything else of that nature.

R.A . Toupin

LECTURE. II

THE AETHER RELATIONS AND THE METRICAL STRUCTURE OF EVENT-SPACE

"As everyone knows, the aether played a great part in the physics of the nineteenth century; but in the first decade of the twentieth, chienfly as a result of the failure of attemps to observe the earth's motion relative to the eather, and the acceptance of the principle that such attempts must always fail, the word 'aether' fell out of favour, and it became customary to refer to the interplanetary spaces as 'vacuous', the vacuum being conceived as mere emptiness, having no properties except that of propagating electromagnetic waves. But with the development of quantum electrodynamics, the vacuum has come to be regarded as the seat of 'zero-point' oscillations of the electromagnetic field, of the 'zero point' fluctations of electric-charge and current , and of a 'polarisation corresponding to a dielectric constant different from unity. It seems absurd to retain the name 'vacuum' for an entily so rich in physical properties, and the historical word 'aether' may fitly be retained. "

E. T. Whittaker

II. 1. INTRODUCTORY REMARKS

In the first lecture it was assumed only that the set of all events was a smooth 4-dimensional manifold. Two physical quantities represented by a 2-cochain and a 3-cochain in called magnetic flux and electric charge, respectively, were introduced along with two unit spaces $\underline{\underline{\Phi}}$ and $\underline{\underline{Q}}$. The distributions of magnetic flux and electric charge were considered independently of one another and were restricted only by Faraday's law of induction and the law of conservation of charge. These two familiar laws of the classical theory of electromagnetism assume the form of very simple assertions in the present formalism, viz., that magnetic flux and electric charge are exact forms (potential forms) in event-space. But anyone familiar with the classical theory of Maxwell and Lorentz knows that the electromagnetic field always determines a unique distribution of charge and current, so that these distributions are not independent. If the aether has a "metrical" structure that can be inferred from the electromagneric equations as was done by Lorentz, then this structure must lie in the relation between the distributions of magnetic flux and electric charge since, as we have seen, it cannot be inferred from the properties of either distribution separately. The aim of this lecture will be to show how the rlation between the distributions of magnetic flux and electric charge assumed by the classical theory actually determines a cone (the light cone) at each event.

II. 2. THE AETHER RELATIONS

The classical relation between the distributions of magnetic flux and electric charge may be characterized as follows.

A6: A certain linear function $K(\phi)$ of the electromagnetic field is a charge-current potential.

More fully now, consider the tangent space $V^4(\xi)$ of $\mathcal{E}$ at an event ξ and the 6-dimensional linear space $V_{[4^2]}(\xi)[\Phi]$ of 2-covectors at ξ having phys. dim. $[\Phi]$, and the 6-dimensional linear space $V_{[4^2]}(\xi)[Q]$ of 2-covectors at ξ having phys. dim. $[Q]$. Then, according to A6 there exists a linear transformation

$$K(\xi) : V_{[4^2]}(\xi)[\Phi] \rightarrow V_{[4^2]}(\xi)[Q] \tag{2.1}$$

such that the field $\eta(Q)$ given by $\eta(\xi, Q) = K(\xi, \varphi(\xi, \Phi))$, where φ is the electromagnetic field is a charge-current potential. Thus

$$C(\mathcal{E}^3, Q) = \oint_{\partial \mathcal{E}^3} K(\xi, \varphi) \tag{2.2}$$

We call the mapping K, the <u>aether tensor.</u> It is clear that we must have

$$\text{phys. dim. } [K] = [Q \Phi^{-1}] \tag{2.3}$$

We postulate now the following properties of the aether tensor which suffice to characterize it uniquely.

A7 : Invariant properties of the aether tensor.

P1: Electric and Magnetic Reciprocity

$$K^2(\xi) = -k^2 I, \tag{2.4}$$

where $k^2 > 0$ is a constant (i.e., is independent of ξ), and I is the identity transformation in $V_{[4^2]}$ $[\Phi]$ phys. dim. $k^2 = [Q^2 \; \Phi^{-2}]$.

P2 : Symmetry. Let $\tilde{D}$ be the duality transformation defined by an arbitrary 4-vector at ξ. Define the bilinear (quadratic) form $\hat{K}(\varphi, \eta)$ in $V_{4^2}(\xi)$ by $\hat{K}(\varphi, \eta) = \tilde{D}(\eta) \cdot K(\varphi)$. The symmetry property is the assertion that

$$\hat{K}(\varphi, \eta) = \hat{K}(\eta, \varphi)$$

for every pair of 2-covectors, φ and η.

P3: Existence of pure electric fields. There exists at least one direction (covector) in $V_4(\xi)$, say, $\epsilon^4(\xi)$, such that for every simple 2-covector having $\epsilon^4(\xi)$ as a factor $\hat{K}(\varphi, \varphi) > 0$ of $\hat{K}(\varphi, \varphi) < 0$, depending on the units of charge and flux and the orientation of the 4-vector which defines the dual transformation $\tilde{D}$.

II 3. LORENTZ FRAMES

Let $e_\alpha(\xi)$ be a basis in the tangent space $V^4(\xi)$ of Then $e_\alpha \vee e_\beta$, $\alpha < \beta$, is a basis in the space $V^{4^2}(\xi)$ of 2-vectors at ξ, and $\epsilon^\alpha \vee \epsilon^\beta (\epsilon^\alpha(e_\beta) = \delta^\alpha_\beta)$ is a basis in the space $V_{4^2}(\xi)$ of 2-covectors at ξ. The components of the aether tensor $\underset{4^2}{K}$ are then defined by

$$K^{\alpha\beta}_{\mu\pi} = K(\epsilon^\alpha \vee \epsilon^\beta, \; \underset{\approx}{Q}, \underset{\approx}{\Phi}) \cdot e_\alpha \vee e_\beta \, . \tag{3.1}$$

The aether tensor has a total of 36 independent components that can be arranged in a square 6 X 6 matrix. The rows and columns can be arranged according to the values of the pairs $(\alpha\beta)$ of the tensor indices. We shall use the conventional ordering, (14), (23), (24), (31), (34), (12).

Theorem II. 1 : There exists a basis $e_\alpha(\xi)$ in the tangent space $V^4(\xi)$ of ξ and units $\underset{\approx}{Q}$ and $\underset{\approx}{\Phi}$ of electric charge and magnetic flux such that, relative to this basis, the components of the aether tensor have the canonical values

$$(3.2) \qquad K^{(\alpha\beta)}_{\mu\tau} = \mathrm{diag}\,(B, B, B),$$

where B is the 2 X 2 matrix $\left\|\begin{matrix} 0 & -1 \\ 1 & 0 \end{matrix}\right\|$.

A basis $e_\alpha(\xi)$ in $V^4(\xi)$, for which the components of K have the values (3.2) is called a <u>Lorentz frame</u> at ξ, and the corresponding units of charge and magnetic flux are called <u>natural units.</u> The vector $e_4(\xi)$ of a Lorentz frame is said to be <u>time-like</u>, and the remaining three elements $e_1(\xi)$, $e_2(\xi)$, and $e_3(\xi)$ are said to be <u>space-like</u>. A simple electromagnetic field with a time-like divisor is a <u>pure electric field</u>.

In order to prove Theorem II. 1 we use the following.

Lemma II. 1. The 2-covectors φ and φ K(φ) are linearly independent.

Proof: Suppose φ is proportional to $K(\varphi)$ so that $\varphi = a\,K(\varphi)$ for some real number a. Then $K(\varphi) = a\,K^2(\varphi)$. By P 1 it then follows that $K(\varphi) = -a\,k^2\,\varphi = -a^2\,k^2$, or

$(1 + a^2 k^2) K(\varphi) = 0$. Since $k^2 > 0$, this has no real solution for a, $K(\varphi) \neq 0$, which proves the lemma.

Lemma II. 2 : If φ is a simple 2-covector, then the 2-covector $K(\varphi)$ is also simple.

Proof : A 2-covector $\varphi \neq 0$ is simple if and only if det $\varphi = E(\varphi, \varphi) = 0$, where $E \neq 0$ is any- 4-vector and we define $E(\varphi, \varphi) = \varphi E \wedge \varphi) = \varphi D \varphi$. But $E(K(\varphi), K(\varphi)) = \hat{K}(\varphi, K(\varphi))$. Using P2 this last expression is equivalent to $K^2(\varphi) \cdot D(\varphi)$. Thus using Pl we have $E(K(\varphi), K(\varphi)) = -k^2 \varphi D(\varphi) = -k^2 E(\varphi; \varphi)$, which proves the lemma .

Lemma II. 3 : If φ and η are linearly independent and have a common divisor, then so also are $K(\varphi)$ and $K(\varphi)$ linearly independent with a common divisor.

Proof. $E(K(\varphi), K(\eta)) = -k^2 E(\varphi, \eta)$ as can be shown using the same methods and properties of K as in the proof of Lemma II. 2 above.

Proof of Theorem II. 1 : First of all, from phys.dim. $k^2 = [\underline{Q}\, \underline{\underline{\Phi}}^{-2}]$, if follows that one may choose the units of electric charge and magnetic flux such that $k^2 = 1$. Then , for such a choice of the units, $K^2 = -I$. From the existence of pure electric fields (property P3) there exists a covector $\epsilon^4(\xi)$ such that $\hat{K}(\overset{1}{\varphi}, \overset{1}{\varphi}) > 0$ for the simple covector $\overset{1}{\epsilon} \vee \overset{4}{\epsilon}$, where $\overset{1}{\epsilon}$ is any covector linearly independent of $\overset{4}{\epsilon}$. Then by Lemmas II.1 and II.2 the 2-covector $K(\overset{1}{\varphi})$ is simple and linearly independent of $\overset{1}{\varphi}$. Set $K(\overset{1}{\varphi}) = \overset{2}{\epsilon} \vee \overset{3}{\epsilon}$. It then follows from $E(\overset{1}{\varphi}, K(\overset{1}{\varphi})) = \hat{K}(\overset{1}{\varphi}, \overset{1}{\varphi}) > 0$ that the covectors $\overset{1}{\epsilon}$, $\overset{2}{\epsilon}$, $\overset{3}{\epsilon}$, and $\overset{4}{\epsilon}$ are linearly independent. Nowset

$\varphi^2 = \epsilon^2 \vee \epsilon^4$. Since φ^2 and φ^1 are linearly independent and have a common factor, it follows from Lemma II.3 that $K(\varphi^2)$ and $K(\varphi^1)$ are linearly independent and have a common factor. By P3 this common factor cannot be ϵ^2 else we would have $\hat{K}(\varphi^2, \varphi^2) = 0$; hence the common factor must be ϵ^3 (i.e. , a covector proportional to ϵ^3). Thus $K(\varphi^2) = \pi \vee \epsilon^3$ for some value of the covector π . But φ^1 and $K(\varphi^2)$ have a common factor also since $E(\varphi^1, K(\varphi^2)) = \hat{K}(\varphi_1, \varphi_2) = E(\varphi^2, K(\varphi^1)) = 0$, where we have used the symmetry property P2 . This common factor cannot be ϵ^4 ; hence it must be proportional to ϵ^1 . Thus $K(\varphi^2) = a\, \epsilon^3 \vee \epsilon^1$ for some value of a. Now set $\varphi^3 = \epsilon^3 \vee \epsilon^4$. Then $K(\varphi^3)$ has a factor in common with $K(\varphi^2)$. This factor cannot be ϵ^3 (P3 again) ; hence we can set $K(\varphi^3) = \pi' \vee \epsilon^1$ for some value of π'. But $\pi' \vee \epsilon^1$ has a factor in common with $\epsilon^2 \vee \epsilon^3$ and this cannot be ϵ^3 ; hence $K(\varphi^3) = b\, \epsilon^1 \vee \epsilon^2$ for some value of b . Choosing now as a basis in $V^4(\xi)$ the four linearly independent vectors $e^\alpha(\xi)$ reciprocal to the four covectors $\epsilon_\alpha(\xi)$ we find that the components of K have the canonical values set forth in (3.2) except that in place of some of the 1's and -1's there appear the unknown scalar values of a and b . But from the condition $K^2 = -I$ we may deduce that the values of a and of b must both +1 , and the theorem is proved.

II. 4. THE LIGHT CONE

Two quadratic forms q and q' in any vector space V^m are <u>conformal</u> if $q = a q'$; i.e. , if one be proportional to the other. A quadratic form with signature (1, 1, ..., 1, - 1) determines a <u>cone</u> . The cone is the set of all vectors such that $q(v, v) = 0$

Every quadratic form in the same conformally equivalent class determines the same cone. Conversely, we can show that the cone of a quadratic form of signature (1, 1, ..., 1, - 1) determines the quadratic form uniquely up to a factor, i.e., determines the conformal class of the form. To see this recall that any symmetric bilinear form $q(u, v)$ is determined by its values $q(w, w)$ because $q(u, v) = \frac{1}{2}(q(+v, u+v) - q(u,u) - q(v,v))$. Now suppose that $q'(u, u) = 0$ whenever $q(u,u)=0$. It is required to show that $q' = aq$. Let e_i, $i=1,..n$ be a basis sich that $q(e_i, e_j)=\mathrm{diag}(1, 1,...1, -1)$. Then $q'(e_A \pm e_n, e_A \pm e_n)=0$ for $A=1, 2,.. n-1$because $e_A \pm e_n$ belongs to the cone of q. From these conditions on q' deduce that $q'(e_A, e_n) = 0$, and that $q'(e_A, e_A 9 + q'(e_n, e_n) = 0$. But $u = \frac{1}{2}(e_A+e_B)+ + e_n$ is also an element of the cone of q for $A \neq B$; hence, $q'(u, u) = 0$. If this condition be expanded, then using the properties of q' already established we get $q'(e_A, e_B) = 0$, for $A \neq B$. Therefore, $q'(e_1, e_j) = \mathrm{diag}(a, a, ..., a, -a)$, which proves the assertion.

The adjoint q of a quadratic form q with signature (1, 1, 1, - 1) determines a cone of convectors in $V_4(E)$ satisfying $q^{\dagger}(\alpha, \alpha) = 0$.

Theorem II.4.1: T here exists a unique cone in $V^4(E)$ such that, for all covectors $\alpha_1, \alpha_2, \beta_1$, and β_2,

$$(4.1) \qquad \hat{K}(\alpha_1 \vee \beta_1, \alpha_2 \vee \beta_2, \underset{\sim}{\Phi}, \underset{\sim}{Q}) = q(\alpha_1, \alpha_2) q(\beta_1, \beta_2) - q(\alpha_1, \beta_2) q(\alpha_2, \beta_1),$$

for one of the quadratic forms $q(v, v)$ in the conformal class which determines the cone.

Proof: Let e_α be a Lorentz frale at ξ abd let e^α be the reciprocal basis. The definition of the quadratic form K in $V_{[4^2]}(\xi)$ depends linearly upon the choice of the 4-vector used to define the duality transformation. Choose for this 4-vector, $E = e_1 \vee e_2 \vee e_3 \vee e_4$. Let natural units be chosen for charge and magnetic flux so that $K^2 = -1$. For this choice of basis, of E, and of the units of charge and flux we have the following values for the components of $\hat{K}$:

$$\hat{K}^{(\alpha\beta)(\mu\tau)} = \hat{K}(\epsilon^\alpha \vee \epsilon^\beta, \epsilon^\mu \vee \epsilon^\tau, \underset{\approx}{\Phi} \underset{\approx}{Q}) = \mathrm{diag}(-1, 1, -1, 1, -1, 1).$$

The rows and columns of the six by six matrix of independent components of $\hat{K}$ are labeled according to the same convention (14), (23), (24), (31), (34), (12) used in (II; 3.2). Now let Latin indices range over the values 1, 2, and 3. One easily verifies that

$$(4.3) \qquad \hat{K}^{ijkl} = q^{+ik} q^{+jl} - q^{+il} q^{+jk} = \delta^{ik} \delta^{jl} - \delta^{il} \delta^{ik}$$

From this equation we may deduce that

$$(4.4) \qquad q_{ij} = \delta_{ij} \det \| q_{rs} \| .$$

On taking the determinant of this equality we find that $\det \| q_{rs} \| = \pm 1$; hence ,

$$(4.5) \qquad q_{rs} = \pm \delta_{rs} , \quad q^{+rs} = \pm \delta^{rs} .$$

Now use

$$(4.6) \qquad K^{ijk4} = q^{+ik} q^{+j4} - q^{+i4} q^{+jk} = 0$$

and (4.5) to deduce that

(4.7) $$q^{\dagger j4} = 0 .$$

Finally, from (4.5) , (4.7), and

(4.8) $$K^{i4j4} = q^{\dagger ij} q^{\dagger 44} - q^{\dagger i4} q^{\dagger j4} = -\delta^{ij}$$

one gets

$$q^{\dagger 44} = -1 \text{ when } q^{\dagger ij} = +\delta^{ij},$$

(4.9) $$q^{\dagger 44} = +1 \text{ when } q^{\dagger ij} = -\delta^{ij} .$$

This proves the theorem and also show a that if $\epsilon_a(\xi)$ is a Lorentz frame at ξ , then the cone determined by K (<u>the light cone</u>) is given by

(4.10) $$q = [a \sum_{i=1}^{3} \epsilon_i \times \epsilon_i - \epsilon_4 \times \epsilon_4],$$

where a is an arbitrary factor of proportionality.

II. 5. THE ELECTROMAGNETIC SYMMETRY GROUP

The aether tensor $K(\xi)$ determines the electromagnetic symmetry group as follows. Every non-singular transformation

(5.1) $$S \; V^4(\xi) + \underline{Q} + \underline{\underline{\Phi}} \rightarrow V^4(\xi) + \underline{\underline{Q}} + \underline{\underline{\Phi}}$$

induces a non-singular linear transformation

(5.2) $$T_S \; W \rightarrow W$$

in the tensor space $W = V_{[4^2]}(\xi) \otimes V^{[4^2]}(\xi) [\underline{\underline{Q}} \; \underline{\underline{\Phi}}]$ of which K is an element. The transformation T_S is defined by the condition

(5.3) $$\bar{K}(\bar{\alpha} \vee \bar{\beta}, \bar{Q}, \bar{\Phi}) . \bar{u} \vee \bar{v} = K(\alpha \vee \beta, Q, \Phi) u \vee v ,$$

where $(\bar{Q}, \bar{\Phi}, \bar{u}, \bar{v}) = (S(Q), S(\Phi), S(u), S(v))$, $\bar{K} = T_S(K)$, and $\bar{\alpha}$ and $\bar{\beta}$ are defined by the corresponding conditions $\bar{\alpha}\ \bar{v} = \alpha\ v$, $\bar{\beta}\ \bar{v} = \beta\ v$. The transformation S is an element of the electromagnetic symmetry group if and only if

(5.4) $$T_S(K) = K ,$$

i.e., if and only if the aether tensor is an invariant tensor under the induced transformation T_S. But if K is invariant under T_S, then so also is the light cone determined by K. Hence the electromagnetic symmetry group is a subgroup of the transformations (5.1) for which

(5.5) $$T'_S(q) = a\ q ,$$

where T'_S is the linear transformation induced by S in the space $V_{\binom{4}{2}}(\xi)$ of symmetric dimensionless 2-cotensors at ξ. The group of transformations S defined by the condition (5.5) is the <u>conformal Lorentz group</u>. The electromagnetic symmetry group is a subgroup of the conformal Lorentz group.

It has been established that the aether tensor has the representation

(5.6) $$K = k\ (e_2 \vee e_3) \otimes (\epsilon^1 \vee \epsilon^4) - e_1 \vee e_4) \otimes (\epsilon^2 \vee \epsilon^3) \\ + \text{cyclic permutations of } 1, 2, 3, . \quad ,$$

where k is a scalar with phys. dim. $[Q\ \Phi^{-1}]$. The condition $k = 1$ defines natural units for Q and Φ. The e_α are the elements of a Lorentz frame at ξ and $\epsilon^\alpha(e_\beta) = \delta^\alpha_\beta$. The transformations S in (5.1) have the form

$$(5.7)\qquad S = \left\| \begin{matrix} s & & \\ & \Phi^{-1} & \\ & & Q^{-1} \end{matrix} \right\|$$

where $s : V^4(\xi) \to V^4(\xi)$ is a non-singular linear transformation of the tangent space at ξ, and Φ and Q are the rations of the new and old units of flux and charge. From the representation (5.7) of K we deduce that a necessary condition for K to be invariant under T_S is that

$$(5.8)\qquad Q\,\Phi^{-1}(\text{sign}\ \det\ s) = +1.$$

In other words, the transformation S must be a proper transformation, and the absolute value of the product $Q\,\Phi^{-1}$ must be 1. One now convinces himself that the two conditions (5.5) and (5.8) are both necessary and sufficient that K be invariant under T_S.

II. 6. DISCUSSION

Taking the relation between the distributions of electric charge and magnetic flux as a starting point, we have shown how this relation determines a unique cone of directions at each event. We have defined an electromagnetic symmetry group as the invariance group of the aether tensor which describes completely how the distributions of charge and flux are related in the classical theory. This symmetry group turns out to be a certain subgroup of the transformations in a

6-dimensional vector space, the direct sum of the tangent space $V^4(\xi)$ of the manifold of events, and the two unit space of electric charge and magnetic flux. If the units of electric charge and magnetic flux are held fixed, then the subgroup of the electromagnetic symmetry group defined by that condition turns out to be the proper conformal Lorentz group of transformations of $V^4(\xi)$. On the other hand, an improper conformal Lorentz transformation of $V^4(\xi)$ when accompanied by an improper transformation of the unit space $\underline{\underline{\Phi}} \otimes \underline{\underline{Q}}^c$ of determinant -1 is an electromagnetic symmetry element. Thus, oddly enough, the statement commonly made that Maxwell's equations are invariant under the Lorentz group of transformations (transformations which leave a quadratic form of signature (1, 1, 1, -1) invariant) is not quite correct for two opposing accounts. First of all, this statement is usually made uned the agreement that the units of charge and flux have been fixed upon and not subject to transformation. In this case, the Lorentz group does not contain the corresponding subgroup of electromagnetic symmetry transformations. Moreover, if the units are held fixed, since only proper transformations of $V^4(\xi)$ are then contained in the corresponding subgroup of electromagnetic symmetry transformations, thys subgroup does not contain the full Lorentz group. On the other hand, if transformations of the units of electric charge and magnetic flux are properly taken into account (they are certainly of equal importance to the transformations of the tangent space) we have seen that Maxwell's equations are invariant under certain improper transformations of $V^4(\xi)$ provided they are accompanied by an appropriate improper transformation of the unit space $\underline{\underline{\Phi}} \otimes \underline{\underline{Q}}$.

LECTURE III
ACTION AND GRAVITY

" These physical hypotheses, however, are entirely alien from the way of looking at things which I adopt, and one object which I have in view is that some of those who wish to study electricity may, by reading this treatise, come to see that there is another way of treating the subject, which is no less fitted to explain the phenomena, and which, though in some parts it may appear less definite, corresponds, as I think, more faithfully with our actual knowledge, both in what it affirms and in what it leaves undecided."

J. C. Maxwell

III. 1. INTRODUCTORY REMARKS

An essential qualitative difference between electric charge and gravitational or inertial mass is that charge appears to occur in Nature with either sign, but mass occurs with but one sign. Thus charge and mass are essentially different qualities of matter requiring essentially different mathematical representations. How can we introduce in a natural way this essential difference within the present framework of concepts and mathematics? We have seen in the first two lectures how Maxwell's equations for the electromagnetic field and the charge-current field can be viewed as conditions upon the distributions of two physical quantities, electric charge and magnetic flux. Each of these physical quantities has been represented mathematically as a linear function of oriented, submanifolds of events; the first by a linear function of 3-dimensional submanifolds, the second by a linear function of 2-dimensional submanifolds. In neither of these cases would it be natural or even possible to introduce a condition that the distributions be positive everywhere (or negative). The situation is different, however, in the case of a linear function of 4-dimensional submanifolds in an orientable embedding space of 4-dimensions. The 4-dimensional oriented curvilinear simplexes in event-space can be divided into two equivalence classes of similarly oriented simplexes. This cannot be done with simplexes of lower dimension. Thus it is natural to seek a theory and representation of mass connected in some way with the simplexes of dimension four in $\mathcal{E}$. If a physical quantity is represented by a 4-cochain in event-space, it is possible to introduce the condition that the value of the cochain, for any fixed choice of the physical unit, has the same sign on every similarly oriented 4-simplex in $\mathcal{E}$. In turns out that mass itself is not the natural quantity which is of uniform sign on such a class, but rather a quantity we shall call action which we may think of as related to the integral of the mass of a 3-simplex over a 1-simplex in a time-like direction. These remarks are inteded only to guide the intuition, since formal definitions of mass and action are given later.

III. 2. ACTION AND GRAVITY

We introduce two more unit spaces $\underline{\underline{A}}$ and $\underline{\underline{G}}$ alongside $\underline{\underline{Q}}$ and $\underline{\underline{\Phi}}$, and two corresponding physical quantities called action and gravity .

A8: Action is a continuous 4-cochain in event-space with phys. dim. $[\underline{\underline{A}}]$.

It follows that there exists a continuous 4-form μ in such that the action $A(\mathcal{E}^4, \underset{\approx}{A})$ of a given oriented 4-dimensionale submanifold $\mathcal{E}^4$ measured in units of action $\underset{\approx}{A}$ is given by

(2.1) $$A(\mathcal{E}^4, \underset{\approx}{A}) = \int_{\mathcal{E}^4} \mu(\underset{\approx}{A}) \quad ,$$

and

(2.2) $$\text{phys. dim. } \mu = [\underline{\underline{A}}].$$

Now by A1 , $\mathcal{E}$ is orientable . Thus it is possible to divide the oriented simplexes of $\mathcal{E}$ into two equivalence classes of similarly oriented 4-simplexes, Denote these two classes by $\mathcal{E}^+$ and $\mathcal{E}^-$.

A8: if S and S' are any two 4-simplexes in the same orientation class $(\mathcal{E}^+$ or $\mathcal{E}^-)$, then action has the property

(2.3) $$A(S, \underset{\approx}{A})\ A(S', \underset{\approx}{A}) \geqslant 0 \quad .$$

Now every continuous 4-cochain in $\mathcal{E}$ has a potential. Thus there exists a regular 3-form π in $\mathcal{E}$ such that

(2.4) $$A(\mathcal{E}^4, \underset{\approx}{A}) = \oint_{\partial\mathcal{E}^4} \pi(\underset{\approx}{A}) \quad .$$

As in the case of the electromagnetic and charge-current potentials, π is not uniquely determined by the distribution of action. If π satisfies (2.4) then so also does

(2.5) $$\pi' = \pi + \operatorname{rot} \beta ,$$

where β is any regular 2-form in $\mathcal{E}$.

<u>Gravity</u> is a physical quantity with phys. dim. $[\underline{G}]$.

A9. Gravity is a continuous 1-cochain in event-space.

It follows from A9 that there exists a unique <u>gravitational field</u> γ, a 1-form in $\mathcal{E}$, such that the gravity $G(\mathcal{E}^1, \underset{\approx}{G})$ of any curve in $\mathcal{E}$ measured in units of gravity $\underset{\approx}{G}$ is given by

(2.6) $$G(\mathcal{E}^1, \underset{\approx}{G}) = \int_{\mathcal{E}^1} \gamma(\underset{\approx}{G})$$

$$\text{phys.dim. } \gamma = [\underline{G}] .$$

A10: The gravitational field is circulation free.

Accordingly, if $\mathcal{C}^1$ is any closed smooth curve in $\mathcal{E}$,

(2.7) $$G(\mathcal{C}^1, \underset{\approx}{G}) = \oint_{\mathcal{C}^1} \gamma(\underset{\approx}{G}) = 0 .$$

It follows from A10 that there exist gravitational potentials, ψ, regular 0-forms (scalar fields) in $\mathcal{E}$, such that

(2.8) $$G(\mathcal{E}^1, \underset{\approx}{G}) = \oint_{\partial\mathcal{E}^1} \psi(\underset{\approx}{G}) = \psi(\overset{+}{\xi}) - \psi(\overset{-}{\xi}) ,$$

and

(2.9) $\quad$ phys.dim. $\psi = [\underline{\underline{G}}]$.

In (2.8) , ξ^+ and ξ^- are the end points of the oriented curve $\mathcal{E}^1$. The gravitational potential is determined by the distribution of gravity only to within an additive constant. What we have called the gravity of a curve is equal to the difference between the values of the gravitational potential at its two end points.

III.3. THE GRAVITATIONAL AETHER RELATION

Guided now by the way that the distributions of electric charge and magnetic flux are related in Maxwell's theory, we shall introduce a connection between the distributions of action and gravity which, up to this point, have been treated as independent.

A11: There exists a linear transformation

(3.1) $$L : V_4(\xi)\,[G] \rightarrow V_{[4^3]}(\xi)\ [\underline{\underline{A}}]$$

of the space of 1-forms in $\mathcal{E}$ with phys.dim. $[\underline{\underline{G}}]$ into the space of 3-forms in $\mathcal{E}$ with phys.dim. $[\underline{\underline{A}}]$ such that $L(\gamma)$ is a potential of action.

We shall call L the <u>gravitational aether tensor</u> so as to distinguish it from the (electromatnetic) aether tensor K which plays the analogous role in relating the distributions of charge and flux.

Let $E \neq 0$ be an arbitrary 4-vector in $V^{[4^4]}(\xi)$ and set $D(\alpha) = \text{dual}\,\alpha = E \wedge \alpha$, where α is any r-covector at ξ. Then $\hat{L}$ defined by

$$(3.2) \qquad \hat{L}(\alpha, \beta, \underset{\sim}{G}, \underset{\approx}{A}) = D(\beta)\; L(\alpha, \underset{\sim}{G}; \underset{\approx}{A})$$

is a bilinear form in $V_4(\xi)$ with phys. dim.[$\underline{\underline{AG}}^{-1}$]

A12: The bilinear form $\hat{L}$ defined by the gravitational aether tensor L is symmetric and has signature $\pm(1, 1, 1, -1)$.

The definition of $\hat{L}$ in terms of L depends on the choice of a 4-vector E. Since all such 4-vectors are proportional, L determines $\hat{L}$ up to a factor. Thus the gravitational aether tensor determines, not a quadratic form, but a cone of directions in $V_4(\xi)$ defined by $\hat{L}(\alpha, \alpha) = 0$. We call this cone the <u>gravitational cone</u>. The connection between gravity and electromagnetism is established in part by

A13: The light cone and the gravitational cone coincide at every event.

III.4. GRAVITATIONAL INVARIANCE GROUP AND THE METRICAL STRUCTURE OF EVENT-SPACE

The gravitational invariance group is defined in the same way as the electromagnetic invariance group with K replaced by L. Thus we consider the set of all transformations

$$(4.1) \qquad S: V^4(\xi) \oplus \underline{\underline{A}} \oplus \underline{\underline{G}} \to V^4(\xi) \oplus \underline{\underline{A}} \oplus \underline{\underline{G}}$$

of the six-dimensional vector space consisting in the direct sum of the tangent space at ξ and the two unit spaces $\underline{\underline{A}}$ and $\underline{\underline{G}}$. Each such transformation induces a linear transformation

(4.2) $$T_{S'} \; W \to W$$

in the tensor space

(4.3) $$W = V_{[4^3]} \otimes V^4(\xi) \quad [\underline{\underline{A}}\underline{\underline{G}}^{-1}]$$

of which L is an element. The gravitational invariance group is the set of all transformations S such that

(4.4) $$T_S(L) = L \; .$$

The transformations S are of the form

(4.5) $$S = \left\| \begin{matrix} s & & \\ & \underline{\underline{A}}^{-1} & \\ & & \underline{\underline{G}}^{-1} \end{matrix} \right\| .$$

Since L defines a unique cone in $V_4(\xi)$, it follows that the transformations s in (4.5) must be a subgroup of the conformal Lorentz transformations. It follows from A12 and A13 that there exists a Lorentz frame e_α at ξ such that $L(\xi)$ is given by

(4.6) $$\begin{aligned} L(\xi) = \ell[&\epsilon^1 \vee \epsilon^2 \vee \epsilon^3 \otimes e_4 + \epsilon^1 \vee \epsilon^2 \vee \epsilon^4 \otimes e_3 \\ &+ \epsilon^2 \vee \epsilon^3 \vee \epsilon^4 \otimes e_1 + \epsilon^3 \vee \epsilon^1 \vee \epsilon^4 \otimes e_2] , \end{aligned}$$

where ℓ is a constant and

(4.7) $$\text{phys.dim.} = [\underline{\underline{A}}\underline{\underline{G}}^{-1}]$$

One sees from this representation of L that necessary conditions for S to be a symmetry element are

(4.8) $(\underline{\underline{A}}\ \underline{\underline{G}}^{-1})\ \mathrm{sign}(\det\ s) \longrightarrow 0,$

and

(4.9) $|\det\ s| = \underline{\underline{A}}^2\ \underline{\underline{G}}^{-2}.$

One then verifies that the three conditions, s a conformal Lorentz transformation, (4.8), and (4.9), are both necessary and sufficient that S be a symmetry element.

Consider now the symmetric tensor $\hat{L}$ defined in (3.2) in terms of L and an arbitrary 4-vector $E \neq 0$. For every choice of E, det $L < 0$. Let $E = e_1 \vee e_2 \vee e_3 \vee e_4$ and $\mathcal{E} = \epsilon^1 \vee \epsilon^2 \vee \epsilon^3 \vee \epsilon^4$, where e_a and ϵ^a are reciprocal sets and e_a is a Lorentz frame at ξ. Now we have

(4.10)
$$L:\ V_4(\xi)\,[\underline{\underline{G}}] \to V^4(\xi)\,[\underline{\underline{A}}],$$
$$\det\ L = \mathcal{E}[\ \hat{L}(\epsilon^1) \vee \hat{L}(\epsilon^2) \vee \hat{L}(\epsilon^3) \vee \hat{L}(\epsilon^4)]$$

and

(4.11) $\text{phys.dim.}\ \det\ \hat{L} = [\underline{\underline{A}}^4\ \underline{\underline{G}}^{-4}]$

If $E' = a\,E$, then $\hat{L}' = a\,\hat{L}$ and we see that

(4.12) $\det\ \hat{L}' = a^4\ a^{-2}\ \det\ \hat{L} = a^2\ \det\ \hat{L}.$

It follows that the symmetric tensor $g^\dagger$ defined by

(4.13) $$g^\dagger = \frac{\hat{L}}{\sqrt{-\det\ \hat{L}}}$$

depends only on orientation of E and

(4.14) phys.dim. $\overset{+}{g} = [\underline{\underline{A}}^{-1} \underline{\underline{G}}]$, det $g < 0$.

Thus we see that the gravitational aether tensor determines a unique (up to sign) symmetric tensor field $\overset{+}{g}$ and cotensor field g ($\overset{+}{g}^{\alpha\beta} g_{\beta\tau} = \delta^{\beta}_{\tau}$) with signature $\pm(1,1,1,-1)$. The gravitational symmetry group can be characterized alternatively as the subgroup of proper transformations (4.5) such that

(4.15) $$T_S(g) = g ,$$

where T_S is the linear transformation induced by S in the space of symmetric 2-cotensors at ξ having phys.dim. $[\underline{\underline{A}}\ \underline{\underline{G}}^{-1}.]$ If g smooth, then it is seem that the gravitational aether relation endows event-space with a smooth (pseudo) Riemannian structure defined by a fundamental <u>metric tensor</u> $\pm g$ with phys. dim. $[\underline{\underline{A}}\underline{\underline{G}}^{-1}]$, det $g < 0$.

R.A. Toupin

LECTURE IV
MOTION OF CONTINUOUS MEDIA

" The idea of motion implies the existence of some means of recognizing again and again the entity that moves. By extending the idea of a mathematical point we have the concept of a moving point which we shall call an electrical point and we may start with the fundamental hypothesis that three independent quantities (α, β, γ) are sufficient to specify an electrical point and distinguish it from others" .

H. Baterman

IV.1. INTRODUCTORY REMARKS

It is worth emphasizing that nothing in the preceding lectures depends in any way upon the concept of motion or the concept of a material medium. Thus it stands independently of whatever is now said about continuous media and motion.

One of the earliest questions raised by the new mechanics of special relativity theory concerned the concept and definition of rigid motion. In classical mechanics, a rigid motion of any set of material points is defined by the condition that the distance between every pair of points in the set at each instant of time remain invariant in time. In the Minkowski manifold of special relativity theory, an instant of time is not defined. We have given , rather, a cone of directions at each event. Born was the first to consider the problem

of extenging ortransferring the classical notion of a rigid motion to the new kinematical setting of Minkowski space. The concept and definition of a rigid motion of a continuos medium is an essential preliminary to a relativistic theory of elastic response.
Essential also to an extension of classical continuum mechanics to the more general space-time manifolds of general relativity is the concept and definition of velocity and acceleration. The objective in this lecture will be to show how the counterparts of these classical kinematical ideas can be defined in a natural way in a manifold of events in which there is given at each event a cone of directions corresponding to a metric field g with

det $g < 0$, phys. dim. $g = [\underline{\underline{A}}\underline{\underline{G}}^{-1}]$.

IV. 2. MOTION OF MATERIAL MEDIA

We shall consider only 3-dimensional material media. A 3-dimensional material medium is an orientable smooth manifold of dimension 3, which we denote by $\mathcal{M}$, together with any additional structure which may be assigned to it. In what follows, the only properties of $\mathcal{M}$ which will be used are those which follow from the definition of a smooth manifold.
The points of $\mathcal{M}$ are called <u>material points</u> and will be denoted by X, X', X'', etc. A motion of $\mathcal{M}$ is a smooth mapping

(2.1) $$f : \mathcal{E}' \to \mathcal{M}$$

of a set of events $\mathcal{E}'$ onto $\mathcal{M}$ such that f has rank three at every point in the domain of f and such that the proper vector of with proper value O is time-like. The orbit of a material point X is the set of events $\{\xi f(\xi) = X\}$ experienced by the point X and is called the world-line of X. As remarked in the preliminaries § 12, from what has been assumed and the implicit, function theorem it follows that locally the world-line of every material point is a smooth parametrizable curve in $\mathcal{E}$. In other words, a motion may be represented locally as a mapping

$$f^{\dagger} : \mathcal{M} \times \mathcal{O} \to \mathcal{E} \tag{2.2}$$

where $\mathcal{O}$ is an open set of real numbers. The gradient of, $f^{\dagger}$ with respect to the parameter, $\nabla_{\tau} f^{\dagger}$ is a proper vector of ∇f with proper value zero; it is tangent to the world-line of the point X at the event $f^{\dagger}(X, \tau)$. The world-velocity field of the motion is the normalized field of time-like tangent vectors

$$v = \frac{\nabla_{\tau} f^{\dagger}}{\sqrt{|g(\nabla_{\tau} f^{\dagger}, \nabla_{\tau} f^{\dagger})|}} \tag{2.3}$$

It should perhaps be pointed out that the direction of v, either forward or backward, depends on the parametrization of the world-lines. As yet we have introduced non assumptions which distinguish the future from the past. Thus one should keep in mind the dependence of v upon the parametrization. Let $v^{\dagger}$ be the covector set in correspondence with v by g; i.e., defined by $v^{\dagger}(u) = g(v, u)$. Then the world-acceleration of the motion is defined by

(2.4) $$a^{\dagger} = (\operatorname{rot} v^{\dagger}) \wedge v.$$

Note that $a^{\dagger}(v) = 0$ Note also that phys. dim. $v = [\underline{T}^{-1}]$, phys dim. $v^{\dagger} = [\underline{A}\underline{G}^{-1}\ \underline{T}^{-1}]$, phys dim. $a^{\dagger} = [\underline{A}\underline{G}^{-1}\ \underline{T}^{-2}]$, where we introduce the notation $\underline{T} = +\sqrt{\underline{A}\ \underline{G}^{-1}}$ for the time dilation which accompanies a transformation $A' = \underline{A}^{-1} A$, $G' = \underline{G}^{-1} G$ of the units of action and gravity. Only the sign of the acceleration is affected by the choice of these units and it is independent of the parametrization of the motion.

In what follows it proves convenient to introduce the symbol s which has the value plus one or minus one such that $sg(u, u) < 0$ if u is a time-like vector. The value of s depends on the units of action and gravity. We may indicate this dependence by writing phys. dim. $s = [\underline{A}\underline{G}^{-1}\ \underline{T}^{-2}]$.

IV. 3. RIGID MOTIONS

The gradient ∇f of a motion (2.1) determines a linear mapping

(3.1) $$\nabla f \cdot W_3(X) \qquad V_4(\xi), \quad X = f(\xi)$$

of the tangent space in the material manifold $\mathcal{M}$ into the space of covectors at ξ. In terms of ∇f and g we define a symmetric bilinear function of material covectors C^{-1} by

(3.2) $$C^{-1}(\Omega, \Lambda) = s\, g^{\dagger}(\nabla f(\Omega), \nabla f(\Lambda))$$

and show that C^{-1} is positive definite. Set $\omega = \nabla f(\Omega)$. By hypothesis, $\nabla f . v = 0$, where v is the velocity vector and is time-like. But then $\omega(v) = g^{\dagger}(\omega, v^{\dagger}) = 0$, where $v^{\dagger} . u = g(v, u)$, and $v^{\dagger}$ is time-like. Hence the set of image vectors $\nabla f(W_3)$ consists in covectors each of which is normal to the time-like covector $v^{\dagger}$. It follows that $s\, g^{\dagger}$ restricted to this subspace of $V_3(\xi)$ is positive definite. Therefore, C^{-1} defined in (3. 2) is positive definite, and phys. dim. $C^{-1} = [\underline{T}^{-2}]$. It is seen that a motion $f\mathcal{M}$ in $\mathcal{E}$ determines a set of positive definite quadratic forms $\{C(\xi) ; f(\xi) = X\}$ in each tangent space $W^3(X)$ of $\mathcal{M}$. A motion of $\mathcal{M}$ is <u>locally rigid</u> at X if and only if this set of forms consists in a single element i. e., if and only if $f(\xi) = f(\xi') \rightarrow C(\xi) = C(\xi')$. The above definition of a locally rigid motion is independent of any parametrization of the motion. If a parametrization is given, more familiar geometric results emerge, but it is important to distinguish those results which depend on the parametrization from those which do not.

Let $\tau : \mathcal{E}' \rightarrow R$ be a regular mapping of the events in a neighborhood of $\xi \in \mathcal{E}'$ into the real numbers such that $\nabla\tau$ has rank one and such that each member of the family of level surfaces $\tau(\xi) = \text{constant}$ is space-like. We may choose τ such that $s \nabla\tau . v = -1$. Then $|\tau(\xi_2) - \tau(\xi_1)|$, where ξ_1 and ξ_2 are events experienced by a given material point X is the interval of propertime measured along the world-line of X between the events ξ_1 and ξ_2.
We may view (2. 2) as a oneparameter family of embeddings $\mathcal{M}$ of $\mathcal{E}$ in . The image

$f^{\dagger}(\mathfrak{M}, \tau)$, for each value of τ, is a smooth 3-dimensional surface in $\mathcal{E}$, and the surfaces $f^{\dagger}(\mathfrak{M}, \tau)$ and $f^{\dagger}(\mathfrak{M}, \tau')$ have no points in common if $\tau \neq \tau'$.

In § 12 of the preliminaries it was shown how a motion of $\mathfrak{M}$ determines a one-parameter family of motions of defined by

$$(3.3) \qquad T_{\tau}(\xi) = f^{\dagger}_{t+\tau} \, f^{\dagger}_{t}{}^{-1}(\xi) .$$

The gradient, ∇T_{τ}, of this points point transformation is a linear transformation of the tangent space $V^4(\xi)$ onto the tangent space $V^4(\xi_{\tau})$, $\xi_{\tau} = T_{\tau}(\xi)$. Let the quadratic form g_{τ} be defined by

$$(3.4) \qquad g_{\tau}(u, v, \xi) = s g(\nabla T_{\tau}(u), \nabla T_{\tau}(v), T_{\tau}(\xi))$$

and suppose that the parameter t is proper time. Then the Lie derivative of the metric field g with respect to the velocity field v of the motion is given by

$$(3.5) \qquad \underset{v}{\mathcal{L}} \, g = s \left. \frac{dg_{\tau}}{d\tau} \right|_{\tau=0}$$

Let a superposed dot denote differentation with respect to t. On verifies by direct calculation that

$$(3.6) \qquad C(U, W, X, \tau) = s \underset{v}{\mathcal{L}} g(P_v(u), P_v(w), f^{\dagger}(X, \tau)),$$

where $P_v = I + s\, v \otimes v^T$ is the projection of $V^4(\xi)$ onto the subspace $V^3(\xi, v)$ of vectors at ξ normal to the velocity, and u is the unique solution of $P\, \nabla f^T\,(U) = P_v \cdot u$. It follows from (3.6) that a necessary and sufficient condition that a motion be locally rigid at X is that the rescriction of the Lie derivative of the metric field with respect to the velocity field of the motion to the subspaces $V^3(\xi, v)$, $f(\xi) = X$ vanish.

The differential equations expressing this condition have the form

(3.6) $$B_{\alpha\beta} = v^T_{(\alpha;\beta)} + v^T_{(\alpha}\, a_{\beta)} = 0\ ,\quad f(\xi) = X.$$

where a semi-colon denotes covariant differentiation, v^T_α is the velocity field (covariant differentation, v is the velocity field (covariant components of v), and the a_α are the components of the acceleration defined in (2.4). The tensor B is called Born's rate of strain tensor.

V. 4. MEASURE OF RELATIVE STRAIN AND ROTATION

The transformation

(4.1) $$F_T(\xi_t) = P_v \cdot \nabla_\eta f^T \cdot \nabla f(\xi_t)$$

is a non-singular linear transformation

(4.2) $$F_\tau(\xi_t) : V^3(\xi_t, v) \rightarrow V^3(\xi_{t+\tau}, v)$$

of the subspace of space-like vectors normal to the velocity at ξ_t onto the corresponding space at the event $\xi_{t+\tau}$. In each of these subspaces we have the induced metric

(4.3) $$g_v = -s(g + s \; \overset{+}{v} \otimes \overset{+}{v}) \; .$$

For brevity, let us denote the inner product $g_v(u, w, \xi_t)$ by $(u, w)_t$, where u and w are elements of $V^3(\xi, v)$. We define the adjoint transformation $F_\tau^\dagger$ by the condition

(4.4) $$(u, F_\tau w)_{t+\tau} = (F_\tau^\dagger u, w) \; .$$

Then

(4.5) $$C_\tau(\xi_t) = F^\dagger_\tau F_\tau(\xi_t)$$

is positive definite and self-adjoint in the sense that

$$(C_\tau u, u)_t = (u, C^\dagger_\tau u)_t = (F^\dagger_\tau F_\tau u, u)_t =$$

(4.6) $$= (F_\tau u, F_\tau u)_t > 0$$

if $u \in V^3(\xi, v)$, $u \neq 0$, and

(4.7) $$C_0(\xi_t) = P_v(\xi_t) \; .$$

We call $C_\tau(\xi_t)$ the <u>relative strain measure.</u> Let $D_\tau(\xi_t)$ be the self-adjoint positive definite square root of $C_\tau(\xi_t)$ so that $D^2_\tau = C_\tau$ and $D_\tau = D^\dagger_\tau$. Set

(4.8) $$F_\tau = R_\tau D_\tau \; ,$$

so that

(4.9) $$R_\tau = F_\tau D_\tau^{-1} .$$

Then we may assert that R_τ is isometric in the sense that

(4.10) $$(R_\tau u,\ R_\tau w)_{t+\tau} = (u,\ w)_t \quad .$$

This follows from the observation that

(4.11) $$R_\tau R_\tau^\dagger = F_\tau D_\tau^{-1} (D_\tau^{-1})^T F_\tau^\dagger = P_v \quad .$$

Thus the transformation R_τ is a measure of the relative rotation of the material about X in the configuration at ξ_t and the configuration at $\xi_{t+\tau} = T_\tau(\xi_t)$.

In terms of components one has the following relation between $C_\tau(\xi_t)$ and the tensor C defined in (3.2) .

(4.12) $$\underset{\tau}{C}{}^{\alpha}{}_{\mu}(\xi_t) = g_v^{\alpha\epsilon}(\xi_t) C_{AB}(\xi_{t+\tau})\, f^A{}_{,\epsilon}(\xi_t) f^B{}_{,\mu}(\xi_t) ,$$

where $g_v^{\alpha\beta} = s(g^{\alpha\beta} + s\, v^\alpha v^\beta)$.

By definition, a motion is if and only if $C_{AB}(\xi_{t+\tau})$ is independent of τ . But this is the only term on the right-hand side of (4.12) which depends on τ ; hence, a necessary and sufficient condition th a motion be locally rigid at X is that the relative strain measure $C_\tau(\xi_t)$ be independent of τ .

LECTURE V

MASS, STRESS, ENERGY, ENTROPY, AND MOMENTUM

" It will be remembered that Faraday, when studying the curvature of lines of force in electrostatic fields, had noticed an apparent tendency of adjacent lines to repel each other, as if each tube of force were inherently disposed to distend laterally; and that in addition to this repellent or diverging force in the trasverse direction, he supposed an attractive or contractile force to be exerted at right angles to it, that is to say, in the direction of the lines of force."

" Of the existence of these pressures and tensions Maxwell was fully persuaded; and he determined analytical expressions suitable to represent them. The tension along the lines of force must be supposed to manitain the pondermotive force which acts on the conductor on which the lines of force terminate; and it may therefore be measured (in the system of units we are now using) by the force which is exerted on unit area of the conductor, i.e., $E^2/8$ or $DE/8$. The pressure at right angles to the lines of force must then be determined so as to satisfy the condition that the aether is to be in equilibrium."

E. T. Whittaker

V.1. INTRODUCTORY REMARKS

Nothing considered thus far depends in any way upon definitions or properties of stress, energy, entropy, or momentum. The electromagnetic field and the gravitational field have been introduced as densities of physical quantities called magnetic flux and gravity. They have not been defined in the traditional way in terms of the force exerted on charge, current, and mass; nor have any other "mechanical" attributes been assigned them. The principles of balance of energy, momentum, and angular momentum, and the elementary concepts of force and inertia which suffice as suitable framework for classical continuum mechanics do not extend in a completely natural way to the more general space-time manifolds considered here. We seek a system of mechanical principles sufficiently general to include Einstein's relation between the curvature of space-time and the distribution of energy and momentum, yet specific enough to embrace theory of motion and entropy production in a continuum interacting with an electromagnetic field.

V.2 INERTIAL MASS

Recall that action is a 4-cochain in $\mathcal{E}$ and that

$$(2.1) \qquad A(\mathcal{E}^4, \underset{\approx}{A}) = \int_{\mathcal{E}_4} \mu\,(\underset{\approx}{A})$$

where μ is the density of action.

Let a motion of a material medium $\mathcal{M}$ be parametrizable so that we are given a congruence of world-lines by a mapping

(2.2) $$f^{\dagger}: \mathcal{M} \times \mathcal{O} \to \mathcal{E}$$

such that the tangent field $\xi = \nabla_t f^{\dagger}$ (where t is the parameter) is everywhere time-like, $sg(\dot{\xi}, \dot{\xi}) < 0$. Define J by

(2.3) $$J = \frac{\det \nabla f^{\dagger}}{\sqrt{-sg(\dot{\xi}, \dot{\xi})}} ,$$

(2.4) $$\text{phys.dim. } J = [T^{-1}] .$$

It follows from the Jacobi identies that

(2.5) $$\operatorname{div}(J^{-1} v) = (J^{-1} v^{\alpha})_{,\alpha} = 0 .$$

Let $\hat{\mu}$ denote the dual of the action density μ . Then we define the density of inertial mass ρ by

(2.6) $$\mu = J^{-1} \rho$$

Let v be the velocity field of a motion of $\mathcal{M}$. A tensor field in the world-tube of $\mathcal{M}$ is said to be invariant under the motion if its Lie derivative with respect to v vanishes. It is said to be absolutely invariant under the motion if its Lie derivative with respect to any field λ v proportional to the velocity vanishes. If φ is an r-form ,

(2.7) $$\mathcal{L}_w \varphi = \operatorname{rot}(w \wedge \varphi) + w \wedge \operatorname{rot} \varphi$$

from which we see that an invariant r-form is absolutely invariant if and only if

(2.8) $$w \wedge \varphi = 0.$$

In particular, if a scalar field is invariant under a motion, it is absolutely invariant under that motion.

Recall that in the previous lecture we showed that a motion of $\mathcal{M}$ was locally rigid about a point X if and only if the induced metric g_V where invariant under the motion. With these remarks in mind we come to the important concept of the consitutive relation for the inertial mass of a continuous medium. We consider a class of material media for which the vlaue of the inertial mass $\rho(\xi)$ at a given event is given by

(2.9) $$\rho(\xi) = U(g(\xi), \nabla \overset{T}{f}(\xi) , \omega_a(\xi) , X(\xi))$$

where ω_a, as =1, 2, ... N is a set of fields in the world-tube of $\mathcal{M}$ called state variables. Different functions U define different materials of the class. We say that the state of the material about X in invariant if the motion is rigid and if $\pounds\, \omega_a = 0$, and absolutely invariant if the motion is rigid and $\underset{\lambda v}{\pounds}\, \omega_a = 0$ for arbitrary .

A 14 . (Principle of Material Indifference) The inertial mass is invariant under the motion whenever the state is invariant under the motion.

Replacement Theorem ; If the constitutive function U for the inertial mass satisfies the principle of material indifference and invariance of each of the state variables ω_a implies its absolute invariance, then

(2.10) $$U(g, \nabla f^{\Gamma}, \omega_a, X) = U(g_v, \nabla f^{\Gamma}, \omega_a, X).$$

Proof. Set $g = g_v + r\, v^{\Gamma} \otimes v^{\Gamma}$ and insert this value for g in U. The replacement theorem is equivalent to the assertion that the resulting value of U is independent of the value of r. Assuming, as we do, that U is differentiable in g, this is equivalent to the condition

(2.11) $$\frac{dU}{dr} = \frac{\partial U}{\partial g_{\alpha\beta}} v^{\Gamma}_{\alpha} v^{\Gamma}_{\beta} = 0 .$$

But if U satisfies the principle of material indifference, then

(2.12) $$\frac{\partial U}{\partial g_{\alpha\beta}} \underset{\lambda v}{\pounds}\, g_{\alpha\beta} = 0$$

for arbitrary λ if v is the velocity field of a locally rigid motion. But for a rigid motion we have

(2.13) $$\underset{\lambda v}{\pounds}\, g_{\alpha\beta} = 2\lambda\, a_{(\alpha} v^{\Gamma}_{\beta)} + 2\lambda_{,(\alpha} v^{\Gamma}_{\beta)} ,$$

where a_α are the components of the acceleration. Choosing $\lambda = 0$ at a point and $\lambda_{,\alpha}$ proportional to v^{Γ}_{α} at that point, one sees that (2.12) is sufficient for the replacement of g by g_v.

But g_v is uniquely determined by the gradient of f^{Γ} and the Born strain measures $C_{AB} = s(g_v)_{\alpha\beta} f^{\Gamma\alpha}{}_{,A} f^{\Gamma\beta}{}_{,B}$ which proves the corollary: Every constitutive relation of the type (2.9) is equivalent to one of the form

(2.18) $$\rho(\xi) = U^{*}(C(\xi), \nabla^{\Gamma}\omega_a(\xi))$$

provided that invariance of the state variables ω_a implies their absolute invariance .

V. 3. ENTROPY AND THE CLAUSIUS-DUHEM INEQUALITY

Anlongside the four basic physical quantities, magnetic flux, electric charge, gravity , and action we now set a fifth and last called <u>entropy.</u>

A 15: Entropy is a continuous 4-cochain in event-space.

$$S(\mathcal{E}^4, \underset{\approx}{S}) = \int_{\mathcal{E}^4} \sigma(\underset{\approx}{S}) , \tag{3.1}$$

where σ is the entropy field and $\underset{\approx}{S}$ is the unit of entropy.

$$\text{phys. dim. } \sigma = [\underset{\approx}{S}] ,$$

and

$$S(\mathcal{E}^4, \underset{\approx}{S}) \quad S(\mathcal{E}'^4, \underset{\approx}{S}) \geqslant 0 \tag{3.2}$$

if $\mathcal{E}^4$ and $\mathcal{E}'^4$ are similarly oriented.

Let a parametrized motion of a conituous medium be given and consider the entropy $S(\mathcal{E}^4(\tau), \underset{\approx}{S})$ of $\mathcal{E}^4(\tau) = T_\tau(\mathcal{E}^4)$ where $\mathcal{E}^4$ is a fixed oriented submanifold in the world-tube of the medium and suppose that the parameter is proper time. Then

$$\left.\frac{dS(\mathcal{E}^4(\tau), S)}{d\tau}\right|_{\tau=0} = \int_{\mathcal{E}^4} \mathop{\pounds}_{v} \sigma = \oint_{\partial\mathcal{E}^4} v \wedge \sigma. \tag{3.3}$$

Define the entropy flux relative to the material, q/θ and the rate of p roduction of entropy Φ/θ by setting

(3.4) $$\oint_{\partial \mathcal{E}^4} v \wedge \sigma = -\oint_{\partial \mathcal{E}^4} h/\theta + \int_{\mathcal{E}^4} \Phi/\theta \quad ,$$

where $\theta \neq 0$ is a factor later to be identified with the absolute temperature. It is assumed that $\hat{h}\ v = 0$, and

(3.5) $$\text{phys.dim.}\left(\frac{h}{\theta}\right) = \text{phys.dim.}\left(\frac{\Phi}{\theta}\right) = [\underline{\underline{S}}\,\underline{\underline{T}}^{-1}] .$$

Thus the sign of the entropy produced in $\mathcal{E}^4$ defined by

(3.6) $$\Delta\,(\mathcal{E}^4, \underset{\sim}{S}T^{-1}) = \int_{\mathcal{E}^4} \Phi(\underset{\approx}{S}T^{-1})/\theta$$

depends on the sign of the unit of entropy, the orientation of $\mathcal{E}^4$, and the <u>direction of time,</u> since the definition of Φ/θ compares the entropy of a set of events $\mathcal{E}^4$ and a set $\mathcal{E}^4(\tau)$ obtained from $\mathcal{E}^4$ by translation along the world-lines of a material medium. The sign of the product

(3.7) $$\nabla^*(\mathcal{E}^4, \underset{\approx}{S}^2\, \underset{\approx}{T}^{-1}) = S(\mathcal{E}^4, \underset{\approx}{S})\,\Delta\,(\mathcal{E}^4, \underset{\approx}{S}\underset{\approx}{T}^{-1})$$

however, is independent of the orientation of $\mathcal{E}^4$ and the unit of entropy.

A 16: (The Clausius-Duhem Inequality) . For every $\mathcal{E}^4$, either

$$\Delta^*(\mathcal{E}^4, \underset{\approx}{S}^2\, \underset{\approx}{T}^{-1}) \geqslant 0,$$

(3.8) or

$$\Delta^*(\mathcal{E}^4, \underset{\approx}{S}^2\, \underset{\approx}{T}^{-1}) \leqslant 0$$

depending only on the direction of the time step $d\tau$ in the definition of the rate of production of entropy.

None of out assumptions before (3.8) allow one to distinguish between past and future events along a given world-line. But hte Clausius-Duhem inequality provides this distinction if the inequality sign in (3.8) holds for at least one $\mathcal{E}^4$. The Clausius-Duhem inequality asserts that the entropy of a set of events $\mathcal{E}^4(\tau)$ obtained by <u>advancing</u> the set $\mathcal{E}^4$ into the future along the world-lines of a material medium is never less that the entropy of $\mathcal{E}^4$ plus whatever entropy has flowed across the lateral boundary or the corresponding segment of the world-tube.

V.4 SOME DIFFERENTIAL INDENTITIES

Before considering equations of motion we digress to record some results which will ease the formal manipulations.

Let $t^{(a)}$, a = 1?2? ... N denote the ordered set of components in some coordinate system of the manifold of a differentiable tensor field, or the components of a set of such fields in the manifold. Then the components of the Lie derivative of the set of fields with respect to the vector field w with components w^α are given in terms of the $t^{(a)}$ and the partial derivatives of the $t^{(a)}$ by

$$(4.1) \qquad \underset{w}{\mathcal{L}} t^{(a)} = w^\alpha t^{(a)}{}_{,\alpha} + F^{(a)\alpha}_{(b)\beta} t^{(b)} w^\alpha{}_{,\beta}$$

where a coma followed by an index denotes partial differentiation with respect to the corresponding coordinate. The coefficients $F^{(a)\alpha}_{(b)\beta}$ are constants independent of the coordinates. All partial derivatives in the formula (4.1) may be replaced by the corresponding covariant derivates and the formula remains true. Thus we also have

$$(4.2)\qquad \underset{w}{\mathcal{L}}\, t^{(a)} = w^{\alpha}\, t^{(a)}{}_{\alpha} + F^{(a)}{}_{(b)}{}^{\beta}{}_{\alpha}\, t^{(b)}\, w^{\alpha}{}_{\beta},$$

where a semi-colon denote covariant differentiation. The equivalence of (4.1) and (4.2) follows from

$$(4.3)\qquad t^{(a)}{}_{\alpha} = t^{(a)}{}_{\alpha} - F^{(a)}{}_{(b)}{}^{\beta}{}_{\mu}\, t^{(b)} \left\{ {}^{\mu}_{\alpha\beta} \right\},$$

$$w^{\alpha}{}_{\beta} = w^{\alpha}{}_{\beta} + \left\{ {}^{\alpha}_{\mu\beta} \right\} w^{\mu}.$$

In particular,

$$(4.4)\qquad \underset{w}{\mathcal{L}}\, g_{\alpha\beta} = {}^{\mu} g_{\alpha\beta;\mu} + 2\, w_{\alpha\beta} = 2\, w_{(\alpha\beta)},$$

since $g_{\alpha\beta\mu} = 0$. The Lie derivative and the partial derivative commute :

$$(4.5)\qquad \underset{w}{\mathcal{L}}\, t^{(a)}{}_{\alpha} = \left(\underset{w}{\mathcal{L}}\, t^{(a)} \right)_{\alpha}$$

Suppose $f^{\dagger}: \mathcal{M} \times \mathcal{C} \to \mathcal{E}$, is a parametrized motion of a continuous medium . Then $\nabla f^{\dagger}$ has components which we now denote by $\xi^{\alpha}{}_{A}$ and $\dot{\xi}^{\alpha}$. Now suppose that

$$(4.6)\qquad f^{\dagger}_{\lambda} : \mathcal{M} \times \mathcal{C} \to \mathcal{E}$$

is a one-parameter family of such motions of $\mathcal{M}$ such that $f^{\dagger}_{\lambda}$ is smooth in all its arguments. Set

(4.7) $$w^{a} = \frac{\partial f^{\tau}}{\partial \lambda} .$$

Then we may assert that

(4.8) $$(\overset{*}{\nabla f}{}^{\tau}_{\lambda}) = \frac{\partial(\nabla f^{\tau})}{\partial \lambda} + \underset{w}{\mathcal{L}}(\nabla f^{\tau}) = 0.$$

or, in terms of components,

$$\overset{*}{\xi}{}^{\lambda}_{,A} = \frac{\partial(\xi^{a}_{,A})}{\partial \lambda} + \underset{w}{\mathcal{L}}\xi^{a}_{,A} = 0.$$

(4.9)
$$\overset{*}{\dot{\xi}} = \frac{\partial \dot{\xi}^{a}}{\partial \lambda} + \underset{w}{\mathcal{L}}\dot{\xi}^{a} = 0 .$$

Consider next a 4-form ν in event-space and its integral over an oriented $\mathcal{E}^{4}$ in $\mathcal{E}$.

(4.10) $$N(\mathcal{E}^{4}) = \int_{\mathcal{E}^{4}} \nu$$

The submanifold may be taken as small as one pleases so that, in particular, let us assume that it lies entirely within a single coordinate patch. Let $\hat{\nu}$ denote the dual of ν. Suppose that the value $\hat{\nu}(\xi)$ in any coordinate system is given

(4.11) $$\hat{\nu}(\xi) = \epsilon\, K(g_{\alpha\beta}(\xi),\, g_{\alpha\beta\gamma}(\xi)\, ,\, g_{\alpha\beta\gamma\mu}(\xi)\, ,\, \dots\ )$$

where K is a smooth function of its arguments independent of the coordinate system and $\epsilon = \pm 1$ depending on the orientation of the basis vectors. The functions $K = \sqrt{-g}\, R$, $K' = \sqrt{-g}\, R^{\alpha\beta} R_{\alpha\beta}$, where $R_{\alpha\beta} = R_{\mu\alpha\beta}{}^{\mu}$, $R = g^{\tau\alpha\beta} R_{\alpha\beta}$ and $R^{\tau}_{\mu\alpha\beta}$ is the curvature tensor of $g_{\alpha\beta}$ are examples of functions K satisfying our hypotheses. We show that for every such function one has the identity

(4.12) $$\left(\frac{\delta \hat{\nu}}{\delta g_{\alpha\beta}}\right)_{;\beta} = 0$$

where

(4.13)

is the Hamiltonian or Lagrangian derivative of ν with respect to g. This well-known and oft quoted theorem may be proved as follows. Let $T_\tau \mathcal{E} \to \mathcal{E}$ be any one-parameter family of point transformations such that $T_0 = 1$ and let $w^\alpha = \partial \bar{\xi}^\alpha(\xi, \tau)/\partial\tau \mid \tau = 0$. Let $T_\tau(\mathcal{E}^4) = \mathcal{E}^4_\tau$. Then

$$N = \frac{d}{d\tau} \int_{\mathcal{E}^4_\tau} \nu \Big|_{\tau = 0} = \int_{\mathcal{E}^4_0} \mathfrak{L}_w \nu$$

$$= \oint_{\partial \mathcal{E}^4_0} w \wedge \nu .$$

Thus N vanishes whenever w = o on the boundary $\partial\mathcal{E}^4$ of the region $\mathcal{E}^4$. But this implies that

$$\int_{\mathcal{E}^4} \frac{\delta\nu}{\delta g_{\alpha\beta}} \mathfrak{L}_w g_{\alpha\beta} = 0$$

if w vanishes on the boundary of $\mathcal{E}^4$ together with its derivatives up to order (m-1), where m is the order of the highest derivative of g appearing in the function K. Using (4.4) we then see that

$$\int_{\mathcal{E}^4} \frac{\delta V}{\delta g_{\alpha\beta}} w_{(\beta,\alpha)} = \int_{\mathcal{E}^4} \left[\left(\frac{\partial V}{\partial g_{\cdot\beta}} w_\alpha \right)_{;\beta} - \left(\frac{\delta V}{\delta g_{\alpha\beta}} \right)_{;\beta} w^\alpha \right] = \int_{\mathcal{E}^4} \left(\frac{\delta V}{\delta g_{\alpha\beta}} \right)_{;\beta} w^\alpha = 0 \tag{4.14}$$

and this must fold for every vector field w in $\mathcal{E}^4$ which vanishes with its derivatives up to order (m-1) on the boundary. We then conclude that we must have (4.12) at each interior point of $\mathcal{E}^4$, which proves the assertion..

V. 5. EQUATIONS OF MOTION OF A CONTINUOUS MEDIUM

Recall that in classical mechanics the equation of motion of systems with a finite number of degrees of freedom with generalized coordinate $q^{(a)}$, a = 1, 2, ... N may be stated in the form

$$\delta \int_{t_1}^{t_2} T \, dt = - Q_{(a)} \, \delta q^{(a)} \tag{5.1}$$

for variations $\delta q^{(a)}$ consistent with the constraints i.e., for all variations $\delta q^{(a)}$ of a specified class. The coefficients $Q_{(a)}$ are the generalized forces. The specification of a particular dynamical system consists in laying down constitutive relations for the kinetic energy T and the generalized forces in terms of the coordinates $q^{(a)}$ and their time derivatives. Here we shall propose Lagrange equations for a continuous medium in a general space-time manifold for which, at each event, we are given a cone of directions $g(\xi)$ that varies smoothly from point to point.

Consider the action of a segment of a world-tube of the material medium $\mathcal{M}$.

(5;1) $$A(\mathcal{E}^4, \underset{\approx}{A}) = \int_{\mathcal{E}^4} \mu(\underset{\approx}{A}) \quad ,$$

and suppose that the motion of $\mathcal{E}_{\mathcal{M}}$ is given by $f_0^{\dagger}: \mathcal{M} \times \mathcal{O} \to \mathcal{E}$ where $f_\lambda^{\dagger}: \mathcal{M} \times \mathcal{O} \to \mathcal{E}$ is a one-parameter family of neighboring comparison motions of $\mathcal{M}$. Let

(5.2) $$w = \operatorname{grad}_\lambda f^{\dagger}$$

Suppose that

(5.3) $$\mu = J^{-1} \rho \, ,$$

where the inertial mass ρ is given by a constitutive relation of the form

(5.4) $$\rho = U(g_{\alpha\beta} , \xi^{\alpha}{}_{A} , \xi^{\alpha} , \omega_{(a)} , X) .$$

The generalization to materials for which U might depend on higher derivatives of $f^{\dagger}$ and the metric tensor is not too different from the special case (5.4) . More definite and special theories follow on specific choices for the state variables $\omega_{(a)}$, but for the moment we are interested in exposing the common structure of all such dynamical systems. Let $g_\lambda(\xi) = g(\xi, \lambda)$ and $\omega_{(a)\lambda} = \omega_{(a)}(\xi, \lambda)$ be fields such that $g(\xi, 0) = g(\xi)$ and $\omega_{(a)}(\xi, 0) = \omega_{(a)}(\xi)$. Let

(5.5) $$\delta g_{\alpha\beta} = \left.\frac{\partial g_{\alpha\beta}}{\partial \lambda}\right|_{\lambda=0} , \quad \delta\omega_{(a)} = \left.\frac{\partial \omega_{(a)}}{\partial \lambda}\right|_{\lambda=0}$$

and call these quantities the variations of g and $\omega_{(a)}$. For each comparizon state and motion of $\mathcal{M}$, the function

$$A_\lambda = A(\mathcal{E}_\lambda, A) = \int_{\mathcal{E}_\lambda} \mu(g_\lambda, \nabla f_\lambda, \omega_{(a)\lambda})$$

is differentiable with respect to the parameter λ. Its derivative is given by

$$\text{(5.7)} \qquad \overset{*}{A}_0 = \int_{\mathcal{E}_0} \overset{*}{\mu} = \int_{\mathcal{E}_0} \left[\frac{\partial \mu}{\partial g_{\alpha\beta}} g_{\alpha\beta} + \frac{\partial \mu}{\partial \omega_{(a)}} \omega_{(a)}\right] + \frac{\partial \mu}{\partial \mu_{a}} \omega_{(a)}]$$

where we have used the identities (4.8), and the superposed star denotes the linear operator $\partial/\partial\lambda + \underset{w}{\mathcal{L}}$.

In view of (5.7) we define the generalized forces $Q^{\alpha\beta}$ and $Q^{(a)}$ by setting

$$\text{(5.8)} \qquad \delta A = \overset{*}{A}_0 = -\int_{\mathcal{E}_c} \left[\frac{1}{2} Q^{\alpha\beta} g_{\alpha\beta} + Q^{(a)} \omega_{(a)}\right].$$

This is the Lagrange equation for the dynamical system with constitutive equation (5.4).

The Lagrange equation (5.8) is a variational equation to besatisfied for some specified class of variations $(w, \delta g, \delta\omega_{(a)})$ of the motion the metric, and the state variables. As in the case of point mechanics, we adopt the view that constitutive equations for the generalized forces $Q^{\alpha\beta}$ and $Q^{(a)}$ expressing these quantities as functions or functionals of the motion and state variables in addition to the constitutive relation for the inertial mass are required to specify the dynamical system. What we have before us is merely a framework into which one can fit most every more definite special case of a theory of motion and deformation of a continuous medium. This remains to be demonstrated by examples. The nature of the system and of the special theory depends on the physical and geometrical properties assigned the state variables $\omega_{(a)}$ and upon the detailed nature of the constitutive relations for the generalized forces. But we shall assume that in every special case, these constitutive relations are consistent with

> A 17 : (The Principle of Local Determinism) The value of each generalized force $(Q^{\alpha\beta}, Q^{(a)})$ at an event ξ is uniquely determined by the values of the fields $g, \varpi f^{\tau}$, and $\omega_{(a)}$ at events ξ' not later that ξ belonging to the same world line.

V.6 THE ELECTROMAGNETIC AND GRAVITATIONAL STRESS-ENERGY-MOMENTUM TENSORS

As in the case of the mechanics of systems with finite degrees of freedom, part or all of the generalized forces may possess a potential. What we do now is essentially to classify the generalized forces into types; a given generalized force is considered the sum of forces of various types. Of these special types of foces, we consider first electromagnetic forces.

Consider the scalar function

$$\hat{\mu}_{(\varphi)} = - \frac{k}{2a} \sqrt{-g} \quad (\varphi, \varphi) \tag{6.1}$$

$$\text{phys.dim. } \hat{\mu}_{(\varphi)} = [\underline{\underline{A}}] ,$$

where (φ, φ) denotes the quadratic form in the space of

2-covectors defined by the metric field g(and $g^{\dagger}$) as in § 5 of the preliminaries. The constant k with phys.dim. $[\underline{\underline{Q}}\,\underline{\underline{\Phi}}^{-1}]$ is the constant appearing in the aether tensor K of (II.2) so that $-k^2 I = K^2$, and a is the fine structure constant.

(6.2) $$\text{phys. dim. } a = [\underline{\underline{Q}}\ \underline{\underline{\Phi}}\ \underline{\underline{A}}^{-1}] .$$

The function $\hat{\mu}_{\varphi}$ of the fields g and φ is the dual of a 4-form μ_{φ} whose integral over an oriented $\mathcal{E}^4$ is the electromagnetic action of that set. The symmetric tensor field defined by

(6.3) $$T^{\alpha\beta}_{(\varphi)} = 2 \frac{\delta\mu}{\delta g_{\alpha\beta}}$$

is the electromagnetic stress-energy-momentum tensor.

The covariant divergence of the electromagnetic stress-energy-momentum tensor yields the electromagnetic Poynting identity

(6.4) $$f^{\alpha}_{(\varphi)} = T^{\alpha\beta}{}_{\beta} = \frac{1}{a} g^{\alpha\beta\lambda}{}_{\gamma}{}^{\gamma} \varphi_{\beta\gamma} .$$

Its value is the generalized Lorentz force $f^{(\beta)}_{(\varphi)}$ which is a measure of the rate at which electromagnetic energy and momentum is converted to other forms of energy and momentum.

In analogy with the above, we define the density of gravitational field action $\hat{\mu}_{(\gamma)}$ by

(6.5) $$\hat{\mu}_{(\gamma)} = \frac{1}{2} a' \sqrt{-g}\ (\gamma, \gamma) .$$

The gravitational constant a' has

(6.6) $$\text{phys. dim. } a' = [\underline{\underline{G}}^{-1}] .$$

The gravitational stress-energy-momentum tensor is defined by

(6.7) $$T^{\alpha\beta}_{(\gamma)} = 2 \frac{\delta \hat{\mu}_{(\gamma)}}{\delta g_{\alpha\beta}} ,$$

and the gravitational Poynting identity is

(6.8) $$f^{\alpha}_{(\gamma)} = T^{\alpha\beta}_{(\gamma)\,\beta} = a' \hat{\mu}_g T^{\alpha\beta}{}_{\gamma\beta} .$$

The generalized gravitational force $f^{\alpha}_{(\gamma)}$ is measure of the rate at which gravitational field energy and momentum is converted into other forms of energy and momentum.

Finally, let

(6.9) $$\hat{\mu}_{(g)} = a''\sqrt{-g}\,K\,(g_{\alpha\beta},\ g_{\alpha\beta\,\gamma},\ g_{\alpha\beta\,\gamma\delta}, \ldots$$

where K is a scalar function of the metric tensor and its derivatives having phys. dim. $[\underline{\underline{A}}^{-1}\ \underline{\underline{G}}]$ and independent of any other field quantities. We call $\hat{\mu}_{(g)}$ the action of curvature. We assume that $\hat{\mu}_{(g)}$ has phys. dim. $[\underline{\underline{A}}]$ so that the relativity constant a'' has

(6.10) $$\text{phys. dim. } a' = [\underline{\underline{G}}] .$$

Later to exhibit Einsten's theory of motion and gravitation as a special case one chooshes K = R , where R is scalar curvature. But whatever choice be made for the function K one has the Bianchi identity (or a special case of that identity)

$$f^{\alpha}_{(g)} = T^{\alpha\beta}_{(g)\ \beta} = 0$$

$$T^{\alpha\beta}_{(g)} = 2 \frac{\delta \hat{\mu}(g)}{\delta g_{\alpha\beta}} ,$$

satisfied by the Einstein stress-energy-momentum tensor $T^{\alpha\beta}_{(g)}$.

V. 7. THE INTRINSIC STRESS-ENERGY-MOMENTUM TENSOR

It has been assumed that the action density $\hat{\mu}$ is of the form

(7.1) $$\hat{\mu} = J^{-1} \rho$$

where the density of inertial mass ρ satisfies the principle of material indifference and J is defined in terms of the metric field and the motion of a material medium by

(7.2) $$J = \frac{\det \nabla f^{T}}{\sqrt{-sg(\nabla f^{T}, \nabla f^{T})}}$$

The intrinsic stress-energy-momentum tensor is defined by

(7.3) $$T^{\alpha\beta}_{(\mu)} = 2 \frac{\delta \hat{\mu}}{\delta g_{\alpha\beta}}$$

Substituting from (7.1) for $\hat{\mu}$ one gets

(7.4) $$T^{\alpha\beta}_{(\mu)} = -s\, \mu\, v^{\alpha} v^{\beta} + 2\, J^{-1} \frac{\partial \rho}{\partial g_{\alpha\beta}}$$
$$= -s\, \mu\, v^{\alpha} v^{\beta} + T^{\alpha\beta}_{(e)}$$

where $T^{\alpha\beta}_{(e)}$ is the <u>elastic stress tensor.</u> If the state variables $\omega_{(a)}$ are such that the replacement theorem applies, then the velocity vector v is a characteristic direction of the elastic stress tensor and

(7.5) $$T^{\alpha\beta}_{(e)} \; v^{\tau}_{\alpha} = 0 \; .$$

Hence , the velocity vector is a characteristic vector of intrinsic stress-energy-momentum tensor,

(7.6) $$T^{\alpha\beta}_{(\mu)} \; v^{\tau}_{\beta} = \hat{\mu} \quad v^{\alpha} \; ,$$

and the proper value is precisely the action density itself.

V.8 SOME FURTHER CONSTITUTIVE RELATIONS AND THE ENTROPY EQUATION

If $T^{\alpha\beta}_{(b)}$, b = (μ, φ , γ , g) is a stress-energy-momentum tensor, we call

(8.1) $$\hat{\epsilon}_{(b)} = s \; T^{\alpha\beta}_{(b)} \; v^{\tau}_{\alpha} \quad v^{\tau}_{\beta}$$

the density of relative energy of the corresponding type. Thus $\hat{\epsilon}_{(\mu)} = \hat{\mu}$ is the intrinsic relative energy, , $\hat{\epsilon}_{(\varphi)}$ is the relative electromagnetic energy density, etc. The adjective "relative" is used because the definition of the energy density depends on the velocity of matter at that event, and such energy (except in the case of the intrinsic energy) is a relative quantity in the sense that it depends upon the motion.

Now set

(8.2) $$Q^{\alpha\beta} = T^{\alpha\beta}_{(\varphi)} + \; T^{\alpha\beta}_{(\gamma)} + \quad T^{\alpha\beta}_{(g)} + Q^{\alpha\beta}_{(d)} \quad ,$$

which serves to define the <u>dissipative stress-momentum tensor,</u>

$Q_{(d)}^{\alpha\beta}$.

A 18 : The relative dissipative energy vanishes .

(8.3) $$Q_{(d)}^{\alpha\beta} \, v_{\alpha}^{\dagger} \, v_{\beta}^{\dagger} = 0 .$$

Assumption A 18 implies that if we define a complete stress-energy-momentum tensor T by

(8.4) $$\begin{aligned} T &= T_{\mu} + Q \\ &= T_{\mu} + T_{\varphi} + T_{\gamma} + T_{(g)} + Q_{(d)} \end{aligned}$$

then the complete relative energy is given by

(8.5) $$\hat{\epsilon} = s\, T^{\alpha\beta} v_{\alpha}^{\dagger} \, v_{\beta}^{\dagger} = \hat{\mu} + \hat{\epsilon}_{(\varphi)} + \hat{\epsilon}_{(\gamma)} + \hat{\epsilon}_{(g)} .$$

Thus according to A 18 the only forms of relative energy are intrinsic, electromagnetic, gravitational, and metrical.

Consider now the case where one of the state variables $\omega_{(a)}$ in the constitutive function for the inertial mass is the entropy field σ . We now write the constitutive relation (2.9) in the form

(8.6) $$\mu = J^{-1} \, U(g \, , \nabla f^{\dagger} \, , \sigma \, , \omega_{(a')}) .$$

Alternatively, and preferably, let $\tilde{\sigma} = \overset{\dagger}{J} \hat{\sigma}$. be the absolute scalar measure of entropy and set

(8.6*) $$\hat{\mu} = J^{-1} \, U^{*}(g, \nabla f^{\dagger} , \tilde{\sigma}, \omega_{(a')}) .$$

The Lagrange equation (5.8) takes the more explicit form

(8.7) $$\delta A = \int_{\mathcal{E}^4} \mu = -\int_{\mathcal{E}^4} [\tfrac{1}{2} Q^{\alpha\beta} \overset{*}{g}_{\alpha\beta} - J^{-1}\theta \overset{*}{\tilde{\sigma}} + Q^{(a')} \overset{*}{\omega}_{(a')}] ,$$

where θ is the absolute temperature and

(8.7) $$\text{phys. dim. } \theta = [\underline{A} \; \underline{S}^{-1}]$$

According to A 18, the dissipative stress-momentum tensor has a unique decomposition of the form

(8.8) $$Q^{\alpha\beta}_{(d)} = S^{\alpha\beta}_{(d)} + s(h^{\alpha} v^{\beta} + h^{\beta} v^{\alpha})$$

where $S^{\alpha\beta}_{(d)} v_{\beta} = 0$ is the dissipative stress, and $h^{\alpha} = Q^{\alpha\beta}_{(d)} v_{\beta}$, $h^{\alpha} v_{\alpha} = 0$, is the heat flux.

Suppose now that the Lagrange equation must hold for arbitrary variations of the class

(8.10) $$\delta g_{\alpha\beta} = 0, \quad \overset{*}{\omega}_{(a')} = 0 = \delta\omega_{(a')} + \underset{w}{\mathcal{L}}\, \omega_{(a')} .$$

Necessary conditions are

(8.11) $$T^{\alpha\beta}{}_{\beta} = 0 , \quad \theta = \frac{\partial \rho}{\partial \tilde{\sigma}} .$$

Substituting for T and $Q_{(d)}$ from (8.3) and (8.9) in (8.11), one gets

(8.12) $$\begin{aligned} &- s\,\mu\, \dot{v}^{\alpha} + s\, J^{-1} \dot{\rho}\, v^{\alpha} + T^{\alpha\beta}_{(e)\,\beta} + f^{\alpha}_{(\varphi)} + f^{\alpha}_{(\nu)} \\ &+ S^{\alpha\beta}_{(d)\,\beta} + s\,(h^{\alpha} v^{\beta} + h^{\beta} v^{\alpha})_{\beta} = 0 , \end{aligned}$$

where a superposed dot denotes $(\;)_{,\alpha} v^{\alpha}$ (i.e., differentiation with respect to proper-time along a world-line of the motion).

The equation

(8.13) $$v_{\alpha}\, T^{\alpha\beta}{}_{\beta} = 0 ,$$

a consequence of (8.11), will be called the <u>entropy equation.</u> Using $v^T_\alpha \dot{v}^\alpha = 0$, $S^{\alpha\beta}_{(d)} v^T_\alpha = T^{\alpha\beta}_{(e)} v^T_\alpha = 0$, we find that the entropy equation can be put in the form

$$(8.14) \qquad (\hat{\sigma} v^\alpha + \frac{h^\alpha}{\theta})_{;\alpha} = \frac{1}{\theta}[S^{\alpha\beta}_{(d)} v^T_{(\alpha,\beta)} - h^\alpha (\ln|\theta|)_{,\alpha} + S h^\alpha \dot{v}^T_\alpha - f^\alpha_{(\varphi)} v^T_\alpha - f^\alpha_{(\gamma)} v^T_\alpha - \frac{\partial\hat{\mu}}{\partial \omega_{(a')}} \overset{\dot{}}{\underset{v}{\pounds}} \omega_{(a')}]$$

Comparing this result with equation (3.4) we see that the dissipation function Φ is given by

$$(8.15) \qquad \Phi = S^{\alpha\beta}_{(d)} v^T_{(\alpha,\beta)} - h^\alpha (\ln|\theta|)_{,\alpha} + s h^\alpha \dot{v}^T_\alpha - f^\alpha_{(\varphi)} v^T_\alpha - f^\alpha_{(\gamma)} v^T_\alpha - \frac{\partial\hat{\mu}}{\partial \omega_{(a')}} \underset{v}{\pounds} \omega_{(a')}$$

The Clausius-Duhem inequality and this expression for the dissipation function serve as guides in the construction of special theories.

The dissipation function contains the term $-f^\alpha_{(\varphi)} v^T_\alpha$, which represents the rate at which electromagnetic field energy is converted to other forms of energy. Depending on the nature of the constitutive relations for the charge-current field, some of this energy appears as heat energy so that this term is related to the so-called Joule heat. The dissipation function (8.15) contains also the analogous but less familiar term

$$(8.16) \qquad -f^\alpha_{(\gamma)} v^T_\alpha = -a' \hat{\mu} \dot{\psi} ,$$

where ψ is a gravitational potential. Thus we may conclude that in the rigid motion of a non-heat conducting, electrically neutral body ($f_{(\varphi)} = h = 0$) entropy is produced at the rate

(8.17) $$\Phi = - a' \hat{\mu} \dot{\psi} - \frac{\partial \hat{\mu}}{\partial \omega_{(a')}} \underset{V}{\mathcal{L}} \omega_{(a')}$$

Thus if the gravitational constant $a' \neq 0$, entropy is produced or absorbed by a particle in every motion with invariant state ($\underset{V}{\mathcal{L}} \omega_{(a')} = 0$) in which that particle does not move on a surface of constant gravitational potential. In Einstein's theory of gravitation $a' = 0$, so that these remarks do not apply. An alternative when it is not assumed that $a' = 0$, is to assume that one of the state variables $\omega_{(a')}$ is the gravitational potential so that the inertial mass of a particle depends on the gravitational potential which it experiences. Moreover, one can adjust this dependence in such a way that no entropy is produced or absorbed in a rigid motion of the medium in which all the other state variables are invariant. This requires that we set

(8.18) $$- a' \hat{\mu} - \frac{\partial \hat{\mu}}{\partial \psi} = 0 ,$$

the general solution of which is

(8.19) $$\hat{\mu} = \hat{\mu}_o e^{-a'\psi} ,$$

where $\hat{\mu}_o$ is independent of the gravitational potential. If (8.19) be assumed, the equations of motion (8.12) with all forces set equal to zero except inertial and gravitational forces assume the special form

(8.20) $$s v^{\alpha} = a'(g^{\dagger\alpha\beta} \psi_{\beta} + s \dot{\psi} v^{\alpha}) .$$

In this case, every material point moves on an orbit independent of its mass. If it be further assumed that $\mathcal{E}$ is an affine space free of curvature, the orbits determined by (8.20) when $a' \psi << 1$ lie very close to the classical Newtonian orbits. The perihelia, rather than advancing as in Einstein's theory, recess slowly at a somewhat lesser amount per revolution. Of course, nothing which has been assumed in the general theory requires either that $a' \neq 0$, or that the curvature of event-space vanish.

V.9 . EINSTEIN'S FIELD EQUATIONS

A necessary and sufficient condition that the Lagrange equation (8.7) hold for unrestricted variations of the metric field is

(9.1) $$T = 0$$

or, equivalently,

(9.2) $$-T^{\alpha\beta}_{(g)} = T^{\alpha\beta}_{(\mu)} + T^{\alpha\beta}_{(\varphi)} + T^{\alpha\beta}_{(\gamma)} + Q^{\alpha\beta}_{(d)}$$

if one chooses for the function $\hat{\mu}_{(g)}$ the function $a''\sqrt{-g}(R+C)$, where C is the cosmical constant, phys.dim. $C = [\underline{\underline{GA}}^{-1}]$, then

(9.3) $$-T^{\alpha\beta}_{(g)} = a'' \sqrt{-g} \; [R^{\alpha\beta} - \frac{1}{2} g^{\alpha\beta} (R + C)]$$

and (9.2) becomes Einstein's relation between the curvature of space-time and the distribution of stress, energy, and momentum. It is of couse assumed in Einstein's theory that $a'' \neq 0$, but it is also assumed that what we have called the gravitational constant a' to which $T_{(\gamma)}$ is proportional is equal zero. Under this assumption , the numerical value of the relativity constant can be so chosen that the geodesics of a singular solution of $T_{(g)} = 0$ lie close to the classical orbits of electrically neutral point particles. In Einstein's theory of gravitation, what we have called the gravitational field γ exerts no force on matter and does not influence its motion in any way.

In the special theory of relativity, it is assumed a priori that event-space is affine and that the components of the metrical field (the light cone) are constant in an affine coordinate system of $\mathcal{E}$. Thus, in special relativity, as in classical mechanics, the metrical and temporal structure of the menifold of events is laid down as a postulate at the outset and not affected in any way by the distribution of stress, energy and momentum. The Lagrange equation holds only for variations consistent with the constraint $\delta g = 0$, which is natural from the viewpoint of the special theory. Thus equation (9. 1) does not apply in the special theory. Both the general and the special theory of relativity are embraced as special cases of our general assumptions thus far.

LECTURE .VI

ELECTRODYBAMICS OF DIELECTRIC MEDIA

" It was the great merit of H. A. Lorentz that the brought about a change here in a convincing fashion in principle, a field exists, according to him, only in empty space. Matter-considered as atoms-is the only seat of electric charges, between the material particles there is empty space, the seat of the electromagnetic field, which is created by the position and velocity of the point charges which are located on the material particles. Dielectricity, conductivity, etc., are determined exclusively by the type of mechanical tie connecting the particles, of which the bodies consist. The particle-charges create the field, which , on the other hand, exrts forces upon the charge of the particles, thus determining the motion of the latter according to Newton's law of motion. If one compares this with Newton's system, the change consists in this : action at a distance is replaced by the field, which thus also describes the radiation. Gravitational is usually not taken into account because of its relative smallness; its consideration, however, was always possible by means of the enrichment of the structure of the field i.e. , expansion of Maxwell's law of the field. The physicist of the present generation regards the point of view achieved by Lorentz as the only possible one; at that time, however, it was a surprizing and audacious step, without which the later development whould not have been possible."

A. Einstein

V.I. INTRODUCTORY REMARKS

It is significant to the historical development of the principle of relativity that Einstein's famous paper of 1905 was entitled, 'Elektrodynamik bewegter Korper" The lectures to this point represent an attempt to summarize by a small number of explicit assumptions, a set of general physical principles common to a large class of more special and specific mathematical theories of the electromathetic field in material media and of the consequent motion and deformation of such media. These embrace and are consistent with a vast variety of different theories of matter, motion, gravitation, and electromagnetism. They represent but a framework into which one can fit more specific theories characterized by different descriptions of the state of a medium and by different constitutive relations relating that state to the inertial mass and the generalized forces. One of the earliest class of problems in the new relativity mechanics to be attacked by Einstein, Minkowski, Abraham, Batema, and many others, was the construction of a theory of the electromagnetic field in a moving and deforming medium characterized in part by linear constitutive relations

$$D = \epsilon \cdot E ,$$

(1.1)

$$B = \mu \cdot H ,$$

for the corresponding medium at rest in some inertial Lorentz frame. Maxwell, Lorentz, Voigt, and others had demonstrated how linear constitutive relations like (1.1) and linear generalizations of them which include the effects of small deformation could account with elegance and simplicity for many of the known optical and electro-mechanical properties of solids and fluids. This early work on the construction of "relativistic" counterpatrts of known classical constitutive relations like (1.1) for stationary media to the case of moving and deformable media, and the controversy surrounding the apparently contradictory results different investigators is described in the excellent article by Pauli. It is difficult to perceive in this early work general physical principles not conditioned by the linearity of the underlying classical constitutive relations under consideration and which could be relied upon to guide the construction of the corresponding theory of motion of dielectric media in which the polarization is not a linear function of the electromagnetic field. Thus we have abandoned these earlier methods of reasoning and in this concluding lecture attempt to sohw how the general principles established thus far can guide the construction and physical interpretation of a more definite special theory of deformable dielectric media which does not rest upon the concept of absolute time and Euclidean space.

V.2. NON- MAGNETIC PERFECT DIELECTRICS

In ordinary terms, by a perfect dielectric one means a perfect electrical insulator. More formally now, we shall define a perfect dielectric as follows. It is a material medium in the sense used in the previous lectures. Let v be the velocity of the medium, and let η be a charge potential in the world-tube of the medium. Thus the charge of any oriented set $\mathcal{E}^3$ of events experienced by points of the dielectric medium is given by

$$C(\mathcal{E}^3, \underset{\sim}{Q}) = \oint_{\partial\mathcal{E}^3} \eta(\underset{\approx}{Q}) \tag{2.1}$$

In a perfect dielectric medium, there exists a charge-potential η such that

$$\eta \wedge v = 0 . \tag{2.2}$$

If $\hat{\eta}$ denotes the dual of η, then (2.2) is equivalent to

$$\hat{\eta} \vee v = 0 . \tag{2.3}$$

This equation asserts that the velocity vector at each event is a divisor of a charge-potential . Thus there exists a field $\hat{P}$ such that

$$\hat{\eta} = \hat{P} \vee v \tag{2.4}$$

Now $\hat{P}$ is not uniquely determined by η and v and the relation (2.4) since one may add to any solution of (2.4) for $\hat{P}$ a term proportional to v and obtain another solution. There is only one solution , however, such that

$$\hat{P} \cdot v^T = 0 ., \tag{2.5}$$

and we shall assume as part of the definition of the <u>polarization field</u> $\hat{P}$ that it satisfies (2.5) .

It should be remarked that according to A6 , $K(\phi)$, where K is the aether tensor and ϕ is the electromagnetic field, is also a charge potential, whether matter experiences the events in question or not. This is the Lorentz viewpoint. Thus we also have

(2.6) $$C(\mathcal{E}^3, \underset{\approx}{Q}) = \oint_{\partial\mathcal{E}^3} K(\phi, \underset{\approx}{Q}) .$$

But the potential of charge, it must be rememebered, is not uniquely determined by the distribution of charge. If this were true, then one would infer that $K(\phi) = \text{dual } (\hat{P} \vee v)$, and this is not implied at all by what has been assumed above. Rather, one can only infer that

(2.7) $$\oint_{\partial\mathcal{E}^3} K(\phi) = \oint_{\partial\mathcal{E}^3} \text{dual } (\hat{P} \vee v) ,$$

for every submanifold of events $\mathcal{E}^3$ experienced by the material points of a perfect dielectric. Assuming that $K(\phi)$ and $\hat{P} \vee v$ are regular forms, (2.7) is equivalent to

(2.8) $$\text{rot } [K(\phi)] = \text{rot } [\text{ dual } (\hat{P} \vee v)],$$

or, taking the dual of this equation,

(2.8') $$\text{div } [K(\phi)] = \text{div } (P \vee v) ,$$

which we can also write in the form

(2.9) $$\text{div } D = 0 , \quad D = K(\phi) + v \vee \hat{P} .$$

Equation (2.9) is one pair of Maxwell's equations relating the charge, current, electric, and magnetic fields in a perfect dielectric medium.

One sees also from the above that the dual , $\hat{\gamma}$, of the charge-current field in a dielectric has the special form

$$(2.10) \qquad \hat{\chi} = -(\operatorname{div} \hat{P})\, v + \overset{\circ}{\hat{P}}$$

where $\overset{\circ}{\hat{P}} = \underset{v}{\mathcal{L}} \hat{P}$. We call, $-\operatorname{div} \hat{P}$, the density of polarization charge or bound charge, and $\overset{c}{\hat{P}}$, the current of polarization. Note that (2.10) does not always correspond to a decomposition of $\hat{\chi}$ into its components along v and normal to v since

$$\overset{\circ}{\hat{P}}\, v^{T} = \overline{\hat{P} \quad v^{T}}^{\circ} - \hat{P} \quad \overset{\circ}{v^{T}}$$

$$(2.12) \qquad = - \hat{P} \cdot a \quad ,$$

where $a = \overset{c}{v}^{T}$ is the acceleration. Thus the current of polarization in an accelerated medium is not always perpendicular to the velocity as is the polarization.

We next assume for this special theory of dielectrics that the inertial mass $\rho = \mu \quad J$ of a simple dielectric medium is a function

$$(2.13) \qquad \rho = U(g, \quad \nabla f^{T} , \sigma \ , \quad P \ , \ X \ ) \ .$$

In words, the inertial mass is a function of the metric field, the deformation gradient, the entropy, and the polarization. By his assumption one excludes any consideration of many other aspects of dielectrics which might be considered in a more general theory such as, for example, strain gradient effects, or diffusion.

The state variables σ and P do not have the property required in the replacement theorem of Lecture V viz. , that their invariance under the motion implies their absolute invariance. For this reason, we replace the state variables σ and P , which appear most natural in the first instance, by the scalar measures of entropy and polarization defined by

(2.14) $$\tilde{\sigma} = J \ \hat{\sigma}$$

(2.15) $$\Pi = J \ \nabla f \ \hat{P} \ .$$

In terms of components relative to an arbitrary system of coordinates about X in $\mathcal{M}$ and ξ in $\mathcal{E}$, this last equation reads

(2.16) $$\Pi^A = J \ X^A_{\ \alpha} \ \hat{P}^\alpha \ .$$

Since $\hat{P}^\alpha v^T_{\ \alpha} = 0$, this equation has a unique solution

(2.17) $$\hat{P}^\alpha = J^{-1} \ P_v{}^\alpha{}_\beta \ \xi^\beta{}_A \ \Pi^A$$

where $P_v = 1 - s \ v \otimes v^T$ is the projection of $V^4(\xi)$ onto $V^3(\xi, v)$. Thus $\hat{P}$ is a function of the polarization measures Π, the metric, and the deformation gradient. Also, $\hat{\sigma}$ is a function of the scalar entropy measure $\tilde{\sigma}$, the metric, and the deformation gradient. Therefore, the constitutive relation (2.13) is equivalent to one of the form

(2.18) $$\rho = U \ (g \ , \nabla f^T \ , \sigma \ , \Pi \ , X)$$

The polarization measure π is the dual of a 2-form in the material manifold $\mathcal{M}$, but with respect to $\mathcal{E}$ it may be viewed as a set of 0-forms, or scalar fields in the world-tube of $\mathcal{M}$. Therefore, each of the state variables in the constitutive relation (2.18) for the inertial mass is absolutely invariant under the motion of $\mathcal{M}$ if it is invariant, and the replacement theorem applies. Thus if the principle of material indifference is assumed, every constitutive relation for the inertial mass is equivalent to one of the form

(2.19) $$\rho = U(C, \nabla f^{\dagger}, \Pi, \tilde{\sigma}, X)$$

It can be shown in several ways now that this function U must be independent of $\nabla f^{\dagger}$. One way is as follows. Since σ, C, II, and ρ are scalar fields in one has

(2.20) $$\underset{w}{\mathcal{L}} q = w^{\alpha} \partial_{\alpha} q , \quad q = \{\sigma , C , \pi \ \rho \}.$$

Consider then the vector fields $w^{\alpha}_{(\Gamma)} = \frac{\partial \xi^{\alpha}}{\partial X^{\Gamma}}$, where $X^{\Gamma} = (X^A, \tau)$, τ the parameter of the motion . One then has

(2.21) $$\underset{w_{\Gamma}}{\mathcal{L}} q = w^{\alpha}_{(\Gamma)} \partial_{\alpha} q = \frac{\partial q}{\partial X^{\Gamma}}$$

But

(2.22) $$\underset{w}{\mathcal{L}} \nabla f^{\dagger} = 0 ,$$

so that computing $\frac{\partial \rho}{\partial X^{\Gamma}}$ in each of two alternative ways one gets .

(2.23)
$$\begin{aligned}
\underset{w_{\Gamma}}{\mathcal{L}} \rho &= \frac{\partial U}{\partial C_{AB}} \underset{w_{\Gamma}}{\mathcal{L}} C_{AB} + \frac{\partial U}{\partial \Pi^A} \underset{w_{\Gamma}}{\mathcal{L}} \Pi^A \\
&+ \frac{\partial U}{\partial \sigma} \underset{w_{\Gamma}}{\mathcal{L}} \tilde{\sigma} + \frac{\partial U}{\partial X^A} \underset{w_{\Gamma}}{\mathcal{L}} X^A \\
&= \frac{\partial U}{\partial C_{AB}} \frac{\partial C_{AB}}{\partial X^{\Gamma}} + \frac{\partial U}{\partial \Pi^A} \frac{\partial \pi^A}{\partial X^{\Gamma}} \\
&+ \frac{\partial U}{\partial \xi^{\alpha}_{\cdot \Omega}} \frac{\partial^2 \xi^{\alpha}}{\partial X^{\Omega} \partial X^{\Gamma}} + \frac{\partial U}{\partial \tilde{\sigma}} \frac{\partial \tilde{\sigma}}{\partial X^{\Gamma}} \\
&+ \frac{\partial U}{\partial X^A} \frac{\partial X^A}{\partial X^{\Gamma}} .
\end{aligned}$$

All terms but one in the last equality cancel each other and we are left with the condition

(2.24) $$\frac{\partial U}{\partial \xi^{a}{}_{\Omega}} \quad \frac{\partial^{2} \xi}{\partial X^{\Omega} \, \partial X^{\Gamma}} ,$$

which must hold for all motions and coordinate systems. But this implies that the function U is independent of ∇ f.[T] Thus

(2.25) $$\rho = U(C , \Pi , \tilde{\sigma} , X) .$$

Consider next the Lagrange equation (8.7) for a perfect dielectric and denote the generalized force conjugate to the polarization measure Π by $\mathcal{E}/\omega$ where a is the fine structure constant. Suppose that the Lagrange equation holds for arbitrary variations of the class

(2.26) $$\overset{*}{g}_{a|} = 0, \qquad \overset{*}{\tilde{\sigma}} = 0 .$$

This will imply that

(2.27) $$\frac{\partial \mu}{\partial \pi^{A}} = - (\frac{1}{a}) \ \mathcal{E}_{A}$$

The electromagnetic energy-momentum vector in a perfect dielectric has the special form

(2.28) $$f_{a} = (1/a) \ \phi_{a\beta} [\ -(\mathrm{div} \ \hat{P}) \ v^{\beta} + \overset{c}{\hat{P}}{}^{\beta}]$$

On taking the Lie derivative of (2.15) with respect to one gets the following relation between the current of polarization and $\dot{\Pi}$.

(2.29) $$\overset{c}{\hat{P}}{}^{a} + s \, v^{a} \ a_{\beta} \ \hat{P}^{\beta} = J^{-1} \, P^{\ a}_{v \ \beta} \ \xi^{\beta}_{\ A} \ \dot{\Pi}^{A} .$$

Consider next the dissipation function (8.15) for the case in hand. Using the results (2.27) , (2.28) , and (2.29) just established, we find that the dissipation function can be expressed in the form

$$\Phi = S^{\alpha\beta} v^{\dagger}_{\alpha\beta} - h^{\alpha}(\ln|\theta|)_{\alpha} + s h^{\alpha} a_{\alpha}$$
(2.30)
$$+ (1/a) \mathcal{E}_{(d)A} \dot{\Pi}^{A},$$

where the dissipative-roatary component $\mathcal{E}_{(d)}$ of the generalized force conjugate to the polarization is given by

(2.31) $$\mathcal{E}_{(d)A} = \mathcal{E}_{A} - J^{-1} \xi^{\alpha}{}_{A} \Phi_{\alpha\beta} v^{\beta}.$$

The electromotive intensity at a point in the world tube of the medium is defined by

(2.32) $$\underset{v}{E}_{\alpha} = \Phi_{\alpha\beta} v^{\beta}.$$

Since

(2.33) $$\dot{\Pi}^{A} = J X^{A}{}_{,\alpha} \overset{c}{\hat{P}}{}^{\alpha}$$

we find that the polarization term in the dissipation function can be expressed in the alternative form

(2.34) $$\mathcal{E}_{(d)A} \dot{\Pi}^{A} = \mathcal{E}_{(d)\alpha} \overset{c}{P}{}^{\alpha},$$

where,

$$\mathcal{E}_{(d)\alpha} = \mathcal{E}_{\alpha} - \underset{v}{E}_{\alpha}$$
(2.35)
$$\mathcal{E}_{\alpha} = J^{-1} X^{A}{}_{,\alpha} \mathcal{E}_{A}.$$

Now let

(2.36) $$\Delta = \{ S_{(d)}, h, \mathcal{E}_{(d)} \}$$

denote the set of fields whose values determine the rate of entropy production in the medium . The constitutive relations for dielectrics consist in the function U whose value determine the inertial mass, and the relations giving the values $\Delta(\mathcal{E})$ at each event in the world-tube of $\mathfrak{M}$ in terms of the motion and state of the medium. According to the principle of local determinism, $\Delta[\mathcal{E}(X)]$ is a functional of the values of the field variables $(g, \nabla f, {}^{T}\sigma, P)$ at events $\mathcal{E}'(X)$ not later than $\mathcal{E}(X)$, Whatever may be the explicit form of these constitutive relations, they must be consistent with the condition $\Phi \geqslant 0$.

The simplest constitutive relation for Δ consistent with this Clausius - Duhem inequality is

$$\Delta = 0 \tag{2.36}$$

These are the constitutive relations for a perfectly elastic, thermal insulator, which is transparent, optically passive, and a perfect electrical insulator. The next simplest class of dielectrics shares all these properties except that they may be optically active. The constitutive relations for this class are of the form

$$S_{(d)} = 0, \quad h = 0, \tag{2.37}$$

$$\mathcal{E}_{(d)} = \Gamma \wedge \overset{\circ}{\hat{P}}, \tag{2.38}$$

where Γ is a 2-form called the <u>gyration coefficient</u> whose value at each event is a function of the values of the deformation gradient and electromagnetic field at that event. This class of media is non-dissipative in the sense that $\Phi = 0$.

If a Lorentz frame exists such that $g_{\alpha\beta} = \mathrm{diag}(1, 1, 1, -1)$ in appropriate units, and if the velocity of the medium <u>relative</u> to this frame as measured by the first three components v^i,

$i = 1, 2, 3$ of the velocity vector is every small compared with unity ; i.e. , $V^2 = \sum_{i=1}^{3} v^i v^i \ll 1$, in the units chosen, and if the rate of change of the inertial mass (internal energy) is everywhere small in the sense that

$$| J^{-1} \dot{\rho} | \, V^2 \ll | \hat{\mu} \, \dot{v} |$$

the theory of motion of dielectrics and of the electromagnetic field in them based on (2.38) is indistinguishable for all practical purposes from the classical theory of such media described in "A dynamical theory of elastic dielectrics" . Thus it is unnecessary to repeat here the way in which simple solutions or approximate solutions of the system of equations proposed here can be constructed and interpreted physically in terms of known qualitative electromagnetic and electromechanical properties of elastic dielectrics. The present treatment is superior to the one given earlier which relies on the concept and properties of absolute time. Whenever it is necessary to treat both inertia and the electromagnetic field side by side as in the case of the dynamics of dielectrics, a classical treatment of the former leads inevitably to the existence of a preferred frame of reference relative to which the aether is at rest. Such a frame is both Galilean (inertial) in the classical sense, and a Lorentz frame from the point of view of the electromagnetic equations. All such frames are at rest relative to one another. While the effects of motion relative to this calss of frames is small in some sense (an assumption which must be made for strict logical interpretation of the classical theory), certainly the relativistic treatment given here which does not rest on such an hypothesis, can claim greater simplicity.

CENTRO INTERNAZIONALE MATEMATICO ESTIVO

(C.I.M.E.)

CHAO-CHENG WANG

SUBFLUIDS

Corso tenuto a Bressanone dal 31 maggio al 9 giugno 1965

SUBFLUIDS

by

Chao-Cheng Wang

This lecture concerns a class of simple materials [1] called simple subfluids, or more briefly subfluids. These are simple materials for which the isotropy group contains a dilatation group, i.e., a group of all unimodular pure stretches and reflections in three linearly independent directions. In other words, a dilatation group, $\underline{\underline{h}}$ can be indexed by a linearly independent set of three vectors, say $\{\underline{e}_1, \underline{e}_2, \underline{e}_3\}$, such that a tensor $\underline{A} \in \underline{\underline{h}} \Longleftrightarrow$ there exist real numbers $\lambda_1, \lambda_2, \lambda_3$ such that

$$|\lambda_1 \lambda_2 \lambda_3| = 1, \qquad \text{and}$$

$$\underline{A}\, \underline{e}_i = \lambda_i \underline{e}_i \qquad \text{for} \quad i = 1, 2, 3 .$$

From this definition it is obvious that without loss of generality we can choose $\underline{e}_i$ to be unit vectors. We call the linearly independent set of unit vectors $\{\underline{e}_i\}$ a set of axes of $\underline{\underline{h}}$. The unit vectors $\underline{e}_i$ are then called axes of $\underline{\underline{h}}$.

In classical crystallography a crystal class is characterized by a crystallographic group. These crystallographic groups are subgroups of the orthogonal group. According to Noll [2] the groups

---

[1] The constitutive equation of a simple material is of the form

$$\underline{T}(t) = \mathcal{F}(\underline{F}^{(t)}, M)$$

Where $\underline{T}(t)$ is the Cauchy stress tensor at time t, M is an arbitrary local reference configuration, and $\underline{F}^{(t)}$ is a function whose value $\underline{F}^{(t)}(s)$, $s \in [0; \infty)$, is the deformation gradient relative to M at time t-s. For more details of the theory of simple materials we refer to [1].

corresponding to the crystal classes define various types of simple solids. It is known that there exist also infinitely many subgroups of the unimodular group that are not conjugate[2] to subgroups of the orthogonal group.

According to Coleman [3] a simple material for which the isotropy group is not conjugate to a subgroup of the orthogonal group is called a simple liquid crystal. For a simple liquid crystal there exists no reference configuration for which the isotropy group is a subgroup of the orthogonal group. From the definition, it is obvious that a conjugate of a dilatation group is again a dilatation group. Since a dilatation group does not preserve the inner product, it is never a subgroup of the orthogonal group. Therefore a subfluid is a simple liquid is a simple liquid crystal.

There are infinitely many types of simple liquid crystals that are not subfluids. It turns out that among all simple liquid crystals a distinguishing feature of subfluids is this : The local configuration of a subfluid is always characterized by the density and the orientation of certain material linear manifolds. For example, a simple material for which the isotropy group is the group of all unimodular

---

[2] Noll [2] proved that the transformation rule of the isotropy group of a simple material relative to change of reference configuration is a conjugation whithin the general linear group, i.e., if M and N are two reference configuration and g_M and g_N are the corresponding isotropy group , then

$$g_N = \underline{G}\, g_M\, G^{-1} ,$$

where $\underline{G}$ maps M onto N.

tensors that leave one plane invariant is a subfluid. The local configuration of that subfluid obviously is characterized by the density and the orientation of that invariant plane. We shall see some more examples later

An interesting problem arises naturally. We wish to know exactly how many distinct types of subfluids there are and what are they? For solid crystals mathematically there exist infinitely many distinct types, although in most books on crystallography only 32 crystal classes[3] are listed. For subfluids, however, it turns out that mathematically there exist exactly 14 types of subfluids. For c convenience, we number them Type 1 to Type 14. Simple fluids[4] constitute Type 1. The remaining type numbers are assigned so as to agree with the partial ordering of subgroups by indusion. Hence the isotropy group of Type 14 is a dilatation group. Roughly speaking, subfluids having a small type number behave more like a simple fluid, while those having a large type number are more like simple solids. Except for the simple fluids themselves, a

[3] In fact , these 32 crystal classes form only 13 distinct types of simple solids, cf. Coleman & Noll [4] .

[4] I. e., simple materials for which the isotropy group is the full unimodular group, cf. Noll [2] .

subfluid is never an isotropic material[5].

Relative to a set of axes, the component matrces of the 14 subgroups which characterize the 14 type of subfluids are listed in the following table.

---

5 According to Noll [2] a simple material is isotropic if there exists a reference configuration for which the isotropy group contains the full orthogonal group. There exist only two types of isotropic materials : simple fluids and isotropic simple solids.

Type number	Component matrices	
1 (Simple fluids)	$\begin{vmatrix} a & b & c \\ d & e & f \\ g & h & i \end{vmatrix}$	$\lvert\det\rvert = 1$, a, b, c, ... are arbitrary real numbers.
2	$\begin{vmatrix} a & b & 0 \\ c & d & 0 \\ e & f & g \end{vmatrix}$	
3	$\begin{vmatrix} a & b & c \\ d & e & f \\ 0 & 0 & g \end{vmatrix}$	
4	$\begin{vmatrix} a & b & c \\ 0 & d & e \\ 0 & 0 & f \end{vmatrix}$	
5	$\begin{vmatrix} a & b & c \\ 0 & d & 0 \\ 0 & 0 & e \end{vmatrix}$ or $\begin{vmatrix} a & b & c \\ 0 & 0 & d \\ 0 & e & 0 \end{vmatrix}$	
6	$\begin{vmatrix} a & b & c \\ 0 & d & 0 \\ 0 & 0 & e \end{vmatrix}$	
7	$\begin{vmatrix} a & b & 0 \\ c & d & 0 \\ 0 & 0 & e \end{vmatrix}$	

8	$\begin{vmatrix} a & 0 & 0 \\ b & c & 0 \\ d & 0 & e \end{vmatrix}$ or $\begin{vmatrix} a & 0 & 0 \\ b & 0 & c \\ d & e & 0 \end{vmatrix}$
9	$\begin{vmatrix} a & 0 & 0 \\ b & c & 0 \\ d & 0 & e \end{vmatrix}$
10	$\begin{vmatrix} a & 0 & 0 \\ b & c & 0 \\ 0 & 0 & d \end{vmatrix}$
11	Generated by the union of types 12-14.
12	$\begin{vmatrix} a & 0 & 0 \\ 0 & b & 0 \\ 0 & 0 & c \end{vmatrix}$, or $\begin{vmatrix} 0 & a & 0 \\ 0 & 0 & b \\ c & 0 & 0 \end{vmatrix}$, or $\begin{vmatrix} 0 & 0 & a \\ b & 0 & 0 \\ 0 & c & 0 \end{vmatrix}$
13	$\begin{vmatrix} a & 0 & 0 \\ 0 & b & 0 \\ 0 & 0 & c \end{vmatrix}$, or $\begin{vmatrix} 0 & a & 0 \\ b & 0 & 0 \\ 0 & 0 & c \end{vmatrix}$
14	$\begin{vmatrix} a & 0 & 0 \\ 0 & b & 0 \\ 0 & 0 & c \end{vmatrix}$

To see that these 14 types are exhaustive, we need a lemma in group theory. I shall state a definition first.

Definition. A set $\sigma = \{ \underline{a}, \underline{b}, \underline{c}, \ldots \}$ of vectors is called independent if every three vectors of σ are linearly independent.

Lemma Let $\{ e_1, e_2, e_3, f_1, f_2, f_3 \}$ be an independent set. Then the union of the dilatation groups h, k with axes $\{e_i\}$, $\{f_i\}$, respectively, generates the full unimodular group. A proof of this lemma is given in [5]. The completeness of the 14 types of subfluids is a direct consequence of this lemma.

The next important problem concerning subfluids is to determine the most general from of the constitutive equation. For an arbitrary simple material the constitutive equation is of the form

$$(1) \qquad T(t) = \mathcal{F}(F^{(t)}, M)$$

where M is the reference configuration. If we choose the present configuration N(t) as reference, equation (1) reduces to

$$(2) \qquad T(t) = \mathcal{F}(F^{(t)}_{(t)}, N(t))$$

where $F^{(t)}_{(t)}$ is the relative deformation history, i.e., $F^{(t)}_{(t)}(s)$ maps N(t) onto N(t-s), $s \in [0, \infty$. From the principle of material frmae-indifference it can be shown [6] that (2) reduces to the form

---

6 For the statement of the principle of material frame-indifference and the proof of (3) we refer to [1].

$$(3) \qquad \underline{T}(t) = \mathcal{G} \left(\underline{C}_{(t)}^{(t)} , \quad N(t) \right)$$

where $\underline{C}_{(t)}^{(t)}$ is the relative Cauchy-Green tensor defined by

$$(4) \qquad \underline{C}_{(t)}^{(t)}(s) = \underline{F}_{(t)}^{(t)}(s)^{T} \underline{F}_{(t)}^{(t)}(s) , \quad s \in [0, \infty).$$

The superscript T of $\underline{F}_{(t)}^{(t)}(s)^{T}$ in (4) denotes transposition. For simple fluids it is easily seen that the local configuration N(t) is characterized by the density $\rho(t)$. Consequently, the constituitive equation reduces to the form

$$(5) \qquad \underline{T}(t) = \mathcal{G} \left(\underline{C}_{(t)}^{(t)} , \quad \rho(t) \right)$$

From the table above we see that the local configuration of a subfluid is characterized by the density and the orientation of certain material linear manifolds. It turns out that for each type of subfluid we can assign a linearly independent set of unit vectors. $\{i_1(t), i_2(t), i_3(t)\}$ which represents the preferred material linear manifolds at time t. We call the set $\{i_i(t)\}$ a <u>preferred</u> basis of the subfluid. The preferred basis is chosen from the sets of axes of the dilatation groups that are contained in $\underline{\underline{g}}_{N(t)}$. Using the preferred basis and the density as the representation of N(t), we can write the constitutive equation of a subfluid as follows:

$$(6) \qquad \underline{T}(t) = \mathcal{G} \left(\underline{C}_{(t)}^{(t)} , \rho(t) , \underline{i}_i(t) \right) .$$

From the principle of material frame-indifference we observe that the functional $\mathcal{G}$ is isotropic, i.e., it satisfies the relation

$$(7) \quad \mathcal{G}(\underline{Q}\underline{C}^{(t)}_{(t)}\underline{Q}^T, \rho(t), \underline{Q}\,\underline{i}_i(t)) = \underline{Q}\,\mathcal{G}(\underline{C}^{(t)}_{(t)}, \rho(t), \underline{i}_i(t))\,\underline{Q}^T$$

for all orthogonal tensors $\underline{Q}$. Hence we can apply the representation theorem of Cauchy and obtain the following reduced form of the constitutive equation

$$(8) \quad \underline{T}(t) = \mathcal{F}^{ij}(C^{(t)kl}_{(t)}, \rho(t), g_{12}(t), g_{23}(t), g_{13}(t))\,\underline{i}_i(t) \otimes \underline{i}_j(t)$$

where

$$(9) \qquad g_{\alpha\beta}(t) = \underline{i}_\alpha(t) \cdot \underline{i}_\beta(t) \qquad \alpha, \beta = 1, 2, 3,$$

are the components of the metric tensor, and

$$(10) \qquad \underline{C}^{(t)}_{(t)}(s) = C^{(t)kl}_{(t)}(s)\;\underline{i}_k(t) \otimes \underline{i}_l(t)$$

is the component representation of the relative Cauchy-Green tensor with respect to the preferred basis $\{\underline{i}_i(t)\}$. The exact constitutive equations of the various types of subfluids are special cases of (8). For more details we refer to [5].

There are several special theories of subfluids. If a subfluid is also an elastic material it is called an <u>elastic subfluid.</u> If an elastic subfluid possesses a stored-energy function then it is called a hyperelastic[7] subfluid. These hyperelastic subfluids have

---

[7] For a general theory of hyperelastic materials we refer to [1].

some very interesting properties. It turns out that the isotropy group of the stored-energy function of a hyperelastic sufluid generally is different from the isotropy group of its mechanical response function[8]. It seems that these hyperelastic subfluids are the first examples ever given that exhibit this 'interesting property'. The constitutive equations of certain types of hyperelastic subfluids are very simple. For example, it can be shown that the constitutive equations of <u>elastic</u> subfluids of Types 1-4 are as follows[9]

(11) $$\underline{T} = a(\rho)\,\underline{1}$$

(12) $$\underline{T} = a(\rho)\,\underline{1} + b(\rho)\,\underline{i}_3 \otimes \underline{i}_3$$

(13) $$\underline{T} = a(\rho)\,\underline{1} + b(\rho)\,\underline{i}_3 \otimes \underline{i}_3$$

(14) $$\underline{T} = a(\rho)\,\underline{1} + b(\rho)\,\underline{i}_1 \otimes \underline{i}_1 + c(\rho)\,\underline{i}_3 \otimes \underline{i}_3$$

where $\{\underline{i}_i\}$ is the preferred basis and $a(\rho)$, $b(\rho)$,

---

[8] That a hyperelastic material may have two distinct isotropy groups one for the stored energy function and one for the response function- was first remarked by Truesdell [6]. However he did not give any explicit example.

[9] Notice (12) and (13) appear to be the same equation, yet really the vector $\underline{i}_3$ in (12) is the unit tangent to an invariant line, while the vector $\underline{i}_3$ in (13) is the unit normal to an invariant plane. For the detail of the preferred basis we refer to [5].

$c(\rho)$ are arbitrary functions of the density, ρ . For hyperelastic subfluids of Types 1-4 , however, I can prove that the functions $b(\rho)$ and $c(\rho)$ must reduce to linear functions in ρ , i.e. ,

(15) $$b(\rho) = b\rho \ , \quad c(\rho) = c\,\rho \ .$$

There is no restriction induced on $a(\rho)$. In particular, every elastic fluid is hyperelastic, (cf. [1]). The proof of (15) is given in [5] .

If the stress in a subfluid is the sum of an elastic part and a linearly viscous part, we call the subfluid a Newtonian subfluid. The constitutive equation for the various types of Newtonian subfluids has also been worked out explicitly in reference [5].

There are certain dynamical problems that can be treated for incompressible subfluids. These problems have been studied previously by Coleman & Noll [7] , [8] for incompressible simple fluids.

REFERENCES

1 Truesdell, C.& Noll, W., The non-linear Field Theories of Mechanics. To appear as Handbuch der Physik, vol. III/3.

2 Noll, W., Arch. Rational Mech. Anal. 2 (1958/59) 197-226.

3 Coleman, B.D., Simple liquid crystals - Arch. Rational Mech. Anal. (1965).

4 Coleman, B.D. & Noll, W., Arch. Rational Mech. Anal. 15 (1964) 87-111.

5 Wang, C. - C., A general theory of subfluids - Arch. Rational Mech. Anal. (1965).

6 Truesdell, C., Proc. Nat. Acad. Sci. 52 (1964), 1081-1083.

7 Coleman, B.D. & Noll, W., Phys. Fluids, 5 (1962) 840-843.

8 Coleman, B.D. & Noll, W., Arch. Rational Mech. Anal. 3 (1959) 289-303.